Primate Ecology and Human Origins

Garland Series in Ethology

Series Editor: Gordon M. Burghardt

Primate Ecology and Human Origins: Ecological Influences on Social Organization
Edited by: Irwin S. Bernstein and Euclid O. Smith

The Development of Behavior: Comparative and Evolutionary Aspects
Edited by: Gordon M. Burghardt and Marc Bekoff

The Behavioral Significance of Color
Edited by: Edward H. Burtt, Jr.

The Behavior and Ecology of Wolves
Edited by: Erich Klinghammer

Handbook of Ethological Methods
By: Philip N. Lehner

Cat Behavior: The Predatory and Social Behavior of Domestic and Wild Cats
By: Paul Leyhausen

Dominance Relations: An Ethological View of Human Conflict and Social Interaction
Edited by: Donald R. Omark, Fred F. Strayer and Daniel G. Freedman

Species Identity and Attachment: A Phylogenetic Evaluation
Edited by: Aaron Roy

Other Garland Books in Ethology

Primate Bio-Social Development: Biological, Social, and Ecological Determinants
Edited by: Suzanne Chevalier-Skolnikoff and Frank E. Poirier

Faces from New Guinea
By: Paul Ekman

The Dog: Its Domestication and Behavior
By: Michael W. Fox

Ethological Dictionary: In English, French, and German
By: Armin Heymer

Primate Ecology and Human Origins:

Ecological Influences on Social Organization

edited by

Irwin S. Bernstein
University of Georgia
and
Yerkes Regional Primate
Research Center of
Emory University

Euclid O. Smith
Yerkes Regional Primate
Research Center of
Emory University

Garland STPM Press
New York & London

15 14 13 12 11 10 9 8 7 6 5 4 3 2 1

Library of Congress Cataloging in Publication Data

Main entry under title:

Primate ecology and human origins: ecological influences on social organization.

"The volume arises from the proceedings of a conference on 'Ecological influences on social organization: evolution and adaptation' held at the European conference center of the Wenner-Gren Foundation for Anthropological Research in August, 1977."

Bibliography: p.

1. Primates--Behavior--Congresses. 2. Primates--Ecology--Congresses. 3. Social behavior in animals--Congresses. 4. Human evolution--Congresses. 5. Social Evolution--Congresses. 6. Mammals--Behavior--Congresses. 7. Mammals--Ecology--Congresses. I. Bernstein, Irwin Samuel, 1933- II. Smith, Euclid O. III. Wenner-Gren Foundation for Anthropological Research.

QL737.P9P67234 599'.8'045 79-11152
ISBN 0-8240-7080-1

Published by Garland STPM Press
545 Madison Avenue, New York, New York 10022

Printed in the United States of America

Contents

Preface

This volume arises from the proceedings of a conference on "Ecological Influences on Social Organization: Evolution and Adaptation" held at the European Conference Center of the Wenner-Gren Foundation for Anthropological Research in August, 1977. Unlike other symposia, Wenner-Gren conferences allow the participants ample opportunity to present, redefine, question, and reformulate particular theoretical positions during the course of the meeting. This conference, in the tradition of Wenner-Gren conferences, allowed the participants to critically evaluate each other's work and to rewrite their presentations in light of these discussions. We have organized these presentations in the present volume in an effort to demonstrate the logical themes which emerged from the intellectual cross fertilization which transpired. We have attempted to guide the reader through the continuity of the various chapters by: (1) writing an introduction setting out the major theoretical goals of the conference, (2) writing individual chapter introductions which will hopefully guide the reader through the development of the conference themes, and (3) writing a summary of the general findings of the conference. We hope to have produced a theoretically and logically complete work. If we have failed, we are the only ones to blame as the contributors offered ample grist for the intellectual mill.

Certainly we must acknowledge the Wenner-Gren Foundation for Anthropological Research, and particularly Ms. Lita Osmundsen, Director of Research, for their enthusiasm over the original idea of the symposium and their continued support to its conclusion. Special thanks to staff at the European Conference Center for their untiring efforts to make every conference unique. Additionally, we should thank our colleagues who agreed to participate in this effort and whose contributions added immeasurably to our knowledge of ecology and social organization through many informal discussions.

Irwin S. Bernstein
Euclid O. Smith

List of Contributors

Stuart A. Altmann and Jeanne Altmann

Department of Biology
University of Chicago
Chicago, IL

John D. Baldwin and Janice I. Baldwin

Department of Sociology
University of California
Santa Barbara, CA

Joseph B. Birdsell

Department of Anthropology
University of California
Los Angeles, CA

François Bourlière

Départment de Physiologie
Faculté de Medicine
Paris, France

Claud A. Bramblett

Department of Anthropology
University of Texas
Austin, TX

C. Loring Brace

Department of Anthropology
University of Michigan
Ann Arbor, MI

Bernard Campbell

Sedgeford Hall
Hunstanton
Norfolk PE36 5LT
England

Timothy H. Clutton-Brock

King's College Research Centre
King's College
Cambridge, England

Anthony M. Coelho, Jr.

Primate Ethology Laboratory
Southwest Foundation for Research and Education
San Antonio, TX

Robin I. M. Dunbar

King's College Research Centre
King's College
Cambridge, England

John F. Eisenberg

National Zoological Park
Smithsonian Institution
Washington, D.C.

Paul H. Harvey

School of Biological Sciences
University of Sussex
Brighton, England

Ueli Nagel

Zoologisches Institut der Universitat Basel
Rheinsprung 9
Basel, Switzerland

Larry B. Quick

Department of Anthropology
University of Texas
Austin, TX

Thelma E. Rowell

Department of Zoology
University of California
Berkeley, CA

Akira Suzuki

Primate Research Institute
Kyoto University
Inuyama City, Aichi, Japan

Editors' Introduction

Knowledge is power. To understand something is to have the potential to control it. If you know what causes an effect, and if you can control the cause, you can produce the effect at will.

Taking this as an article of faith, human beings have studied the world around them, and that world includes other human beings. Even with only modest ability to remember and learn, an examination of experience reveals certain regularities. The recognition and identification of correlational sequences then permits the application of inductive logic to hypothesize causal relationships; identification of causes allows prediction of specified outcomes. The appearance of a specific event of the class which has in the past always preceded an outcome allows for the prediction of another such outcome.

We may recognize this process and formalize it by calling it the Scientific Method, but there is nothing terribly mysterious or complex about the process itself. It may be recognized as a description of a learning process used by every child who reaches the age of reason. The first portion of this process may be found in every creature capable of learning and merely assumes that the past is a predictor of the future, that is, what has happened regularly in the past will continue to happen in the future. If two events have always occurred together in our past experience, then whenever we see the first, we look for the second.

The next portion of the process requires an ability to manipulate the environment so that the effect can be produced by gaining control over the cause. When we formalize the process, we call the testing of the relationship between two events, by producing one at will and looking for the second, an experiment.

The original collection of experience, or data, has been called empiricism and the induction of explanations, which permit the generation of specific hypotheses, is theory building. Only those theories which can be used to deduce specific future outcomes, in situations which can be produced and replicated at will, are considered in the domain of Science. It is not enough that the predictions be logically true as derived from the postulated premises, reality testing (or experimentation) looks to see if the specified outcomes occur. If the experimental outcome is as predicted, we feel that there is some support for the relationships postulated in the theory, whereas if we fail to find the predicted outcome, we have cast doubt on the premises leading to the prediction.

Inasmuch as we can never exactly duplicate our past experiences, and inasmuch as particular outcomes may result from more than one preceding set of events, our tests cannot produce absolute proof and we can only generate probability statements concerning various alternative explanations. Since experimental outcomes contain an element of uncertainty, certain risks are inherent in testing presumed relationships. Where outcomes may prove dangerous, we proceed with some caution. When we are concerned with the effect of a variable upon some important process in ourselves, and are fearful of the potential risks, then we seek a substitute, or a model for ourselves. Such a substitute should be as much like ourselves as possible so that the influence of the original manipulation will not be modified by interaction with elements not found in our own systems.

This then accounts for the development of animal models in research, and for our particular fascination with nonhuman primates. Problems which are impossible to examine with human subjects, or where the risk to humans would be too great, are examined using a substitute species. We have considerable latitude in the selection of subjects when the processes are basic and general, but when the processes reflect specializations absent in many lineages, then we turn to the primates as animals which are likely to share common features with ourselves, due to common origins and similar development in the face of similar selective pressures.

In the study of social behavior we may ask questions about the basis of common social patterns, when the origins of such patterns are buried in the history of human origins and are no longer directly observable. An animal model may be examined to search for common basic elements and to ask how the primitive condition could have evolved into the present expression.

Many people from diverse disciplines have addressed themselves to the study of social behavior and social organization. They have examined particular environmental conditions in which alternative forms of organization and behavior exist in order to try to understand what factors account for the observed variance in expression. Some have viewed societies as immediate responses to ecological conditions, while others have tried to derive modern social systems from a presumably more primitive state responding to ecological pressures in time. Many have marshalled comparative data, available fossil and paleoecological data, and argued for one interpretation or another of early hominid society. These explanations account for data in logically acceptable fashions and build cases arguing that a particular sequence should have logically come about.

The difficulty with such models is that they are all logical and satisfactorily "explain" the same data, and yet may be contradictory. It is apparently all too easy to argue for the rationality of any logical model. It is far more difficult to design a test for such models, especially if the events being described are buried in the past.

We have, therefore, turned to the nonhuman primates in our search for understanding of human social organizations and social behavior. First, we asked what was basic and general in the order. Next we asked if there were particular conditions which produced particular variations. Then we asked if there were suitable primate subjects living under conditions sufficiently similar to the circumstances of early hominid groups such that we might argue that the organization of the nonhuman primate group would be similar to that of early hominid groups.

The first models were proposed after examining our accumulated knowledge of different kinds of primates (and other animals) under captive conditions and in their natural habitats. The pioneers in this field included Sir Solly Zuckerman (1932), C. R. Carpenter (1939), and M. F. Ashley Montagu (1943). They found themselves faced with a paucity of data which would allow only very general statements. It was clear that much more information would be required for proper theory building. The pioneering field work of such people as C. R. Carpenter revealed the inadequacies of much of the available information, which consisted of little more than anecdotes and hasty generalizations, many of which proved unreliable. A whole generation of investigators then returned to the field to collect the basic data required. This did not prove an easy task, for not only were working conditions difficult, but the task itself was difficult to define. Neither social organization nor ecology can be placed on a unidimensional scale for measurement, and we are still identifying the component variables of each.

As values for identified variables became available, new theories were generated. Washburn and DeVore clearly exemplify the interest in theory building which motivated some of the first modern primate field studies (DeVore, 1963). It, nonetheless, took nearly ten years before a sufficient diversity of studies had been completed to permit correlational approaches. Crook and Gartlan (1966) tried to relate selected aspects of social organization to five conceptual habitat types. They reviewed the new data on primate social organization and tried to relate it to the descriptions of field study locations. They recognized this as only a beginning and have continued to review new data and revise their original hypotheses.

The rate of data acquisition has not diminished, but the fundamental attributes of ecology and social organization are still not fully identified. Model building has become more sophisticated and correlational models have suggested more fundamental relationships. Denham (1971) tried to define ecology more explicitly in terms of two attributes of food distribution: quantity and dispersion. Although dealing with only a limited aspect of ecology, this model allowed for measurement and prediction. If food were the primary selective pressure operating on social organization, then this model would account for the greatest portion of the variance in social organization.

Populations are surely shaped primarily by the most intensive selective pressures operating. Evolutionary processes produce adaptive changes which relieve the pressures, but any time a pressure is relieved, it becomes less important relative to other pressures, which in their turn now become the primary pressures operating. Populations are therefore shaped first by one pressure and then by another, not only because the pressure itself may change, but also because adaptive processes may reduce the strength of the pressures. Morphological and behavioral responses will therefore accumulate according to the history of their adequacy in dealing with past and present situations, and at any one time will be most influenced by the factors which most limit genetic fitness. The history of such adjustments lies in phylogeny, and, to the extent that these adjustments are now carried in the genetic system, we may expect certain attributes of individuals to reflect adjustments to past "primary pressures" rather than the paramount pressures of today. This is no more than to say that present behavior is a result of the interaction of the individual's responses to present situations and his genetic inheritance.

Many social scientists trained in various behaviorist traditions may find this statement too strong. Perhaps it should be restated, then, to say that since morphology is clearly at least partially an outcome of genetic coding, and since morphology includes: (1) the receptor system which delimits stimuli which an organism may receive, (2) the motor system which delimits the responses which an organism may make, and, (3) the nervous system which acts to link and modify the connections between the receptor system and the motor system, it follows that the range of behavioral response is clearly limited (rather than determined) by genetic inheritance. This syllogism need be carried but one step further to demonstrate that social organization, a reflection of the behavioral interactions of individuals, is itself subject to evolutionary processes.

In searching for the means by which evolution could modify social organization, Goss-Custard *et al.* (1972) presented a description of the theoretical mechanisms which might apply. Accepting the idea of phylogenetic input to social organization, Eisenberg *et al.* (1972) developed a model of evolutionary stages in primate grouping patterns and categorized existing knowledge in accordance with their stage model. Altmann (1974) later reviewed some of the multiple parameters which must be considered in identifying ecological pressures on social organization. Emlen and Oring (1977), in their model of avian social organization, further make the point that ecology must be considered as permissive rather than determinant in the development of certain social systems. This perhaps only emphasizes the broader principle that selection is for the tolerable and not the optimal. Hans Kummer (1971:90) reminds us of this when he states "Discussions of adaptiveness sometimes leave us with the impression that every trait observed in a species must by definition be ideally adaptive, whereas all we can say with certainty is that it must be tolerable, since it did not lead to extinction. Evolution, after all, is not sorcery."

This book, therefore, represents, not an attempt to prove how various social organizations must have necessarily resulted in response to the ecological settings in which they are found, but rather, a further refinement of our thinking about the relationships between social organization and ecology. We will indicate some of the considerations which must go into theories relating these two complexes and we will reexamine selected hypotheses relating variables within the complex of social organization to factors of ecology. Although the extent to which such hypotheses account for the available data gives them some credence, the ultimate test is always the ability of a theory to accurately predict the outcome of studies not yet undertaken. Ideally, these studies will examine: (1) the same species in multiple habitats (habitat influence), (2) multiple species in the same habitat (phylogenetic influence), (3) multiple populations of the same species in similar habitats (to test for variability of expression), and (4) multiple species in multiple habitats (to determine the range of possible adaptations). As we gain confidence in our understanding of the processes which account for the social organizations of living primates, we may gain confidence in our models of the evolution of social organization in our own species.

The order Primates is noted not only for its sociality, but also for behavioral flexibility, coupled with long periods of immaturity and biological dependence. Such a combination surely produces many responses which are acquired during the lifetime of each individual, more specifically, during the long developmental period. This acquired behavior will develop in response to the experiences accumulated in an environment which is often largely social. Humans, with the maximum potential to adjust behavior as a function of experience, should therefore be most responsive to immediate ecological situations. However, with increasing emphasis on the transmission of information from generation to generation rather than individual problem solving, it is also true that human adaptations will be retained so long as they are tolerable solutions to the problem of life (rather than optimal solutions). Traditions and acquired behavior will then follow some of the same principles as evolving genetic patterns. Variability of solution as well as conservatism in change may thus occur in both cases, although the potential time courses may differ.

In the chapters that follow we will consider:

1. What do we mean by an "adaptive" social pattern
2. What is the "quality" of the habitat
3. How selection operates on life history processes, and the implications thereof
4. The influence of random drift factors and local ecology on demographic processes
5. The significance of acquired behavior
6. A specific human social organization and its adaptation to a range of habitats
7. The significance of variability in nonhuman primate social organizations
8. The variability possible in sympatric species
9. Some correlations among population characteristics
10. Ecological, phylogenetic, distributional factors, and their interrelationships

11. Some factors which marked major shifts in human adaptation
12. The impact of major habitat changes on the development of human social organization
13. A theoretical approach showing how social behavior is organized and relates to the ecology.

The reader is then asked to decide to what extent we can have confidence in models of human social organization and its origins given the available data base and the limits of the methodology so far employed. This approach may not produce any revolutionary or exciting new theories of human social origins, but we hope that the conservative statements which we can make may endure the tests of new data somewhat longer than some of the flashier speculations which sometimes attract both serious students and the public alike.

References

Altmann, S. A. 1974. Baboons, space, time and energy. *Amer. Zool.* 14: 221–248.

Carpenter, C. R. 1939. Behavior and social relations of free ranging primates. *Sci. Mon. N. Y.* 48:319–325.

Crook, J. H., and Gartlan, J . S. 1966. Evolution of primate societies. *Nature* 210:1200–1203.

Denham, W. W. 1971. Energy relations and some basic properties of primate social organization. *Amer. Anthrop.* 73:77–95.

DeVore, I. 1963. A comparison of the ecology and behavior of monkeys and apes. Pages 301–319 in *Classification and human evolution.* Ed. S. L. Washburn. Chicago: Aldine.

Eisenberg, J. F., Muckenhirn, N. A., and Rudran, R. 1972. The relation between ecology and social structure in primates. *Science* 176: 863–874.

Emlen, S. T., and Oring, L. W. 1977. Ecology, sexual selection, and the evolution of mating systems. *Science* 197:215–223.

Goss-Custard, J. D., Dunbar, R. I. M., and Aldrich-Blake, F. P. G. 1972. Survival, mating and rearing strategies in the evolution of primate social structure. *Folia Primat.* 17:1–19.

Kummer, H. 1971. *Primate societies: Group techniques of ecological adaptation.* Chicago: Aldine-Atherton.

Montagu, M. F. A. 1943. On the relation between body size, waking activity and the origin of social life in the primates. *Nature, Lond.* 152:573–574.

Zuckerman, S. 1932. *The social life of monkeys and apes.* London: Rutledge and Kegan Paul.

Chapter 1

How Would We Know if Social Organization Were Not Adaptive?

Thelma E. Rowell

Thelma Rowell introduces us to the mood of the conference by asking basic questions concerning many premises which we have always accepted as articles of faith. We have perhaps all too readily believed that social organizations developed in response to various ecological pressures and that variability in one would produce variability in the other. We have speculated at length as to the adaptive significance of many observed aspects of social organization, but have never seriously considered whether we had grounds to reject the null hypothesis; that is, that social organization, or some social behavior might not be adaptive.

Is it possible that the recognized random processes which influence genetic changes might also influence behavioral expressions? If genetic drift is a factor to be reckoned with in small populations, then surely similar drift phenomena can operate to influence behavior in the small social groups typical of primates. Traditions may also "drift" insofar as the changes observed remain tolerable solutions to environmental demands. To prove that natural selection rather than random processes will account for the data collected will take more than an argument that the observed phenomena are "adaptive."

Ecological factors will surely influence demography and kinship proliferation; predation may exert pressures favoring group living and the size of troops, but such factors as the number of troops into which a population is divided may relate more to "tradition" than to some optimal strategy of predator avoidance or resource exploitation. The "tradition" may continue because it is a tolerable adaptation rather than an optimal one. The concept of tolerable versus optimal will be a recurrent theme in this volume, as will consideration of random processes which may be expected to produce variability in small populations.

Dr. Rowell also introduces us to the idea that different primate taxa may have different basic strategies. Whereas all primates may be regarded as K *selection strategists due to their longevity and reproductive rates, within the order some may lean more to the* r *strategist end of the spectrum. These different basic strategies may influence adaptation to particular habitats and modify the direct influence of ecological variables upon social organization.*

Dr. Rowell also reminds us that the learning capacities of primates permit considerable flexibility of behavior. Primates may have to learn a complex diet each generation, and some drift and variability is to be expected. Social organization itself consists of a series of relationships, many of which are learned. The variability of behavior introduced by the ability to modify behavior thus may result in considerable variability of social organization as individuals learn specific relationships with one another. Primate field workers, limited to what is essentially a case history approach to the study of social organization, have perhaps searched too hard for particular environmental variables which would account for differences observed between two troops studied in different areas by different observers. In fact we should expect to find variation even in the same group studied in the same location by the same observer, if studies are temporally separated. Random processes will influence demography, traditions may drift, and life history processes are expressed on a time scale much longer than any study.

Introduction

It is an article of faith among primatologists that social organization is determined by ecological influences. The view has been most persuasively expressed by Crook (1970), and it represents a very respectable Darwinian position that observed characteristics have evolved as a result of natural selection—the agents of natural selection are environmental pressures which should be identified by ecological study. In this paper I should like to take a critical look at this view and the evidence we have to support it.

First, I should like to establish some basic premises about the effect of natural selection on animals as we see them. A population cannot be uniform if natural selection is to act. There is a range of forms which are viable at a given time. Some of the forms may turn out to have a present advantage, and be in the process of increasing their representation in the population. Others may actually be at a slight disadvantage now, and are about to be eliminated, through natural selection. Still other variations will be equally viable under present conditions, but when new selection pressures begin to affect the population some of these forms may turn out to be preadapted to survive well under the new conditions, others may be unable to survive. If we assume that natural selection as envisaged by Darwin is a continuing process, we must accept that not all the characteristics we observe in a population of animals are the result of natural selection, nor can they all be optimally adapted to present conditions.

The genetic characteristics of a population may change without selection having occurred, through various manifestations of chance. Mutations may become fixed or eliminated in isolated small breeding populations by the process of genetic drift. A species may be differentiated by geographical isolation of a small population from the parent stock, followed by the accumulation of mutations eventually leading to reproductive incompatibility between the stocks. The species would differentiate even if the selection pressures acting on the two stocks were identical, and the characteristics which distinguish the new species would be within the range permitting viability in their identical niches.

We observe variation in phenotypes, not genotypes. Although all variation has a genetic base, the genotype determines the phenotype in the sense that it limits the range of possible interaction with the environment that permits survival and successful reproduction. A good example of genotype allowing wide latitude to environmental effects is human height. How tall people grow is greatly influenced by their diet and activity during growth, as well as by their genotype. Height has thus rather low heritability, in that, without knowing the way of life of parent and child, the height of the parent is only a moderately useful predicter of the height of the child. Much of the behavior of an animal is similarly greatly influenced by its interaction with its environment, by the process of learning among others; behavior remains modifiable throughout life as well, in a way that height does not. The behavior of an animal is determined by its genotype in the same way as its height is, but the heritability of particular behavioral characteristics is likely to be low.

It is clear that the environment can affect behavior, both in its interaction with individual genotypes to produce the observed phenotypes, and, through natural selection, to alter the genotypes found in a population over time. It is also clear that not all differences between populations necessarily have a heritable basis, and not all differences are necessarily the result of ecological influences. In examining the idea that environmental pressures determine social organization, then, we must work from the null hypothesis that the variation we observe in social organization is due to chance. What sorts of observations would be required to reject the null hypothesis, and do we have these data?

Two aspects of the problem can be identified. The first is how the environment might determine social organization, by what mechanism it could produce changes. The "selective tools" of the environment can also be described as the problems the animal needs to overcome in order to survive to reproduce successfully. Although infinitely varied in detail, the problems fall into three types: All animals must get food, of appropriate quantity and quality; they must avoid predation, with which we may include disease; and they must escape the physical limitations of the their habitat, which might range from temperature or humidity extremes to a shortage of sleeping places. It may sometimes be convenient to separate from this last category the problem of acquiring materials required for reproduction. In real life these problems are not independent, but will exacerbate each other (Alexander, 1974).

The second aspect of the problem is *what* will be affected by the environment. The distinction must be drawn between a labile responsiveness to the environment of the behavioral phenotype, and the production, by natural selection, of a new genotype, setting new limits to behavioral variation. It will also be necessary to distinguish clearly between primary effects of the environment on the determinants of behavior, and secondary effects, where, for example, a change in the animal occurs as a result of natural selecton which has, as a side effect, repercussions on the animal's behavior, producing changes there which themselves are selectively neutral. Social organization, being a high-order, integrative phenomenon, might be expected to be especially susceptible to such secondary effects.

The first requirement is to describe social organizations in terms that will allow the detection of differences and changes. Here I should like to use a dichotomy I have suggested elsewhere (Rowell, 1972a), and separate social *structure*, meaning such measures as population density, group size, and demographic character, from social *organization* which I should like to limit to the pattern of interactions between individuals, a description of behavior.

Ideally, a hypothesis would be testable by experiment. Phenotypic lability of behavior in response to environment can indeed be tested experimentally, but the effects of selection on long-lived animals can only be inferred. The method must be comparative, and the basis of comparison must be a large enough sample of populations to be amenable to statistical treatment. A single association of a behavioral trait with an environmental condition may be entertaining and thought-provoking, but it may also be due to chance.

My hope is that by firmly keeping before us the question "how would the environment affect what," we might generate testable hypotheses which would allow us to reject the null hypothesis of random drift in social behavior.

Group Living

All mammals which have been studied have a social organization, in that individuals recognize other individuals and show differentiated patterns of interaction with them. Most primates are gregarious and move around in groups larger and more permanent than the mother-offspring association that lactation imposes as minimal sociality on mammals. Discussion of primate social organization has concentrated on interaction within groups, with some interest on interaction between groups. We should also note that the equivalence of the group, a unit of social organization, to the deme, the population within which interbreeding occurs, is unlikely. The extent of the deme needs to be established in each case and the information is essential to any evolutionary interpretation of social behavior.

Our first question, then, must be why do monkeys live in groups. It is obvious to anyone watching monkeys that it is subjectively very important to the individual to be part of a group. There is a measurable "social drive" in that a monkey will work a lever or undergo considerable discomfort to join or stay in a group. We have interpreted this to imply a strong selection pressure for group living among primates, but I do not think the implication has been subject to rigorous analysis for any primate, in the way it was recently done for bank swallows (*Riparia riparia*) by Hoogland and Sherman (1976) for example. In that study experimental manipulations and comparisons between 54 colonies of different sizes were used to test hypotheses about the advantageous and disadvantageous consequences of colonial nesting, and the probable selection pressure for group living was identified as predation on eggs and young.

Like bank swallows, primates rarely hunt cooperatively. If food is evenly distributed, the individual should be better off foraging alone. If food is in large clumps, group foraging is not deleterious: In the special case of widely separated, difficult-to-find, abundant sources of food, group foraging may be advantageous in that it reduces the risk of finding no food at all, rather than maximizing feeding rate (Thompson, Vertinsky, and Krebs, 1974). This condition is difficult to match to most known primate feeding patterns, but may be relevant for those primates which exploit poorer habitats, such as the hamadryas baboon (*Papio hamadryas*, Sigg, personal communication). A case can be made for the primate group as an educational institution in which learned information about feeding in an exceedingly complex environment can be shared to general advantage (Rowell, 1975). This could provide a food-related mechanism for extending the family group into adulthood, but is difficult to stretch to the maintenance of larger groups.

Protection from predation seems to me the most probable advantage of group living, although very little is known about predation on primates. In slow-reproducing animals loss of a very few individuals to predators will be sufficient to control population. Bank swallows protect themselves from predators by vigilance and by mobbing; fieldworkers, whose behavior closely resembles that of a stalking predator, are well aware of the effectiveness of both these behavior patterns in monkeys. Mobbing of other predators has been described in talapoins (*Miopithecus talapoin*) (Rowell, 1972b) and vervets (Lancaster, 1971 and film).

To assess the value of group living, we would like to compare the behavior of animals living in groups with solitary individuals of the same species. Nearly always, solitary monkeys of gregarious species are adult males. Since adult males are larger and more experienced than the average monkey in a troop, the observation that only they seem to be able to survive alone perhaps in itself supports the view of the group as a protection against predators. The adult male is likely to be at the upper end of the size range of prey that a regular monkey predator can manage. Slatkin and Hausfater (1976) have described the behavior of one solitary male baboon (*Papio cynocephalus*) from the time he left a group until his return to society. They noted that he fed well, confirming the idea that the group is not necessary to foraging. The education hypothesis is not, however, excluded by the foraging competence of solitary adult males, since they may be presumed to have already profited from the system.

Social Structure

Group Size

The environment must determine an upper limit on population density, which is equivalent to determining population size if a limited area (island) of habitat is available. The proximal limiting factor of a plant-eating species is likely to be predation in a stable environment—and the majority of primate species are native to tropical forest, the most stable of terrestrial ecotypes. Food supply may become critical when the environment changes, as seems to have been the case at Amboseli for baboon and vervet (*Cercopithecus aethiops*) populations (Struhsaker, 1976). Two studies have suggested water as a limiting factor at the edge of primate range, in these cases acting as a density independent population control. At Gilgil in Kenya, a water supply system put in for cattle allowed baboons to expand into new areas and their population has rapidly increased (Harding, 1976); and at the Waza Game reserve in Cameroon waters holes dried up in the drought, and baboons and other species died in large numbers (Gartlan, 1975).

Population density must interact with group size in a general way. We may imagine an optimal range of group sizes, with an upper limit set by the number of individuals that can move around together maintaining contact and find enough food. This in turn must be determined by aspects of the structure of the environment, such as its density, or opacity, and the size and spacing of food sources. The lower limit would be set by whatever promotes gregariousness, as discussed above. Within this range, we should assume that group size has no effect on the survival and reproduction of group members unless we can demonstrate otherwise.

There has been much discussion of the possible mechanisms for group splitting. In fact it has rarely been observed other than in rapidly expanding artificially maintained populations, and combinations of groups have been reported even more rarely (Altmann and Altmann, this volume, Chapter 3). On the contrary, the evidence, especially from expanding populations (Gilgil) and contracting ones (Amboseli) is that groups can maintain their identity over large size ranges. Observations of different-sized adjacent troops in more or less homogeneous environments also suggest wide latitude in viable group size. If there are environmental pressures towards an optimal group size, they are not strong enough to counteract tradition in the short term and must produce their effect very slowly.

Some observations, all made at the extreme edge of habitable range, show that it is possible for environment to limit quite strictly some aspects of group size: Aldrich-Blake *et al.* (1971) found that, at the edge of their range in Ethiopia, anubis baboon (*Papio anubis*) troops were of normal size (44) morning and evening, but broke up into small foraging parties during the day. This recalls the group structure of hamadryas baboons in similar or still poorer habitat. Kummer and Kurt (1963), who described the separation of hamadryas baboons into small foraging parties during the day, suggested that the sparse food resources limited foraging group size, and this explanation was also accepted by Aldrich-Blake *et al.* (1971) who pointed out, however, that the composition of the anubis foraging parties was different from that of the hamadryas parties. Hall (1963) described baboons stranded on islands in the flooding Kariba dam as foraging through their nearly exhausted habitat in small groups or singly.

These observations can all be taken to suggest that, for baboons, the large group is not the most efficient foraging unit: in terms of feeding behavior, it is something of a luxury to be indulged in, in relatively affluent circumstances. It is possible that the foraging disadvantage of the large group is counteracted by increased protection from predation or some other advantage. It is also possible that the large group is a byproduct of a period of good survival and recruitment coupled with a tendency for families to stay together whose survival value (educational or protective) is fulfilled at quite a small group size. To test these two alternative hypotheses we require data on rate and cause of mortality in troops of different sizes. Since neither hypothesis predicts that food would be a significant variable, troops from a fairly wide range of nonextreme habitats could usefully be compared.

On present evidence, then, the best guess is that while the overall population density is environmentally controlled, the number of troops into which a local population is divided is largely a matter of tradition. We would therefore expect *average* group size in a local population to vary as a function of overall density, with the variance being rather large. This again is a hypothesis that could be tested relatively easily.

Demography

The remaining aspect of social structure is the demography of the population. Clearly this will be affected by environmental effects which produce age or sex specific mortality rates. Southwick and Siddiqi (1968) described highly eccentric population structures in rhesus macaque (*Macaca mulatta*) groups subjected to heavy trapping pressure which selectively removed juveniles. Dittus's study (1975) of a toque macaque (*Macaca sinica*) population provides the most complete analysis available of the effects of age-sex specific mortality rates. He found that female mortality exceeded male from birth until old age. Males showed a sharp increase in death rate as they approached maturity, and their mortality remained higher than that of the females throughout adult life. Although all males left their natal troop as they reached maturity this had little effect on troop composition in the long run since they were replaced by young males from neighboring troops. Sex ratios thus varied with age; overall, recruitment exactly balanced losses over the three-year study. Compared with the demography of expanding populations of other macaques (provisioned populations), Dittus found that the main difference was in the survival rate of juvenile females, though the birth rate was also somewhat higher in the expanding populations.

Baboon females show flexibility in their breeding, with birth intervals reported from different populations varying from just over a year to over two years (Ransom and Rowell, 1972). Similarly, whereas age at maturity in a rich habitat seemed to be the same as that observed in captivity, with first conception occurring in the fourth year, females in the deteriorating environment of Amboseli have been reaching six or seven years before entering the breeding population (Altmann *et al.*, 1977). I am assuming that this variation represents a phenotypic response to the quality of available nutrition rather than a genetically based difference in the population, since females from a variety of habitats undergo puberty at about 3½ years when in captivity. An analysis of breeding records, at for instance, the Southwest Foundation colonies in Texas, where baboons of different origin have been bred for many years, should allow any genetic differences in the populations to be detected.

Thus we have evidence to infer that some primate species will show changes in their demography in response to variation in their environment, but that the demographic structure observed at a particular field site is unlikely to be a fixed species characteristic.

If the known flexibility of baboons were general, a problem would be raised whenever we wished to compare two species with differing social structure that also have different habitats. I was recently able to approach this problem by comparing the breeding records of seven species of African monkeys at the Kenyan Institute of African Primatology at Tigoni, near Limuru. All the species had been housed and fed similarly and had been allowed to breed in small family groups, relatively undisturbed, for up to 15 years. In this situation, the actual figures obtained are possibly unrepresentative of the animals' behavior in the wild. Differences between the species, however, presumably represent the interaction of different genotypes with this particular environment (Rowell and Richards, 1979). Data were derived from four *Cercopithecus* species, two *Cercocebus*, and *Colobus guereza*. We studied timing of births in the lifetime of females of each species, considering age at first birth and birth intervals, and also such indications as we had on longevity; we also considered the timing of births relative to climatic changes (the animals were housed outdoors, very near the equator), and also their timing relative to each other, looking for possible effects of social facilitation. There were significant differences between the birth intervals shown by the different species, and there were also large differences in age at first conception.

There was a general correlation between fast maturation and short birth intervals. Two species, the vervet (*Cercopithecus aethiops*) and the patas (*Erythrocebus patas*), emerged as fast breeders, much faster than the laboratory rhesus monkey, the other two guenons the Sykes and DeBrazza monkeys (*C. mitis* and *C. neglectus*). The black and grey mangabeys (*Cercocebus albigena* and *C. torquatus*) were much slower at both maturation and having successive infants than the laboratory rhesus. The colobus matured slowly, but once adult they gave birth annually. The fast breeding vervet seemed to be short-lived: no known-aged females survived beyond 12. In contrast, the slower DeBrazzas and Sykes monkeys were still breeding in their teens, and the senior female mangabey was at least 27 and still breeding regularly. Within a primate context, one might think in terms of *r* and *K* selected species, and as would be expected, the *r* selected species are native to the more varying and unpredictable habitats: vervets and patas live mainly at the edge of forest and in the open grasslands, respectively, while the other species are all forest dwellers. It is reasonable to guess that different genotypes for reproductive strategies have evolved in response to different sets of environmental pressures.

Two species, the vervets and the Sykes monkeys, showed significant correlation of conceptions with climate: vervets conceived in dry months, Sykes during periods of diminishing rainfall. Both species are reported to breed seasonally in the wild. The rest bred aseasonally: breeding is reported to be aseasonal in the wild in colobus and the black mangabey; there is no information of the grey mangabey nor on DeBrazza's. The patas is known to be highly seasonal in the wild, but at Tigoni bred all the year around. We suggested that the breeding season in wild patas is determined by a direct response to changes in food supply—conceptions occur a few weeks after the onset of the rains, at the same time as the new growth of grass. Such a mechanism would allow patas to avoid breeding in a bad year, when low rainfall in the wet season would lead to an inadequate supply of food in the following dry season. In conditions of continuously abundant food, the same mechanism would allow continuous breeding. Seasonal breeding in vervets and Sykes is apparently not determined by changes in food supply, but cued directly by climatic events. These climatic events had no effect on the breeding of other species. It seems that the species differ in their mechanisms of timing reproduction in relation to environmental fluctuations.

Births of vervets occurred in clusters; births of colobus were dispersed in time, while those of the other species occurred at random intervals. These findings, though they need to be confirmed with larger samples over longer periods, suggest that social facilitation of sexual activity may help to limit breeding seasons in some species but not in others; in some conditions social inhibition of sexual activity seems to be possible. Another explanation would be that social facilitation (or inhibition) have different thresholds or require different conditions in different species, and that these conditions were only fulfilled at Tigoni for two species.

In summary, then, we have evidence of both phenotypic and genotypic variation in reproductive rate and timing among African monkeys. Phenotypic variation may be possible for some species but not others; genotypic differences are surprisingly large within one closely related genus, and seem to be related to habitat differences. Thus we can see how the environment could affect demographic characteristics in both the short and the long term, other than through differential mortality rates. It is easy to imagine ways in which selection might act to produce species differences in reproductive strategies by acting on the reproductive success of the individual female. If the selection pressure is still in action, it could be detected by an examination of individual reproductive histories. Females whose reproductive history differed greatly from the modal pattern would be expected to leave fewer grandchildren than those close to the mode. Adequate data might be available to test some of the simpler hypotheses about such selection pressure from the records of the provisioned macaque troops in Japan and Puerto Rico, where breeding records have been maintained on large numbers of animals for many years.

Social Organization

There remains the question of how the environment might affect social organization in the narrower sense as defined on page 5. We are now looking for something beyond constraints on group size and effects on demography. We are asking for patterns of social interaction among individuals that will enhance the chances of the individuals involved in getting food and avoiding predators in the short term, and help them to leave more offspring in the long term (we may add the possibility that they may be enabled to assist close relatives to the same ends). Some interaction patterns, such as the protection of infants by mothers and allowing them to share the mother's food, have obvious immediate rewards and are not confined to gregarious mammals.

We might regard the development of matrifocal clans as a paedomorphic extension of the advantages of parental care. Can we go beyond this banal level of description?

If the behavior of each individual in a group is generally predictable by its fellows, communication about the environment can be achieved by very slight changes in behavior which are perceived as anomalous by comparison with that which is expected. In contrast to ritualized signals, such as alarm calls, such communication involves no extra expenditure of energy and is unlikely to be detected by the outside observer (such as a predator) which is not so finely attuned to the expectation of behavior, thus carrying minimal additional risk. For example, a talapoin monkey whose attention is caught by the blink of a resting owl stops eating and peers at the bird. These actions, in the context of a foraging group, are sufficient to alert its neighbor and direct its attention to the predator, causing similar changes in behavior. Soon the whole group has been alerted and begins to mob the owl, but the first stages of communication leading to this group cooperative activity have been achieved without any overt signals being exchanged. Similarly, a sudden movement into hiding announces the arrival of a hunting hawk as effectively as an alarm call, and a change in feeding behavior directs the attention of the group to a new prey item. All the postulated advantages of group living will be enhanced if the general predictability of each individual's behavior is increased. Since the movements of one monkey are much influenced by the behavior of others, the predictability of his behavior is much enhanced if he and his neighbors are conforming to some generally accepted rules of conduct. For this reason, there must be a selective advantage to developing and observing such rules. But do the actual details of the rules matter?

We have the common observation that species behave differently—indeed it is the "special flavor" of each species' behavior that provides the pleasure of comparative studies. Part of the difference lies in the different signals used, or the different frequency of similar signals in each species. Such differences need not result in differences in social organization at all. But there are also differences in "tone" that make one feel justified in describing species with such adjectives as quarrelsome or relaxed; differences in the roles of age/sex classes; differences in the amount and predictability of subordinacy observed—all of which derive from quantifiable differences in the pattern of interaction between individuals.

First, some differences have been shown to be directly due to the immediate environment: a single troop can be induced to change its behavior immediately by altering its circumstances; for example, by changing the size of its living space (Southwick, 1967). Gartlan found very different social organization in troops of vervet monkeys living in different habitats (Gartlan and Brain, 1968), and these differences are probably again the direct result of interaction with these different environments.

Second, differences in social organizations between troops of the same or different species could be a direct result of differences in their demographic structure and kinship structure, which might in turn be the result of either immediate environmental pressures or of genetic differences in reproductive rates and life histories. The possibility has not generally been considered, mainly I think, because primatologists always have an inadequate sample of groups to study and because most studies are of short duration relative to the generation time of a primate. Altmann and Altmann (this volume, Chapter 3), explore the consequences for social behavior of demographic variation. Their attention has been drawn to these effects by working with a baboon population through a period considerably longer than their generation time; my own enthusiasm for the importance of demographic effects derives from keeping monkeys in captivity, and observing the great changes in social organization which occur as a number of strange adults are placed together, settle down, acquire infants, then a cohort of juveniles, and at last a second generation. (The change when a group acquires a playgroup of juveniles is particularly striking.) The most obvious and quantifiable change is a shift from a social organization which can be largely characterized by some simple hierarchical model to one in which a hierarchy becomes of little use as a description, being replaced by a system of kinship and cohort alliances.

Third, differences in social organization between species may arise by a process of random drift, either cultural or genetic, if the important factor is simply to have regularity to provide predictable behavior, and not the actual rules themselves. To distinguish between differences which are truly adaptive and those produced by random drift would be extremely difficult, and it therefore behooves us not to accept all differences as having selective advantage without bearing the alternative in mind. It is even very difficult to find out whether particular differences are cultural or genetic. Kaufman (1976) has been trying to find out whether differences in mothering behavior between pigtail and bonnet macaques are simply due to early experience with a conspecific mother or result from inborn differences in contact behavior; the technical problems are almost overwhelming.

With these alternative hypotheses in mind, we may look again more critically at some of the widely quoted examples of adaptiveness of social behavior. A useful line of reasoning begins with observed differences between troops of the same species. For example, in some troops of Japanese macaques, (*Macaca fuscata*) males show protective behavior towards juveniles which have been replaced at the breast by new infants when they are a year old. It is easy to invent a possible adaptive significance for such behavior, yet in other troops males show parental behavior rarely or not at all (Itani, 1959). No difference in survival of juveniles is reported for these troops, and Itani described it as being an example of cultural drift with no adaptive significance under present conditions. Roughly similar interactions between adult males and infants have been reported for various other populations of macaques and baboons (Mitchell and Brandt, 1972, review) and a selective advantage to either infant or male has often been suggested by the describer. By analogy with the paternal Japanese macaques, cultural drift would seem at least equally probable. If a selective advantage is postulated, the argument is incomplete without some explanation as to why the trait should be advantageous to this species and not to a closely related species which lacks it.

The dominance-subordinacy relationship has often been cited as having adaptive significance. Insofar as such a relationship represents an increase in the predictability of behavior of group-members, we have seen that it must increase the advantageousness of group living; however, the generally observed correlation between clearcut hierarchies and high frequencies of aggression somewhat counteracts that idea since fighting is conspicuous and uses a lot of energy. The relationship has also been assumed to confer a selective advantage on the dominant animal because it gains priority of access to limited resources—this assumption usually implies a heritable component to dominance. Outside the most restricted laboratory conditions, it is not clear that high-ranking animals do inevitably obtain priority of access: for example, Duvall, Bernstein, and Gordon (1976) found that male social rank in a corralled rhesus monkey group was not correlated with the number of infants sired. Hausfater (1975) did find a correlation between rank and frequency of copulation in wild baboon groups, but rank-order changed so frequently that it seems unlikely that individuals had characteristic lifetime dominance ratings—dominance was not a heritable trait. Some claims to demonstrate an effect of rank on reproductive success do not adequately control for correlated variables; for example, age and rank are inextricably entwined in adult female rhesus monkeys studied by Drickamer (1974). There has been a great deal of confused thinking in this area, and it is only recently that pioneer studies like those of Duvall, Drickamer, and Hausfater have provided data to test the assumptions.

A more likely candidate for an environmentally determined social organization is the pattern of interactions that leads to young males leaving the breeding group and forming all-male groups. The all-male groups of geladas (*Theropithecus gelada*) forage differently from the breeding groups at some times of the year and Crook (1966) suggested that competition was thereby reduced to the advantage of breeding females; patas breeding groups have priority at waterholes over all-male groups (though this could be because they are simply more numerous rather than because of their composition) and Gartlan (1974) suggested this might be critical for survival of breeding females in drought years. In both cases the supposed advantage of the breeding groups is only inferred, and on the other hand, the all-male groups of langurs (*Presybtis spp.*) (Sugiyama, Yoshiba, and Pathasarathy, 1965) use the same resources as the breeding groups, on a different time schedule. Perhaps we should think rather carefully about exactly what behavior we think is being selected; it is difficult to imagine the young males putting themselves at a disadvantage, or even parents driving their own offspring away and so reducing their chances of survival. It might be better to think of young males being able to exploit alternative resources if they can take advantage of their greater mobility by deserting the less active mothers with infants. If we could develop a more precise formulation of the supposedly adaptive behavior, we could then propose hypotheses generating predictions which could be tested by further observation. At present we have no way of rejecting the null hypothesis (of no environmental selection pressure for all-male groups).

While primary effects of the environment on the interaction patterns between individuals may be difficult to imagine, there is also the possibility of secondary effects, the by-product of a more direct interaction with the environment. One source of such secondary effects is the inevitable sloppiness in selecting for a polygenic effect, for which the experience of sheep breeders provides a useful model. Sheep are bred to thrive under a variety of circumstances from foraging on poor rangeland to being handfed in folds. A shepherd in the Scottish highlands likes his ewes to produce one lamb a year; twins are unlikely to survive the harsh spring, and the effort to rear them severely weakens the ewe. On gentler pastures, twins can be reared easily, but triplets may present problems, while in intensive folds, triplets or even larger litters are economical. The shepherd tries to select for the appropriate level of fertility, but is never entirely successful. Most of the highland sheep produce one lamb, but there is always a significant proportion of yeld (infertile) animals, and a few which twin. The pasture ewes mainly produce twins, yeld ewes are rare, but single lambs and triplets are fairly common. In the folded breeds the mean litter size is higher, but again many births produce litters just larger or just smaller than the optimum.

The following two examples of observed behavior might have been produced by such a selection for an optimum of a polygenic character. In both cases a by-product effect on social interaction is probably neutral under normal circumstances, but becomes potentially pathological in captivity.

1. Adult males of most monkey species will move towards a potential predator and feign an attack or even attack it, while females and juveniles withdraw. In some, but not all, species adult males will also behave aggressively towards smaller conspecifics, and will thus dominate them, an effect especially noticeable in captivity. If aggressiveness towards predators is under the same control as aggressiveness towards conspecifics, it could be worth putting up with a domineering male, as it were, in return for effective defense. The male's intraspecific behavior is not itself advantageous, but is a by-product of a more efficient defense which is not disadvantageous under normal circumstances, although it may be disruptive in captivity (Rowell, 1974).
2. Female patas monkeys have a very strong tendency to pick up small infants they see near them. In their native habitat, where they normally forage scattered over a wide area, we can imagine that this behavior might function to rescue closely related infants misplaced during an alarm, at little

cost to the rescuer. Females are often observed to be carrying infants which do not belong to them. In captivity, perhaps because the mother is always in sight of her groupmates, this urge to pick up infants results in a great deal of baby-stealing which is clearly very distressful to the infant. In small gang cages the baby-stealing can reach such a high level that mothers are unable to regain their infants, which die of starvation. (Field observations, Gartlan, personal communication; captive studies, Chism, 1978; and Taub, personal communication.)

A Working Example

Finally, I should like to present an outline of some recent field observations of my own, as an example of one relatively simple situation for which a theory of environmental influences determining differences in social organization would have to be able to provide testable explanatory hypotheses.

Last year I spent six weeks watching monkeys in the Kakamega Forest Nature Reserve, in western Kenya. The study area was a small part of the forest about 1.5 by .75 km bounded on one side by grassland and the other by cultivated land, with the two ends contiguous with more forest. A rather uneven grid of paths spaced at 50 by 200 m is maintained in the forest, which is much used as a research area by Kenyan biologists. The forest is a montane semideciduous type which was selectively felled in the 1940s, so that visibility is unusually good for forest monkey habitats. The monkeys are unmolested and somewhat accustomed to people with binoculars, so that once I had learned some of their habitual routes it was fairly easy to stay with a selected group for six hours or more in a day.

The area was the home of 80 to 100 black and white colobus (*Colobus polykomos*), about the same number of blue monkeys (*Cercopithecus mitis stuhlmanni*) and 40 redtail monkeys (*Cercopithecus ascanius*).

The colobus monkeys lived in groups of up to 13, typically including an adult male, 3 adult females most of which were carrying white or transitional infants during the study period, and some juveniles of both sexes. Sometimes these groups seemed to break up into subgroups, and they occasionally approached a neighboring troop to within 50 m or so, probably within sight of each other. Roaring by one group elicited roaring by neighbors, as has been described by Marler (1969). A troop could usually be found within an area about 100 m^2 if one searched thoroughly enough. Clearly, however, the colobus participated in social interactions beyond the 13-animal troop.

The redtails lived in two troops, of 22 and 18 animals. The 22-animal redtail troop was the main focus of the study. It included one very large adult male and two younger males, seven adult females one of which had apparently never bred and one which was probably pregnant for the first time. There was a cohort of five juveniles under one year old; at the very start of the study one of them was still occasionally carried by its mother. Copulations were seen at the beginning of the study (in early November), and this was probably the end of a fairly short breeding season. The redtails moved about the whole area in a fairly compact group, though on a few occasions they divided into two subgroups. On two occasions the two troops met—one went out of its way to find the other—and there were prolonged exchanges of threats and lunges, mainly by the adult females, backed by juveniles. The adult males sat to one side. The encounters occurred in the same area, which possibly was a territorial boundary. However, after an evening encounter the troops slept close together; it is probable that a female and her infant joined the other troop for a while after one encounter, and the young adult males may also have spent part of a day with the "wrong" troop.

The blue monkeys had the most enigmatic social organization. They could be found in groups of from 6 to 60 animals; sometimes, but not always, smaller groups seemed to coalesce in the evening and break up again in the morning. The composition of the parties was not stable, and larger parties seemed to include a higher proportion of small juveniles than smaller ones. Few adult females had infants being carried or at foot, and there was no sign of like-aged cohorts of juveniles, which might indicate the existence of a sharply defined breeding season.

The two guenons would sometimes travel together for several hours, but they did not form stable associations as Gautier-Hion and Gautier (1974) report for their sibling species (*C. nictitans* and *C. cephus*) in Gabon. Even the colobus would move a short way in the same direction when a large group of guenons passed them. Infants of all three species played together, though since the colobus were much larger than redtails of the same age the latter seemed rather overwhelmed and usually left soon. Blues could supplant the smaller redtails but interactions between the adults were seldom seen. When an adult male blue made his loud call, the redtail male in the same association always followed immediately with his loud call, but the redtails did not respond to loud calls from blues they were not accompanying.

Both blues and redtails ate mainly fruits, the same fruits. The diets largely overlapped, though redtails spent more time hunting for arthropods and the blues were more often seen to eat young leaves. The only predator seemed to be a monkey-eating eagle, which was seen to take a juvenile colobus and make unsuccessful attacks on both blues and redtails. The species differed in size: an adult male redtail weighs perhaps 5 kg, an adult male blue about 9 kg, and an adult male colobus 13 kg. They differed in their population density, demography, and reproductive strategies. The composition and stability of subgroups differed, and this was perhaps the most obvious and easily measured difference between them in social organization.

It is the nature of our discipline to try to suggest adaptive significance to the observed differences, and we can invent explanations in those terms, even on the basis of the very incomplete data I have just presented. If this is to be more than an entertaining parlor game, we must formulate such explanations as hypotheses which can be tested and potentially disproved by further observations. The null hypothesis which must be accepted, until an alternative can be demonstrated to fit the observations better, is that the differences in social organization are the result of random drift, by which I understand that each species could use the same food sources and avoid predation as successfully if it were organized like one of the others. In the Kakamega forest situation, it might be possible to use the variation of behavior within a species over time and between groups, to test hypotheses about adaptation. The general prediction would be, in effect, that blue monkeys will be less successful when they behave more like redtails, and vice versa. Useful data might be differences in feeding rates or predator attack-frequency in subgroups of different sizes; it would be significant if grouping patterns were changed following a series of predator attacks. Limiting factors might act only briefly during the annual cycle of reproduction or food availability so that variation in grouping might be sharply reduced for short periods.

Considering these possibilities, it is clear that even in a relatively simple situation such as I have described, sustained and detailed investigation is required before the null hypothesis can be confidently rejected. No wonder that the rigor of scientific method is so often abandoned in favor of the theoretician's armchair. It seems to me we are cursed with our own mental agility: even as I write I can envisage an adaptive explanation for any variation I can imagine I might observe, or for changes in either direction.

References

Aldrich-Blake, F. P. G., Bunn, T. K., Dunbar, R. I. M., and Headley, P. M. 1971. Observations on baboons, *Papio anubis*, in an arid region in Ethiopia. *Folia Primat.* 15:1–35.

Alexander, R. D. 1974. The evolution of social behavior. *Ann. Rev. Ecol. Syst.* 5:325–383.

Altmann, J., Altmann, S. A., Hausfater, G., and McCluskey, S. 1977. Life history of yellow baboons: Physical development, reproductive parameters, and infant mortality. *Primates* 18:315–330.

Chism, J. 1978. Relations between patas infants and group members other than the mother. Pages 173–176 in *Recent advances in primatology*, vol. 1. Eds. D. J . Chivers and J. Herbert. New York: Academic Press.

Crook, J . H . 1966. Gelada baboon herd structure and movement: A comparative report. *Symp. Zool. Soc. Lond.* 18:237–258.

Crook, J. H . 1970. The socio-ecology of primates. Pages 103–166 in *Social behaviour in birds and mammals*. Ed. J. H . Crook. London: Academic Press.

Dittus, W. P. J. 1975. Population dynamics of the toque monkey (*Macaca sinica*). Pages 125–152 in *Socioecology and psychology of primates*. Ed. R. Tuttle. The Hague: Mouton.

Drickamer, L. C. 1974. A ten-year summary of reproductive data for free-ranging *Macaca mulatta*. *Folia Primat.* 21:61–80.

Duvall, S. W., Bernstein, I. S., and Gordon, I. 1976. Paternity and status in a rhesus monkey group. *J. Reprod. Fertil.* 47:25–31.

Gartlan, J . S. 1975. Adaptive aspects of social structure in *Erythrocebus patas*. Pages 161–171 in *Proc. symp. 5th cong. int'l. primat. soc. (1974)*. Eds. S. Kondo, M. Kawai, A. Ehara, and S. Kawamura. Tokyo: Japan Science Press.

Gartlan, J. S., and Brain, C. K. 1968. Ecology and social variability in *Cercopithecus aethiops* and *C. mitis*. Pages 253–292 in *Primates: Studies in adaptation and variability*. Ed. P. Jay. New York: Holt, Rinehart, Winston.

Gautier-Hion, A., and Gautier, J. P. 1974. Les associations polyspecifiques de Cercopitheques du Plateau de M'passa (Gabon). *Folia Primat.* 22:134–177.

Hall, K. R. L. 1963. Variations in the ecology of the chacma baboon, *Papio ursinus*. *Symp. Zoo. Soc. Lond.* 10:1–28.

Harding, R. S. O. 1976. Ranging patterns of a troop of baboons (*Papio anubis*) in Kenya. *Folia Primat.* 25:143–185.

Hausfater, G. 1975. Dominance and reproduction in baboons (*Papio cynocephalus*). *Contrib. Primatol.* 7:1–150.

Hoogland, J . L., and Sherman, P. W. 1976. Advantages and disadvantages of bank swallow (*Riparia riparia*) coloniality. *Ecol. Monogr.* 46:33–58.

Itani, J. 1959. Paternal care in the wild Japanese monkey (*Macaca fuscata fuscata*). *Primates* 2:61–93.

Kaufman, I. C. 1976. Learning what comes naturally: The role of life experience in the establishment of species typical behaviour. Pages 37–50 in *Socialization as cultural communication*. Ed. T. Schwartz. Berkeley: University of California Press.

Kummer, H., and Kurt, F. 1963. Social units of a free-living population of hamadryas baboons. *Folia Primat.* 1:4–19.

Lancaster, J. B. 1971. Play-mothering: The relations between juvenile females and young infants among free-ranging vervet monkeys. *Folia Primat.* 15:161–183.

Marler, P. 1969. *Colobus guereza*: Territoriality and group composition. *Science* 163:93–95.

Mitchell, G., and Brandt, E. M. 1972. Paternal behaviour in primates. Pages 173–206 in *Primate socialization*. Ed. F. Poirier. New York: Random House.

Ransom, T. W., and Rowell, T. E. 1972. Early social development of feral baboons. Pages 105–144 in *Primate socialization*. Ed. F. Poirier. New York: Random House.

Rowell, T. E. 1972a. *Social behaviour of monkeys*. London: Penguin.

Rowell, T. E. 1972b. Toward a natural history of the talapoin monkey in Cameroon. *Ann. Fac. Sci. Cameroun* 10:121–134.

Rowell, T. E. 1974. Contrasting male roles in different species of nonhuman primates. *Arch. Sex. Behav.* 3:143–149.

Rowell, T. E. 1975. Growing up in a monkey group. *Ethos* 3:113–128.

Rowell, T. E., and Richards, S. I. (In press). Reproductive strategies of some African monkeys. *J. Mammal.* 60:58–69.

Slatkin, M., and Hausfater, G. 1976. A note on the activities of a solitary male baboon. *Primates* 17:311–322.

Southwick, C. H . 1967. An experimental study of intragroup agonistic behaviour in rhesus monkeys (*Macaca mulatta*). *Behaviour* 28: 182–209.

Southwick, C. H., and Siddiqi, M. R. 1968. Population of rhesus monkeys in villages and towns of Northern India, 1959–65. *J. Anim. Ecol.* 37:199–204.

Struhsaker, T. S. 1976. Further decline in numbers of Amboseli vervet monkeys. *Biotropica* 8:211–214.

Sugiyama, Y., Yoshiba, K., and Pathasarathy, M. D. 1965. Home range mating season, male group and inter-group relations in Hanuman langurs (*Presbytis entellus*). *Primates* 6:73–106.

Thompson, W. A., Vertinsky, I., and Krebs, J . R. 1974. The survival value of flocking in birds—a simulation model. *J. Anim. Ecol.* 43: 785–820.

Chapter 2

Significant Parameters of Environmental Quality for Nonhuman Primates

François Bourlière

Given that the question of "adaptiveness" of a particular social system centers on the identification of key environmental variables, Bourlière then invites us to examine the qualities of the ecology. We may be able to identify minimum conditions of food availability, temperature and cover, but the quality of an environment must be judged on more than the bare essentials. The energy budget of an individual should be considered both in terms of minimum requirements and maximum potential use. The quality of an environment can, therefore, be examined quantitatively.

Bourlière describes the econiche as the situation in which a species can reproduce indefinitely, but then indicates that individuals select habitats on the basis of qualities beyond the bare minimums. Behavior is viewed as being oriented toward energy savings and relates to the abundance and distribution of food in the habitat. Among animals, progressive behavioral refinement increases abilities for active locomotion, and this increases the flexibility of the organism and the number of habitats in which it can make a living. By modification of its life-style, a species thus has the possibility of occupying an alternative habitat; thus, we may find populations of the same species exhibiting different life-styles in different habitats. The ecology may be expected to directly influence diet, to produce changes in demography and, through these efforts, influence social behavior.

Evaluation of the habitat will, therefore, be no easy task, for the life-style of the individual will define the quality of the habitat. We nonetheless may strive for objective criteria in assessing the quality of a habitat for a particular individual and the social unit in which it lives.

Each environment must provide each living organism with two basic requirements: first, the proper set of physical conditions required for the adequate functioning of the category of organisms concerned (for example, a range of temperatures compatible with life, adequate support and cover, and so on) and, second, the necessary amount of assimilable energy, vitamins, minerals, and trace elements to keep them alive and able to reproduce. However, once these requisites are met other conditions are usually needed to make the environment optimal for a species or a "guild" of species sharing the same life-style. Taken as a whole, these physical and dietary requirements make up the "ecological niche" here defined, following Hutchinson (1957), as an *n*-dimensional hypervolume in the environment, the perimeter of which circumscribes the area in which a species can reproduce indefinitely. The parameters of a niche are many, and present-day ecology is largely concerned with their identification and measurement. Furthermore, the relative importance and ecological significance of environmental variables vary from one taxonomic group, and even species, to another. For instance, to a frog or a toad the seasonal availability of a body of water in which to lay its eggs and where the development of its tadpoles will take place is far more important than the floristic structure of the plant cover. Conversely, the right kind of Eucalypt tree is more vital for a Koala than the availability of water; out of the 600 species of genus *Eucalyptus* growing in Australia, no more than 8 can apparently be used as food by this folivorous marsupial which is able to spend its whole life without drinking.

It is therefore extremely difficult to assess the quality of a given environment,[1] all the more so when one deals with animals whose habits are poorly known. For primates, another difficulty arises. Because of our biological affinities with monkeys and apes, we are far too often prone to extend our own preferences and value judgments to the other members of this order. Yet we should not forget how difficult it is for us to assess the quality of our own living conditions when cultural and individual factors play such a great role in environmental perception. In many cases we cannot do better than to define an optimal environment in terms of public health, leaving aside all behavioral variables.

Any attempt to evaluate the quality of a nonhuman primate's environment must consequently be preceded by (1) the choice of objective criteria of environmental quality, and (2) a careful appraisal of the relative importance of known environmental parameters for the species concerned.

The Selection of Objective Criteria of Environmental Quality

Though we cannot rule out the hypothesis that for a monkey or an ape some environments are more pleasant to live in than others, we will probably never be able to prove it by experiment. We have to select, therefore, criteria based upon definite physiological and evolutionary advantages. To help our choice we have first to consider briefly some basic biological constraints which underlie most of our daily activities.

The Allocation of Assimilated Energy

Every organism during its lifetime has a *limited* amount of energy to devote to its physiological maintenance, its various activities, its growth, and the production of offspring. Among vertebrates, and for a given species this amount depends upon the ability of that species to regulate its temperature, and to control its level of energy metabolism. Furthermore, every kind of active warm-blooded animal has both a minimum metabolic potential (roughly equivalent to its Basal Metabolic Rate in a thermo-neutral environment) and a maximum metabolic potential [which in a deer, for instance, approximates 6 times the BMR value on a short term basis, and from 1.24 to 1.45 the BMR value on a daily basis (Moen, 1973)]. Within these lower and upper limits (the so-called scope for activity) the animal has to allocate the available calories to various *conflicting* demands. As Pianka (1974) puts it, this apportionment determines the ways in which the organism can conform to many aspects of its environment and thus indicates a great deal about its ecological niche.

Broadly speaking, three major categories of metabolic expenditures must be covered: maintenance, reproduction, and defense. All three are largely mediated through locomotion and behavior.

Maintenance includes the energy cost of the resting animal, that is, the energy required by the "routine work" of its various organs, from chemical synthesis and conversion of food into active metabolites to the cost of maintaining postural tonus and internal temperature. Such a baseline value of energy metabolism is most often equated to the BMR in human primates, although metabolism is lowered below the basal level during sleep. For other mammals it is ecologically more significant to adopt as a baseline the "standard metabolism," that is, that which includes the cost of routine minimal activities in a restricted environment (usually a respiration chamber for small species). A useful variant is the "average daily metabolic rate" (ADMR) of Morrison and Grodzinski (1975) which is, so far, the best measure of metabolism available to us for constructing daily energy budgets for small mammals. The ADMR is the mean of measurements made over 24 hours at a temperature similar to that of the animal's natural habitat. This measure contains the BMR, the metabolic equivalents of thermo-regulation and minimal activity, as well as the energy of specific dynamic action of food (SDA). ADMR obviously underestimates, even in small mammals, the actual cost of foraging which cannot be measured in a respirometer.

The energy cost of reproduction includes all the physiological as well as behavioral expenditures, imputable to gestation, parturition, lactation, and care of offspring. Whereas the additional energy requirement due to gestation in medium-sized mammals remains small from conception through the first two-thirds of gestation, it increases towards the end, and again during lactation. As an example one can quote Moen's (1973) figures for a 60 kg deer: its energy expenditure, expressed as a multiple of basal metabolism, rises to 1.53 at the end of gestation, and to 2.30 at the peak of lactation. Reproduction can also represent a definite increase in energy expenditures among males. During the rut, a 3½ year old male elk (Harem bull) increases its metabolism to 1.74 times its basal value (Moen, 1973).

We can place under the heading of "defense-cost" all the energy expenses due to protection against predators or parasites and active competition with conspecific individuals. Most of these situations involve some kind of agonistic behavior, which is always very costly, energetically speaking. Catlett (1963) mentions, for instance, a 2.29 increase in oxygen consumption among house mice during aggressive encounters.

Most of the extra-activity expenses of an animal, that is, those which are above the minimal activity level of the ADMR, are by and large due to an increased muscular activity whatever the motivation of the animal (or the goal of its actions). However the energy cost of these movements is very dissimilar. For a 100 kg ungulate for instance, the activity cost (as multiple of BMR) is 0.1 for standing, 0.26 for ruminating, 0.59 for foraging, 0.64 for walking 1 km on the level, 1.35 for walking the same distance on a 10% gradient, 2.0 for playing, and 7.0 for running. The energy cost of lifting the body on a vertical gradient for sheep, horse, and man is over 10 times greater than that for walking on the level (Moen, 1973). Unfortunately we do not yet have comparable figures for nonhuman primates; it is probable, however, that their energy cost of climbing is lower. Another variable has to be taken into account at this stage: body size. Schmidt-Nielsen (1972, 1975) has convincingly demonstrated that the cost of level locomotion, per unit of weight, is lower for larger animals. Furthermore walking and running are far less economical than flying, because their cost of transport (the ratio of power input to the product of speed of transport and body weight) are higher (Tucker, 1970).

Given the limited amount of energy available to cover all the necessary activity expenses and the cost of reproduction, the organism cannot afford to face all these energy expenditures at the same time. It must find the best possible compromise between conflicting demands, taking into account the ever-changing environmental constraints. An increase in reproductive effort, for instance, must be accompanied by a decrease in other activities. As R. A. Fisher pointed out long ago (1930), the apportionment of energy into reproductive versus nonreproductive organs and activities is indeed a pervasive necessity. It is here that behavior comes into the picture. From an ecological viewpoint, its major role appears to be the saving of energy. Being always oriented, behavior becomes the major "anti-chance" mechanism. It reduces random activities to a minimum and thus decreases considerably the cost of locomotion. Furthermore, mental effort by itself requires an insignificant additional expenditure of energy, even in man (Passmore and Durnin, 1955). It seems highly significant that, within the animal kingdom, progressive refinement of behavior always develops on a par with increased ability for active locomotion. Sessile or passively drifting invertebrates most often behave in a way which is not much more elaborate than, say, a carnivorous plant (Bourlière, 1975, 1976). The same relationship between active locomotion and behavior exists even within the limits of the same taxonomic group. Among mollusks, Piaget (1976) very rightly contrasts the "progressive" behavior and learning abilities of the highly mobile cephalopods with the poor performances of most bivalves which spend much of their lives burrowing in sand, or attached to a substrate, filtering the surrounding water. Herbivorous or carnivorous gastropods are in an intermediate position.

Very little work has been done so far on the energy cost of the various activity patterns of nonhuman primates. Estimates of these costs are generally based upon the energy expenditure of similar activities for humans performing actions involving similar muscular movements, and standard metabolism calculated on the basis of Kleiber's formula (Coelho, 1974). It should not, however, be forgotten that recent studies (Hildwein, 1972; Hildwein and Goffart, 1975) have shown that the standard energy metabolism of a number of small tropical forest mammals is 30 to 45% lower than that of temperate species of similar size. According to Grand (1976), arboreal leaf-eaters such as howler monkeys or sloths have a lower percentage of muscle per total body weight (about 25%) than high-speed terrestrial folivores (over 50%). Consequently their metabolic rates and energy needs have to be lower for the same total body weight. Furthermore, some prosimians like the mouse-lemurs and the potto may enter into a state of torpor on a seasonally or daily basis when their oxygen consumption, body temperature, and activities drop to very low levels. However, larger monkeys and apes, whose metabolism has been measured under laboratory conditions on acclimatized individuals, do conform to the value predicted by the Kleiber equation (*see* References and Figure 6 in Hildwein and Goffart, 1975). Similar measurements should be made on wild animals within the range of temperatures they encounter in their natural habitat. As for the energy cost of their various activities, it might be predicted in field conditions by combining "activity budgets" and radio-telemetry of physiological parameters, such as heart rate or respiration rate, which are correlated with metabolic level. Encouraging results along these lines have recently been obtained with white-tailed deer (Holter *et al.*, 1976).

A Basis for the Selection of Environmental Quality Criteria

As the amount of assimilated energy is limited by physiological constraints (the minimum and maximum metabolic potentials), the best possible use must be made of it for the benefit of both the parent-individual and its progeny. An environment will be considered optimal when it provides the animal with the best and most economic level of nutrition, the largest scope for activity, and the best number and quality of offspring. In other words, the quality of a given habitat will increase when environmental conditions will: (1) minimize the cost of maintenance of an individual, (2) minimize predation (by carnivores and pathogens) and competition, and (3) maximize the contribution of the individual to the next generation.

However, within the set of parameters which describe the environment, we must single out the more influential ones. To achieve this goal, a cost-benefit evaluation[2] of the importance of every parameter for each sex, age, and social category will have to be undertaken for each species. This task can be done in two ways; either by demonstrating experimentally that a particular living condition is likely to affect the well-being of the individual and the fitness of the population, or by comparing the behavior and the breeding success of groups living in environments differing only in the variable under study—a very hard condition to fulfill.

We must also remember that an environment never remains stable for long. Even in the absence of human interference, the local climate, the vegetation, and the biotic community will change, at least seasonally. Every animal species, even the most conservative, has to maintain a certain "margin of safety," behavioral as well as physiological, if it is to achieve a successful adaptation to its environment. For mammals with developed learning abilities, some degree of novelty and unpredictability of the surroundings might even confer long-term advantages, enabling certain individuals to explore new possibilities and test new life-styles, which might eventually give rise to new social traditions. It has often been noticed, for instance, that monkeys and apes are more prone to develop new skills in artificial conditions (semicaptive or captive) than in their normal habitat. Beck (1975) points out that the use of tools is more common in captive than in wild primates because exploratory object manipulation is more probable in this context. Peripheral populations living in marginal habitats on the outskirts of a species range also very often develop unusual life-styles. In Senegal, for instance, green monkeys (*Cercopithecus aethiops sabaeus*) not only live in savanna woodlands like grivets or vervets, but enter coastal mangroves and Sahelian thornbush. Such diverse environments strongly influence their troop size, social behavior, and diet. In the Saloum mangroves they feed mostly upon fiddler crabs and the flowers, fruits, shoots, and young leaves of *Rhizophora.* In the Sahelian *Acacia* forest green monkeys, though mostly vegetarian, become seasonally much more opportunistic, preying heavily on rodents and birds (Galat and Galat-Luong, 1976, 1977).

The Various Environmental Parameters: Their Significance for a Nonhuman Primate

Physical Parameters

Climate integrates most of the important physical environmental parameters for terrestrial vertebrates. Generally speaking, its role is far more important for poikilotherms than for homeotherms. For the latter its limiting action is mostly indirect, through a change in food abundance or availability. Provided that they are supplied with sufficient daily rations, most large tropical mammals in captivity can successfully withstand very cold temperatures. During the harsh 1941–42 winter, when temperatures dropped down to a low of -22° C, one escaped female black gibbon *Hylobates concolor,* in Clerès, Normandy, gave birth to an infant and raised it for two months (Olivier, 1952). Near Moscow, Chernyshev (1966) succeeded in maintaining rhesus macaques and vervets in outdoor cages, even in winter.The tails of the vervets were amputated, however, before severe frosts set in. Some populations of two macaque species, *Macaca fuscata* and *M. sylvana,* are known to spend the winter in snowclad evergreen forests. At that time the Japanese macaques of the Shimokita peninsula sustain themselves with tree bark (Izawa and Nishida, 1963). Many barbary macaque troops permanently live in cedar forests in the mountains of Morocco and Algeria, where snow is abundant and temperatures remain below freezing point for up to five months of the year, falling to at least -18° C (Deag, 1974). A provisioned colony of this species has been translocated successfully to the Vosges mountains in northeast France where it breeds normally. In subtropical and tropical mountains, macaques may be found up to an altitude of 4,400 m in southwest China (Hill, 1974) and mountain gorillas up to 4,300 m in eastern Zaire (Schaller, 1963). Obviously climatic factors, particularly temperature, are far less important for primates than diet and social conditions (Bourlière, 1952). But sustained low temperatures anywhere mean increased cost of maintenance; besides a higher cost of thermo-regulation, the energy expenditures for foraging are also increased as a result of the decreased availability of food.

Climate by itself is seldom a limiting factor for primates, but some of its elements play an important role as synchronizers, if not determinants of their reproductive cycle. The respective roles of temperature, photoperiod, and rainfall are not clear, fewer studies having been carried out on primates than on other mammalian orders (Sadleir, 1969). A relationship between time of conception and declining temperatures has however been found in many Asiatic and African species by Lancaster and Lee (1965) and Bourlière, Hunkeler, and Bertrand (1970). A quasi-experimental case is that of the talapoin. In northeast Gabon copulations occur from June to August, during the long dry season when temperatures are the lowest, and births in December and January (Gautier-Hion, 1971). In Mbalmayo, southern Cameroon, only some 300 km northwest of Makokou, the seasons are reversed; here the talapoins mate from December to February during the local dry season, and give birth from June to August (Rowell and Dixson, 1975). As pointed out by these authors, such associated reversals very strongly suggest that the mating season is triggered by stimuli produced by the dry season itself rather than by sidereal clues. Seasonality of reproduction is certainly adaptive, since to be born and raised at the time of the year when food is the most abundant has an obvious survival value for the young.

It can thus be safely predicted that the optimal climate for a nonhuman primate will neither be one with prolonged cold periods nor a uniformly warm one—small, short, and predictable drops in temperature acting as synchronizers of the breeding cycle being useful in increasing the probability of survival of the offspring.

SUPPORT AND SHELTER The support of locomotion is a feature of the physical environment, the importance of which has often been minimized. Although some species of primates are adapted to ground locomotion, most of them remain predominantly arboreal and the nature, size, and orientation of the vegetal support are very important. The high energy cost of lifting the body in man has already been mentioned. Although this expenditure might be smaller in nonhuman primates, it certainly remains higher than that of locomotion on a level plane. Any physical characteristic of the support which will greatly increase muscular work might well render uneconomic the resource exploitation of some forest types. Indeed, some architectural patterns of trees (Hallé and Oldeman, 1970) are very seldom if ever climbed by present-day primates: most Gymnosperms, bamboos, and palms fall in this category. Schultz (1969) points out that no monkey or ape can climb a nearly perpendicular, smooth and branchless tree trunk too thick to be securely grasped with its digits or even hugged between its arms. The only exceptions are the small marmosets which can hold their light weight adequately with their sharp claws. A good example of the role that supports of locomotion play in apportioning the various forest layers to sympatric species of primates is given by Charles-Dominique (1971) in his study of Gabonese prosimians. The stratification of the five different sympatric species is not due to some mysterious height preference but mostly to their differing adaptations to various types of support. The weight of the adult animal is also of importance in this respect. The largest ape, the mountain gorilla, climbs trees the least readily; when danger threatens it descends and flees on the ground, a very unexpected reaction for a monkey or ape (Schaller, 1963).

Weight, body-build, and climbing ability greatly influence the degree to which different species can make use of the same available resources of the arboreal environment. Grand (1972) and Mittermeier and Fleagle (1976) have shown how forelimb suspension in gibbons and spider monkeys can increase their feeding sphere by 100% over that available to a macaque or a black-and-white colobus monkey in a sitting position. This represents an important energy-saving adaptation.

Trees not only provide food for most primates, they represent also a haven of rest and safety, and a locus for sociality where the predation pressure of both terrestrial and aerial predators is minimized. Terrestrial species very often use cliffs as sleeping sites when trees are not available. Such social traditions can become so well established that Kummer (1971) goes so far as to say, when speaking of the hamadryas and anubis baboons living in the Awash National Park, that trees but no cliffs mean anubis, cliffs but no trees mean hamadryas.[3] That the amount of cover might affect agonistic behavior in the wild as well as in captivity has also been suggested by Rowell (1972).

To put it briefly, the nature and orientation of the support of locomotion is important for primates because it facilitates or precludes access to food sources and therefore influences maintenance expenditures. However, shelter in a broad sense is also necessary to decrease predation pressure during the day, and adequate sleeping sites fulfill the same role at night.

Trophic Parameters

Whatever the habitat may be, the existence of a year-round supply of food adequate to cover maintenance, activity, and production expenditures is a basic environmental requirement for any species. In other words, the carrying capacity of the environment must be large enough to satisfy all the population's nutritional requirements. Some studies have been made which relate food production, standing crop, and productivity of a number of wild ungulates and rodents (*see*, for instance, Sinclair, 1975). Unfortunately nothing as elaborate has yet been done for nonhuman primates, their more or less omnivorous diet making it difficult to evaluate both food production and food consumption during the yearly cycle. At least three different aspects of the problem need to be considered: the quantity, the quality, and the availability of food.

THE QUANTITY OF FOOD Seasonal patterns of production are more important than total amount of food produced per year for most animal consumers, particularly for those that feed upon fresh plant or animal tissues and cannot store food. Basically two patterns of food production can be observed in the tropics. In rain forests fresh leaves, fruits, and small animals were supposed to be produced more or less evenly throughout the year. Recent research has shown however, that a certain degree of seasonality is the rule (Taylor, 1960; Smythe, 1970; Charles-Dominique, 1971; Klinge, 1974; Ogawa, 1974; Huttel and Bernhard-Reversat, 1975; Clutton-Brock, 1975; Leigh, 1975). The Hladiks (personal communication) give the following seasonal low and peak values for two major components of the litterfall correlated with food for primates. On Barro Colorado Island, Panama, the leaffall ranges from 1 g/m^2/day during the dry season to 4 g/m^2/day during the rains; similar values for the fruitfall are 0.1 and 0.5 g/m²/day. In Ipassa, Gabon, the leaffall fluctuates from 1.5g to 3 g/m²/day and the fruitfall from 0.1 to 0.6 g/m²/day. In the semideciduous forest of Polannaruwa, Ceylon, the leaffall ranges from 0.5 to 1.5 g/m²/day and the fruitfall from nil to 0.05 g/m²/day. Furthermore, important year to year variations have been observed in Gabon. In tropical savannas, seasonality is strong and universal, becoming more and more important as one travels farther away from the equator (Bourlière and Hadley, 1970; Sinclair, 1975; Lamotte, 1978). An extreme case, where most of the primary and secondary production takes place within two consecutive months, is characteristic of desert steppes like the African Sahel (Bourlière, 1978). In such a harsh environment the maximum possible biomass of sedentary small mammal consumers is determined by the amount of food still available at the end of the long dry season. At that time large species have to scatter more or less widely over large areas and/or migrate seasonally to more fertile regions.

When compared to the plant instantaneous biomass, the total amount of *potential* primate food produced per year and per unit surface area in major tropical habitats is always small. Whereas the yearly leaffall in all tropical rain forests so far studied on three continents ranges from 5.8 to 10 tons/ha (dry weight), the yearly fruitfall is only 0.5–1.4 tons/ha on Barro Colorado (Leigh, 1975), 0.3–1.2 in Ivory Coast (Bernhard, 1970), 0.3–0.8 in Amazonia (Klinge, 1974), and 0.39 in Pasoh, West Malaysia (Ogawa, 1974). In savannas the fruit production is much smaller: 0.023 t/ha/yr in the Sahelian savanna of Fété Olé (Bourlière, 1978). No estimate has ever been made of the yearly production of rhizomes and edible roots. The grass yield varies with rainfall: from 0.0 to 1.2 t/ha/yr in Fété Olé (Bourlière, 1978), from 0.6 to 11.5 t/ha/yr in the Serengeti (Braun, 1973), and from 5.7 to 8.0 t/ha/yr in Nigeria (Hopkins, 1968).

THE QUALITY OF FOOD It is unrealistic to estimate the dietary needs of an adult animal on the basis of its caloric intake alone. Primates are no exception to this rule. True, the bulk of organic material used for food consists of proteins, carbohydrates, and fats, compounds which are, within wide limits, interchangeable in the energy metabolism. Nevertheless a diet may furnish adequate food, measured in calories, yet be utterly inadequate as regards raw materials necessary for growth and for maintenance of the cellular and metabolic machineries. Monkeys and apes, like other animals, also need a supply of certain amino acids vitamins, various minerals, and trace elements for growth and upkeep.

Our present knowledge of the nutritional value of the various kinds of plant and animal foods used by primates in the wild is unfortunately very scarce. Mere extrapolation from values of similar temperate food items is dangerous: Jordan (1971) has shown, for instance, that the average energy content is lower both in tropical leaves and entire tropical plants than in their temperate equivalents. The best data we have so far are those published by Struhsaker (1975) for Uganda and C. M. Hladik (1977) for Barro Colorado, Ceylon, and northeast Gabon. On the whole, leaves contain much more protein (24.8% dry weight on average) than fruits (5.81%), but fruits are richer in soluble carbohydrates (31.1%) and fats (6.5%) than leaves (14.9% and 2.3%, respectively). Shoots contain more protein (35.5%) than either young leaves (29.2%) or mature ones (15.9%). Such essential amino acids as histidine, isoleucine, leucine, lysine, methionine, phenylalanine, threonine, and valine, as well as cystine and tyrosine, are found in leaves and stems of plants eaten by chimpanzees in Gabon. Insects are obviously very rich in protein and (sometimes) fats. The clay eaten by most of the leaf-eating primates may act as an absorbant for tannin.

Whatever the results obtained in the bomb calorimeter or during chemical determinations, the food value of any plant or animal material depends also on its palatability. This in turn is conditioned by the presence or absence of some secondary substances such as tannin, phenolic compounds, alkaloids, aromatic substances, and so on. These compounds have recently attracted a great deal of attention (Freeland and Janzen, 1974; Janzen, 1975; Levin, 1976; Hladik and Hladik, 1977). Suffice it to say here that tannins are most abundant in very old leaves, phenolic compounds are rarely found in shoot tips, lateral cambium, and seeds, whereas alkaloids are more frequent in second growth and marginal habitats than in mature forest, and particularly so in shrubs and woody climbers (lianas).

Another parameter of food quality to be taken into consideration is the capacity for preservation of the edible plant parts. Most of the fruits and young leaves do not last long in a tropical rain forest. They must be consumed quickly, otherwise the combination of heat and humidity, and the abundance of decomposers, lead to their quick decay. The situation is just the opposite in most savannas. Seeds, rhizomes or tubers, all of them containing a very nutritious meal neatly packaged for potential consumers, may store energy for weeks or even months on end. Suzuki (1969) has pointed out that in the savanna woodland of the Kasakati basin, Tanzania, more than half of the 61 species of fruits identified as chimpanzee food are "half-dried fruits," the others also being "more or less dried."

THE AVAILABILITY OF FOOD A major characteristic of many plant and animal populations in the humid tropics is that of "patchiness" (McArthur, 1972). Patchy distribution is indeed the obvious rule in rain forest environments, being less so in more open and seasonal habitats where species diversity is smaller. An optimal foraging strategy is defined as that which maximizes the difference between foraging profits and their costs (Pianka, 1974). It is therefore clearly of advantage for a social and mostly vegetarian consumer like a monkey not to have an overspecialized diet; in doing so it can best profit from the heterogeneity of the habitat.

The observations of Crook and Aldrich-Blake (1968) on sympatric species of baboons have made clear that the percentage of time spent foraging and feeding, and therefore the energy cost, can vary greatly between two species with different food preferences living in the same area. At Debre Libanos, geladas spent 35 to 70% of their time feeding, whereas the anubis spent only 20%. In this case, it is obvious that gelada food requires more effort to assure proper nutrition. Similar differences have been found between the activity budgets of populations of the same species exploiting different habitats. In Senegal, *Cercopithecus aethiops sabaeus* living in the rather rich mangrove environment spent only 8.1% of their time feeding, and 20.6% moving, whereas those dwelling in the impoverished Sahelian gallery forest spent 23 to 32% of their time feeding and 35 to 48% moving (Galat and Galat-Luong, 1976).

The relationships between social dynamics (particularly the temporary splitting of a band into smaller foraging parties), and the spatial distribution of food have been given much attention during the past decade (Crook, 1966, 1967; Kummer, 1971; Altmann, 1974) and thus do not need to be emphasized here.

DIET AND BODY SIZE A much neglected trophic parameter is the size of the animal. Small, short-lived mammals which have a relatively high ratio of body surface to body volume generally also have higher metabolic rates and energy requirements per unit of body weight than do larger ones. Therefore the trophic impact of populations of larger, long-lived mammals upon their environment will be smaller than that of small animals: a biomass of 3,000 kg of herbivores will consume about 2,000 kg of grass per day if made up of voles, and 200 kg if composed only of two hippos! Increased body size can therefore by considered as an energy-saving adaptation, both for the animal and its habitat. A large size means also a lower daily ration of proteins, per unit of body weight, as the minimum protein requirement is also closely related to the basal metabolic rate (Portman, 1970). This explains why small monkeys are more insectivorous than larger ones (Gautier-Hion, in press; Hladik, 1977).

Larger mammals can also withstand starvation more easily than smaller ones; if the amount of food is temporarily insufficient, the remaining energy requirements are covered by the consumption of body substance, primarily stored fat. Nonhuman primates, however, cannot rely much upon this safety device in natural conditions (no more than on food caches): their fat deposits are always small. Schultz (1969) points out that no noteworthy well-distributed amount of fat has ever been found by him in any of the many monkeys and apes he has skinned, except in some overfed captive specimens which had had too little exercise.

This section devoted to the trophic parameters of the environment can be summarized by stating that in a rich environment one may expect (1) an adequate amount of palatable and nutritionally well balanced food, and (2) a steady production of these foodstuffs throughout the year, without prolonged periods of serious food shortage.

A patchy distribution of food supplies is not necessarily detrimental in itself, as long as the food is energetically rich enough to maximize the foraging profits and as long as adequate foraging strategies can minimize its costs.

Whether or not the amount of food is a limiting factor for primate populations under natural conditions is presently a matter of disagreement among primatologists. Food is a limiting factor in the most extreme environments and for species like Japanese and barbary macaques, and probably also for baboons, patas, and guenons of the *Cercopithecus aethiops* group. On the contrary, Coelho *et al.* (1976) feel that the population of howlers and spider monkeys in the quasi-rain forest of Tikal, Guatemala, have an excess of food at their disposal throughout the year. The Hladiks (personal communication) feel that primate populations might well be limited by the amount of adequate food available at the end of the dry season on Barro Colorado and in northeast Gabon, as well as in eastern Ceylon and western Madagascar. The year to year variations in food production might also play an important limiting role. Obviously many more studies of food production versus food consumption are needed before we can draw any conclusion on this matter.

Other Biological Parameters

PREDATION PRESSURE Predation, except by man, does not appear to play an important role as a mortality factor in adult monkeys or apes. Some successful kills of macaques and baboons by large felids have been reported by Altmann and Altmann (1970) and Lindburg (1971). Jouventin (1975) has also seen an adult male mandrill killed by a crowned hawk eagle. But such incidents occur very rarely. Neither Schaller (1963), nor van Lawick-Goodall (1968) and her group have ever seen any gorilla or chimpanzee seriously hurt by a predator. The role of predation in the high infant mortality of baboons (Rowell, 1969), macaques (Dittus, 1975) or chimpanzees (Teleki *et al.*, 1976) remains unknown, however. The fear of snakes, or even of birds of prey, displayed by so many species might be an indication of their predatory role. In this respect, the environmental quality of most tropical habitats is poor.

PARASITE LOAD Most species of wild primates harbor a variety of parasites in their digestive tract and blood, and this heavy parasite load represents for them an energy drain of unknown, yet not negligible magnitude. This rich fauna uses for its own profit part of the nutrients ingested by the host. Parasitic helminths are particularly numerous in kind and abundant in numbers (Dunn *et al.*, 1968; Kuntz and Moore, 1973; File *et al.*, 1976) even in wild Japanese macaques despite their temperate environment (Nigi and Tanaka, 1973). Blood parasites and viral agents are also present. Some taxonomic groups are found more frequently in certain habitats than in others. In the Malayan rain forest, for instance, microfilariae are more common in monkeys living in the canopy, trematodes, and red-cell protozoa in those preferring the middle layers, and cestodes in ground-living species. Nematodes are very common in monkeys from all levels. Thus a heavy parasite load is common in wild primates in all environments, though apparently stronger in rain forests. The more insectivorous species are the most exposed to heavy infections of parasitic helminths of all kinds.

INTERSPECIFIC COMPETITION Primates make up an impressive biomass of primary consumers in undisturbed tropical forest environments. On Barro Colorado Island, their minimum standing crop (fresh weight) ranges from 420 to 504 kg/km^2 (Eisenberg and Thorington, 1973). This biomass rises to 2,370 kg/km^2 at Polonnaruwa, Ceylon (Eisenberg, Muckenhirn, and Rudran, 1972) and to 2,317–3,578 kg/km^2 in the Kibale forest, Uganda (Struhsaker, 1975). The trophic impact of such populations on foliage, fruit, and insect resources must consequently be very strong, and some sort of competition with other leaf- and fruit-eaters is to be expected. That such competition does exist is suggested by the figures we have just quoted. In Panama, primates share the forest resources with another successful group of folivores, the sloths, whose biomass is much higher than theirs (2,781 kg/km^2). There are no sloths in Asia or Africa, so primates have no serious leaf-eating competitors here, and their biomasses are interestingly larger. Other possible competitors are the tree hyraxes in Africa (arboreal browsers), the fruit bats (Pteropidae in the Old World and Stenoderminae in the Neotropics), the kinkajou, and a few other American arboreal Carnivores. Outside forests, competition for food with other mammals is far less likely to take place. However, it can occur in exceptional circumstances, like a pullulation of rodents; Galat and Galat-Luong (1977) have documented such an instance during an *Arvicanthis niloticus* outbreak in northern Senegal.

Competition among sympatric species of primates obviously does occur in certain situations, as shown by the mixed troops of forest monkeys (Gautier and Gautier-Hion, 1969; Gartlan and Struhsaker, 1972; Gautier-Hion and Gautier, 1974; Gautier-Hion, in press). Some are temporary but not random, some species occurring together more frequently than others; others are very stable and the adult male of the dominant species can, in this case, play the role of an "interspecific" leader. In these mixed troops the associating species have more or less the same diet, each species retaining, however, its preferences for food and foraging techniques. This "mixed group paradox" can be explained by the advantages of cooperation. Not only can these associations increase the efficiency of predator detection, but they also multiply the chances of discovering new food patches without increasing intraspecific competition for food or mate, as would be the case in a monospecific troop of similar size. Polyspecific troops do not occur in more open environments, niche separation between sympatric species being apparently more marked among terrestrial open country primates than among forest species (Dunbar and Dunbar, 1974).

INTRASPECIFIC COMPETITION There are two situations in which intraspecific competition can be contemplated. Any increase in population density when the food resources remain stable will lead to intensification of agonistic behavior, increased cost of foraging, as well as to the use of unusual foods and/or marginal habitats. Ultimately such a condition will become detrimental to the fitness of the population concerned, and a decline in numbers can be expected.

A more frequent state of affairs occurs when the size of a population remains constant while its food resources fluctuate seasonally or from year to year. In this instance one might expect a troop to prevent access to its major sources of energy by adjacent troops of conspecifics when resources are diminishing. Recent research has shown that variations in territorial behavior of some birds can indeed be explained in this way. After having compared the nectar production of flowers and studied its consumption by sunbirds Gill and Wolf (1975) demonstrated that these birds formed territories only when the flowers contained slightly more nectar than was necessary to support one bird for 24 hours. There were no territories in very sparse or very rich habitats. In the first case, nectar was so sparsely distributed that it would require an inordinate amount of energy for a male to defend a territory with enough food for himself, while in the second situation nectar was so densely distributed that it would be a waste of energy to defend any feeding area. Hummingbirds apparently behave in the same way. Similar cost-benefit analyses have not yet been made in monkeys. However, the observations of Galat and Galat-Luong (1976, 1977, personal communication) on *Cercopithecus aethiops sabaeus* in Senegal indicate that these monkeys do not defend territories in the impoverished Sahelian environment where they do not react to a playback of the adult male loud call, whereas they display territorial behavior in richer habitats of the Bandia forest and the Sine Saloum mangroves further south. Where food is locally or temporarily superabundant, they again do not defend territories.

Summary

An environment is often said to be better or worse, richer or poorer than another on a purely intuitive basis. An attempt is made here to find more objective criteria of environmental quality for nonhuman primates using a bioenergetic as well as an evolutionary approach.

An optimal environment will be that which provides the animal with the best and most economic level of nutrition, the broadest scope of activity, and the largest number and best quality of offspring. This is achieved by a reduction of the maintenance expenditures, predation pressure, and competition, as well as by an increased contribution of the parent individuals to the next generation.

The amount of assimilated energy to be spent during a certain timespan being limited by physiological constraints, and the quantity of food obtainable being restricted by environmental constraints, the individual organism cannot afford to face all its possible energy expenditures at the same time. The best compromise has to be found between conflicting demands. In this context, behavior may be considered as a major energy-saving device.

The evaluation of the quality of various environmental variables can be made on a cost-benefit basis, both at the level of the individual and that of its social group.

The demographic and social structure of a stable population in its usual undisturbed environment, probably represents the best possible compromise between the phylogenetic capabilities of the species and its normal habitat. However, the roles of the various environmental parameters are difficult to estimate in such climax conditions. The study of populations living in marginal environments can help elucidate the mechanism of action of many ecological factors.

Notes

1. Environmental quality is usually defined as those properties of the environment which ensure the comfort and well-being of the individual and maximize its survival and that of its offspring.
2. Quite recently a great deal of attention has been given to cost benefit analysis in socio-ecology (see for instance Pulliam, 1976; Clutton-Brock and Harvey, 1976, 1977; Emlen and Oring, 1977).

3. This is obviously an oversimplification, as shown by more recent observations by Nagel (1973).

References

Altmann, S. A. 1974. Baboons, space, time, and energy. *Amer. Zool.* 14: 221–248.

Altmann, S. A., and Altmann, J. 1970. Baboon ecology. African field research. *Bibl. Primatol.* 12:1–220.

Beck, B. B. 1975. Primate tool behavior. Pages 413–447 in *Socioecology and psychology of primates.* Ed. R. H. Tuttle. The Hague: Mouton.

Bernhard, F. 1970. Etude de la litière et de sa contribution au cycle des éléments minéraux en forêt ombrophile de Côte d'Ivoire. *Oecol. Plant.* 5:247–266.

Bourlière, F. 1952. Quelque remarques sur l'écologie comparée des primates. *Biol. Med.* 41:1–15.

Bourlière, F. 1975. Review of E. O. Wilson's *Sociobiology. La Terre et la Vie* 29:624–628.

Bourlière, F. 1976. Review of J. L. Brown's *The evolution of behavior. La Terre et la Vie* 30:485–486.

Bourlière, F. 1978. La savane sahélienee de Fété-Olé, Sénégal. In *Structure et fonctionnement des ecosystèmes terrestres.* Ed. M. Lamotte and F. Bourliere. Paris: Masson.

Bourlière, F., and Hadley, M. 1970. The ecology of tropical savannas. *Ann. Rev. Ecol. Syst.* 1:125-152.

Bourlière, F., Hunkeler, C., and Bertrand, M. 1970. Ecology and behaviour of Lowe's guenon (*Cercopithecus campbelli lowei*) in the Ivory Coast. Pages 297–363 in *The old world monkeys: Evolution, systematics, and behavior.* Eds. J. R. and P. H. Napier. New York: Academic Press.

Braun, H. M. H . 1973. Primary production in the Serengeti; purpose, methods and some results of research. *Ann. Univ. Abidjan* E, 6: 171–188.

Catlett, R. A. 1963. Page 255 in *Principles in Mammalogy.* Eds. D. E. Davis and F. B. Golley. New York: Reinhold.

Charles-Dominique, P. 1971. Eco-éthologie des prosimiens du Gabon. *Biol. Gabonica* 7:121–228.

Chernyshev, V. 1966. Acclimatisation of tropical species of monkeys in the suburbs of the Moscow area. *Int. Zool. Yearbook* 6:247–251.

Clutton-Brock, T. H. 1975. Feeding behaviour of red colobus and black and white colobus in East Africa. *Folia Primat.* 23:165–207.

Clutton-Brock, T. H ., and Harvey, P. H. 1976. Evolutionary rules and primate societies. Pages 195–237 in *Growing points in ethology.* Eds. P. P. G. Bateson and R. A. Hinde. Cambridge: Cambridge University Press.

Clutton-Brock, T. H., and Harvey, P. H. 1977. Primate ecology and social organization. *J. Zool. Lond.* 183:1–39.

Coelho, A. M., Jr. 1974. Socio-bioenergetics and sexual dimorphism in primates. *Primates* 15:263–269.

Coelho, A. M., Jr., Bramblett, C. A., Quick, L. B., and Bramblett, S. S. 1976. Resource availability and population density in primates: A soci-bioenergetic analysis of the energy budgets of Guatemalan howler and spider monkeys. *Primates* 17:63–80.

Crook, J . H. 1966. Gelada baboon herd structure and movement: A comparative report. *Symp. Zool. Soc. Lond.* 18:237–258.

Crook, J . H. 1967. Evolutionary change in primate societies. *Sci. J.* 3 (6): 66–72.

Crook, J . H ., and Aldrich-Blake, P. 1968. Ecological and behavioural contrasts between sympatric ground-dwelling primates in Ethiopia. *Folia Primat.* 8:192–227.

Deag, J. M. 1974. A study of the social behaviour and ecology of the wild Barbary macaque, *Macaca sylvanus.* Ph.D. thesis, University of Bristol.

Dittus, W. P. J. 1975. Population dynamics of the toque monkey *Macaca sinica.* Pages 125–151 in *Socioecology and psychology of primates.* Ed. R. H . Tuttle. The Hague: Mouton.

Dunbar, R. I. M., and Dunbar, E. P. 1974. Ecological relations and niche separation between sympatric terrestrial primates in Ethiopia. *Folia Primat.* 21:36–60.

Dunn, F. L., Lim, B. L., and Yap, L. F. 1968. Endoparasite patterns in mammals of the malayan rain forest. *Ecology* 49:1179–1184.

Eisenberg, J . F., Muckenhirn, J ., and Rudran, R. 1972. The relationship between ecology and social structure in primates. *Science* 176: 863–874.

Eisenberg, J. F., and Thorington, R. W., Jr. 1973. A preliminary analysis of a neotropical mammal fauna. *Biotropica* 5:150–161.

Emlen, S. T., and Oring, L. W. 1977. Ecology, sexual selection, and the evolution of mating systems. *Science* 197:215–223.

File, S. K., McGrew, W. C., and Tutin, C. E. G. 1976. The intestinal parasites of a community of feral chimpanzees, *Pan troglodytes schweinfurthii. J. Parasitol.* 62:259–261.

Fisher, R. A. 1930. *The genetical theory of natural selection.* Oxford: Clarendon Press.

Freeland, W. J., and Janzen, D. H . 1974. Strategies in herbivory by mammals: The role of plant secondary compounds. *Amer. Nat.* 108:269–289.

Galat, G., and Galat-Luong, A. 1976. La colonisation de la mangrove par *Cercopithecus aethiops sabaeus* au Sénégal. *La Terre et la Vie* 30:3–30.

Galat, G., and Galat-Luong, A. 1977. Démographie et régime alimentaire d'une troupe de *Cercopithecus aethiops sabaeus* en habitat marginal au Nord Sénégal. *La Terre et la Vie* 31:557–577.

Gartlan, J. S., and Struhsaker, T. T. 1972. Polyspecific associations and niche separation of rain-forest anthropoids in Cameroon, West Africa. *J. Zool. Lond.* 168:221–266.

Gautier, J. P., and Gautier-Hion, A. 1969. Les associations polyspécifiques chez les Cercopithecidae du Gabon. *La Terre et la Vie* 23: 164–201.

Gautier-Hion, A. 1971. L'écologie du Talapoin du Gabon. *La Terre et la Vie* 25:427–490.

Gautier-Hion, A. In press. Food niche and co-existence in sympatric primates in Gabon. In *Primate feeding behaviour in relation to food availability and composition.* Eds. C. M. Hladik and D. J. Chivers.

Gautier-Hion, A., and Gautier, J . P . 1974. Les associations polyspécifiques de Cercopitheques du plateau de M'Passa (Gabon). *Folia Primat.* 22:134–177.

Gill, F. B., and Wolf, L. L. 1975. Economics of feeding territoriality in the golden-winged sunbird. *Ecology* 56:333–345.

Grand, T. I. 1972. A mechanical interpretation of terminal branch feeding. *J. Mammal.* 53:198–201.

Grand, T. I. 1976. Body weight, diet and locomotion. *Primate News* 14 (5):16–17.

Halle, F., and Oldeman, R. A. A. 1970. *Essai sur l'architecture et la dynamique des arbres tropicaux.* Paris: Masson.

Hildwein, G. 1972. Métabolisme énergétique de quelques mammiferes et oiseaux de la forêt équatoriale. *Arch. Sci. Physiol.* 26:387–400.

Hildwein, G., and Goffart, M. 1975. Standard metabolism and thermoregulation in a prosimian *Perodictitus potto. Comp. Biochem. Physiol.* 50A:201–213.

Hill, W. C. O. 1974. *Primates: Comparative anatomy and taxonomy,* vol. 7. Edinburgh: Edinburgh University Press.

Hladik, A., and Hladik, C. M. 1977. Signification écologique des teneurs en alcaloîdes des végétaux de la forêt dense: Resultats de tests préliminaires effectués au Gabon. *La Terre et la Vie* 31:515–555.

Hladik, C. M. 1977. Adaptive strategies of primates in relation to leaf eating. In *The ecology of arboreal folivores.* Ed. G. G. Montgomery. Washington, D. C.: Smithsonian Institution Press.

Holter, J . B., Urban, W. E., Hayes, H . H., and Silver, H . 1976. Predicting metabolic rate from telemetered heart rate in white-tailed deer. *J. Wildl. Manage.* 40:626–629.

Hopkins, B. 1968. Vegetation of the Olokemeji Forest Reserve, Nigeria. V. The vegetation of the savanna site with special reference to its seasonal changes. *J. Ecol.* 56:97–115.

Hutchinson, G. E. 1957. Concluding remarks. *Cold Spring Harbor Symp. Quant. Biol.* 22:415–427.

Huttel, C., and Bernhard-Reversat, F. 1975. Recherches sur l'écosysteme de la forêt subéquatoriale de basse Côte d'Ivoire. V. Biomasse végétale et productivité primaire: cycle de la matiere organique. *La Terre et la Vie* 29:203–228.

Izawa, K., and Nishida, T. 1963. Monkeys living in the northern limits

of their distribution. *Primates* 4:67–88.
Janzen, D. H. 1975. *Ecology of plants in the tropics.* London: Arnold.
Jordan, C. F. 1971. A world pattern in plant energetics. *Amer. Sci.* 59: 425–433.
Jouventin, P. 1975. Observations sur la socio-écologie du Mandrill. *La Terre et la Vie* 29:493–532.
Kleiber, M. 1961. *The fire of life.* New York: Wiley.
Klinge, H . 1974. Litter production in tropical ecosystems. Paper presented at the I.B.P. Synthesis Meeting, Kuala Lumpur, 12–18th August 1974.
Kummer, H . 1971. *Primate societies group techniques of ecological adaptation.* Chicago: Aldine.
Kuntz, R. E., and Moore, J. A. 1973. Commensals and parasites of African baboons captured in Rift valley province of central Kenya. *J. Med. Prim.* 2:236–241.
Lamotte, M. 1978. La savane préforestiere de Lamto (Côte d'Ivoire). In *Structure et fonctionnement des ecosystemes terrestres.* Eds. M. Lamotte and F. Bourlière. Paris: Masson.
Lancaster, J . B., and Lee, R. B. 1965. The annual reproductive cycle in monkeys and apes. Pages 486–513 in *Primate behavior: Field studies of monkeys and apes.* Ed. I. DeVore. New York: Holt, Rinehart and Winston.
Leigh, E. G., Jr. 1975. Structure and climate in tropical rain forest. *Ann. Rev. Ecol. Syst.* 6:67–86.
Levin, D. A. 1976. Alcaloïd bearing plants: An ecogeographic perspective. *Amer. Nat.* 110:261–284.
Lindburg, D. G. 1971. The rhesus monkey in North India: An ecological and behavioral study. *Primate Behavior* 2:1–106.
MacArthur, R. H. 1972. *Geographical ecology: Patterns in the distribution of species.* New York: Harper & Row.
Mittermeier, R. A., and Fleagle, J. G. 1976. The locomotor and postural repertoires of *Ateles geoffroyi* and *Colobus guereza*, and a reevaluation of the locomotor category semibrachiation. *Am. J. Phys. Anthrop.* 45:235–256.
Moen, A. N. 1973. *Wildlife ecology: An analytical approach.* San Francisco: Freeman.
Morrison, P., and Grodzinski, W. 1975. Morrison respirometer and determination of ADMR. Pages 300–309 in *Methods for ecological bioenergetics.* Eds. W. Grodzinski, R. Z. Klekowski and A. Duncan. *IBP Handbook* no. 24. Oxford: Blackwell.
Nagel, U. 1973. A comparison of anubis babbons, hamadryas baboons and their hybrids at a species border in Ethiopia. *Folia Primat.* 19: 104–165.
Nigi, H., and Tanaka, T. 1973. Clinical examinations on the japanese monkey (*Macaca fuscata*). *Exp. Animals* 22:461–470.
Ogawa, H. 1974. Litter production and carbon cycling in Pasoh forest. Paper presented at the I.B.P. Synthesis Meeting, Kuala Lumpur, 12–18th August 1974.

Olivier, G. 1952. Note sur la reproduction du Gibbon *Hylobates concolor. La Terre et la Vie* 6:139–141.

Passmore, R., and Durnin, J. V. G. A. 1955. Human energy expenditure. *Physiol. Rev.* 35:801–840.

Piaget, J. 1976. *Le comportement, moteur de l'évolution.* Paris: Gallimard.

Pianka, E. R. 1974. *Evolutionary ecology.* New York: Harper & Row.

Portman, O. W. 1970. Nutritional requirements (NRC) of nonhuman primates. Pages 87–115 in *Feeding and nutrition of nonhuman primates.* Ed. R. S. Harris. New York: Academic Press.

Pulliam, H. R. 1976. The principle of optimal behavior and the theory of communities. Pages 311–332 in *Perspectives in ethology.* Eds. P. P. G. Bateson and P. H . Klopfer. New York: Plenum Press.

Rowell, T. 1969. Long-term changes in a population of Ugandan baboons. *Folia Primat.* 11:241–254.

Rowell, T. 1972. *The social behaviour of monkeys.* London: Penguin Books.

Rowell, T. E., and Dixson, A. F. 1975. Changes in social organization during the breeding season of wild talapoin monkeys. *J. Reprod. Fertil.* 43:419–434.

Sadleir, R. M. F. S. 1969. *The ecology of reproduction in wild and domestic mammals.* London: Methuen.

Schaller, G. 1963.. *The mountain gorilla: Ecology and behavior.* Chicago: University of Chicago Press.

Schmidt-Nielsen, K. 1972. Locomotion: Energy cost of swimming, flying, and running. *Science* 177:222–228.

Schmidt-Nielsen, K. 1975. *Animal physiology: Adaptation and environment.* London: Cambridge University Press.

Schultz, A. H. 1969. *The life of primates.* London: Weidenfeld and Nicolson.

Sinclair, A. R. E. 1975. The resource limitation of trophic levels in tropical grassland ecosystems. *J. Anim. Ecol.* 44:497–520.

Smythe, N. 1970. Relationships between seasons and seed dispersal methods in a neotropical forest. *Amer. Nat.* 104:25–35.

Struhsaker, T. T. 1975. *The red colobus monkey.* Chicago: University of Chicago Press.

Suzuki, A. 1969. An ecological study of chimpanzees in a savanna woodland. *Primates* 10:103–148.

Taylor, C. F. 1960. *Synecology and sylviculture in Ghana.* Edinburgh: Nelson. Quoted on p. 117 of K. A. Longman and J . Jenik *Tropical forest and its environment.* 1974. London: Longman.

Teleki, G., Hunt, E. E., Jr., and Pfifferling, J . H . 1976. Demographic observations (1963–1973) on the chimpanzees of Gombe National Park, Tanzania. *J. Hum. Evol.* 5:559–598.

Tucker, V. A. 1970. Energetic cost of locomotion in animals. *Comp. Biochem. Physiol.* 34:841–846.

van Lawick-Goodall, J. 1968. The behaviour of free-living chimpanzees in the Gombe Stream Reserve. *Anim. Behav. Mon.* 1:161–311.

Chapter 3

Demographic Constraints on Behavior and Social Organization

Stuart A. Altmann
Jeanne Altmann

The quality of the habitat may influence population densities and the sizes of groups, but factors such as birth intervals, longevity, and the duration of the period prior to sexual maturity will also influence the demographic composition of a group. Altmann and Altmann call our attention to the feedback mechanisms that relate life history processes to demography to social organization to ecological pressures and back to life history processes. If selection is considered to be operating on life history processes, then these parameters will influence demographic rate processes, which, in turn, set the limits on social structure.

The reproductive performance of an individual in any one year must not be considered to be a direct reflection of evolutionary selective forces in animals such as the primates. It should be clear that it will be reproductive success over the individual's entire lifetime that will determine genetic fitness. Short-term cross-sectional studies may inadvertently focus our attention on short-term processes, causing us to neglect the more significant long-term consequences.

Crucial examples are provided by the few longitudinal studies of nonhuman primates that are available. Such studies revealed the importance of kinship to social organization and our attention was then riveted to this parameter by recent advances in population genetics theory and the resulting attention to kin selection. Altmann and Altmann remind us, however, that most of the longitudinal studies of primates have taken place at provisioned sites and that, given unlimited food and protection from various sources of mortality, even K strategists like the primates can show r strategy type expansion. When this occurs, matrilines proliferate and large numbers of surviving kin surround each new infant.

The Altmanns contrast this with their own long-term observations of baboons at Amboseli, living in a more undisturbed state, and point out the consequences of provisioning to natural life history processes. When populations are at equilibrium, what kinds of group structure result? How many kin surround the infant under these conditions? How do life history processes influence the effects of kinship on playmate selection, infant association and care patterns, female association and agonistic aiding patterns? A small change in factors influencing demography may thus have profound consequences on social organization.

Introduction

The thesis that we shall develop can be stated very simply. First, the size and composition of social groups, in terms of age, sex, and kinship, affect behavior and social relationships. Second, demographic processes provide delayed feedback on behavior because they affect group size and composition and are altered, in turn, by the effects of behavior on demographic parameters.

This cycle of effects—of behavior on demographic processes, of the latter on group size and composition, and of these last, in turn, on behavior—is loosely coupled because behavior and other life history processes are affected not only by demographic and behavioral factors, respectively, but by other environmental factors as well (Figure 1). Consequently, an environmental change may have both short-term effects on behavior, through direct responses of individuals to it, and long-term effects, through responses of these individuals or their descendants to the altered demography of the local population.

First we shall consider briefly the two better-known links in the cycle, traditionally regarded as part of a unidirectional chain of influence. We then turn to the effects of group composition on behavior, thereby completing the cycle. The existence of two-way causal relationships between demography and behavior has no doubt been tacitly assumed by many people, but has rarely been discussed in the literature (cf. Mason, 1978). The ramifications of this duality are virtually unexplored.

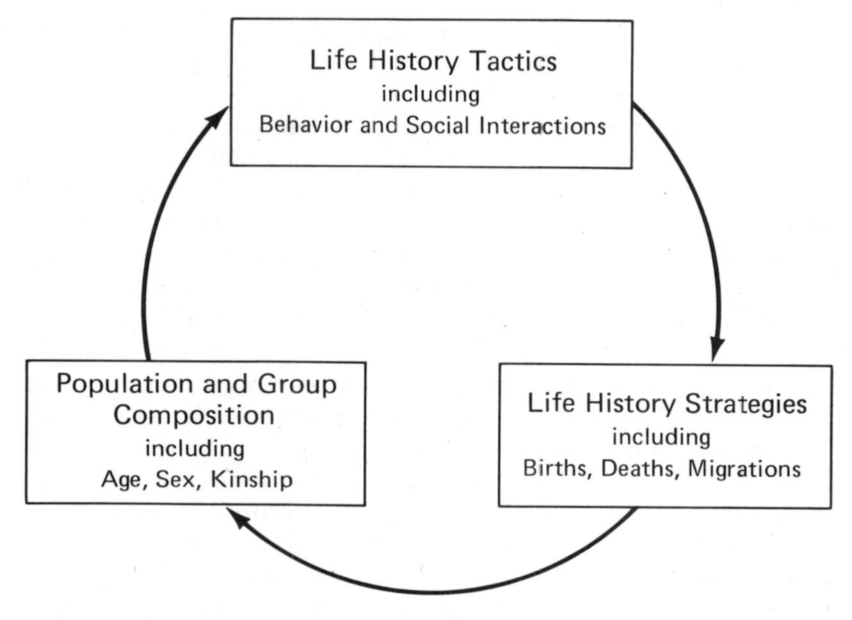

Figure 1.

The Cycle

Life History Processes Affect Demography

In 1954, Lamont Cole published a remarkable paper that succinctly laid out many demographic consequences of life history processes. The significance of that paper was largely unrecognized at the time, and for over a decade, it seldom received more than token citations in the literature. Matters changed quickly thereafter. Publications by Lewontin (1965), Gadgil and Bossert (1970), and Emlen (1970) were among the first substantial contributions to life history theory after Cole. In the last few years the blossoming of this holistic approach to biology has resulted in the publication of several theoretical papers, as well as extensive research on life history phenomena in many species.

Cole's approach was to measure the evolutionary impact of a life history phenomenon by its effect on the intrinsic rate of increase of the population, and that approach has been followed by most subsequent authors. An alternative approach, measuring the effect on reproductive value (Fisher, 1930), was proposed by Medawar (1952), but Charlesworth's recent analysis (1973) suggests that Cole's approach is preferable.

The life history concept has come to include two closely

related sets of phenomena: the basic demographic variables (primarily natality, mortality, and dispersion rates) and the biological means by which the values of these variables are altered or maintained. This distinction between means and ends is significant. For example, it might be biologically important to know whether a decline in fertility with age results from reduced access to mating partners, from decreased frequency of ovulation, reduced litter size, higher rates of spontaneous or induced abortion, or, in humans, from increased use of contraceptives. For most demographic purposes, however, only the resulting age-specific birth rates matter. To distinguish between life history means and ends, we shall call them life history tactics and strategies, respectively. (Demographers regard the latter as demographic indices or parameters.) Sheps and Menken (1973) have explored some of the relations between the two: the reaction of natality indices to variation in reproductive physiology. Stearns (1976) reviewed literature on the adaptive significance of life history tactics, which he defines as sets of coadapted traits that are designed, by natural selection, to solve particular ecological problems.

Evolutionary biologists analyze the adaptive significance of traits at various levels of organization, for example, biochemical, histological, anatomical, and behavioral. The development of life history concepts has meant recognition that selection may act on life history components as well. The life history approach has a much greater significance, however. Differential reproduction between genotypes—that is, natural selection—depends on differences between genotypes in one or more components of a species' life history strategy, and this, in turn, requires a difference in the underlying life history tactics. For some of the latter, such as litter size and age at menopause, a change will, in the absence of compensating changes in other components, directly affect one or more of the demographic variables. Others, including most forms of behavior, will affect life history strategies less directly. In either case a change in life history tactics is a prerequisite to natural selection.

As a result, natural selection acts simultaneously on life history parameters and on the heritable component of any traits that affect them. Suppose, for example, that in a population of animals some individuals are, for genetic reasons, more aggressive, that they thereby get more to eat, and thus have larger litters and raise more offspring to maturity. The result is simultaneous selection for greater aggressiveness and for larger litter size. Changes in life history parameters are the channel through which selective evolutionary changes are brought about, regardless of level of organization.

The implications for behavioral biology are clear: the selective advantage of one mode of behavior over another depends on its effects on life history strategies. Thus, it is not surprising that in recent years behavioral biologists have placed increasing emphasis on the life history consequences of behavior and social organization. King (1973) and Gauthreaux (1978) have summarized what is now known about the effects of agonistic behavior on age at sexual maturity, fertility, age-specific mortality, and patterns of dispersion. Unfortunately, no comparable surveys are available for other modes of behavior.

Demographic Processes Affect Group Size and Composition

The number and distribution, according to age, sex, and kinship, of potential social partners are produced by demographic processes through effects of the latter on group size and group composition. Changes in the size and composition of a group are brought about by just seven processes: birth, death, emigration, immigration, maturation, group fusion, and group fission, each of which is susceptible to small-sample fluctuations (demographic "drift"). At present, we do not have a formal model for predicting group size and composition from all of these processes. A model that predicts the equilibrium distribution of group size (but not composition) from the first four of these processes, that is, birth, immigration, death, and emigration (hence, the BIDE model), was developed by Cohen (1969, 1972), based on earlier work by Kendall (1949) and others. The BIDE model predictions are close to those observed in several populations of wild primates, including colobus (*Colobus guereza*), langurs (*Presbytis entellus*), howlers (*Alouatta palliata*), gibbons (*Hylobates lar* and *agilis*), and baboons (*Papio cynocephalus, ursinus,* and *anubis*) (Cohen, 1969; Keiding, 1977). Distributions predicted by the BIDE model are those of the negative binomial and poisson distributions. According to the BIDE model a population in which the per capita annual rates of birth, death, and emigration are 0.177, 0.173 and 0.55, respectively, and in which the immigration rate is 0.548 individuals per group—all reasonable values for Amboseli baboons (*Papio cynocepha-*

lus)—will have an average group size of about 50. However, in such a population about 9% of all individuals will be in groups smaller than 30, and 11% will be in groups larger than 140. Therefore, any effects on behavior and social relations of such very large and very small groups will be recurring phenomena, a point to which we shall return. The literature on primate social groups contains numerous speculations about the adaptive significance of group size. In the BIDE model the chances that an individual gives birth, emigrates, or dies in any time period are assumed to be constant, regardless of group size, and thus the group rates for these processes are proportional to the number of individuals in the group. Additionally, immigrants are assumed to ignore group size, so that all groups in a population acquire immigrants at the same rate. Thus none of these processes would correct, except fortuitously, for any deviations of a group's size from any optimum, if any optimum exists. The equilibrium distribution in the BIDE model is not a dynamic equilibrium, with compensating feedback, but a long-term, large-sample, statistical equilibrium, the result of many replications of groups, each developing independently. That is, in the BIDE model, group size is a consequence of autonomous birth, death, and migration processes. If the BIDE model is correct, natural selection must act on the birth, death, and migration rates rather than on group size per se, and thus it is these rates, not group size, for which it is appropriate to seek evolutionary explanations. Although Cohen's modelling of group demography needs to be extended in several ways (Altmann, 1972), it is by far the most elegant attempt yet made to account for the size distribution of social groups on the basis of the underlying demographic processes.

Perhaps more important than group size as a determinant of behavior is group composition, that is, the age-sex distribution and kinship relationships within social groups. The general problem of how natality, mortality, and dispersion determine the distribution of surviving kin of each class, by age and sex, in a system of groups has not yet been solved. However, recent developments in demographic theory (Keyfitz, 1977, Chapter 10) make it possible to estimate the expected number of surviving kin of each type in a population, for an individual of specified age and sex.

A few rough estimates, calculated under simplifying assumptions, will serve to illustrate the great potential for large differences in social milieu. We look first at the kin composition of a social group. The expected numbers of any class of relatives available in a group depends on that group's recent demographic history. In a population in which births greatly exceed deaths and dispersal from natal groups is low, each individual will grow up surrounded by relatives. That is exactly what happens in rapidly expanding primate populations. By contrast, consider a group of primates at or near a stationary condition, Alto's baboon group in Amboseli as of 1975. In such a group, what is the chance that a liveborn neonatal infant has a living next-older sibling? From our data (J. Altmann *et al.*, 1977 and in preparation) on mean interbirth interval and female life expectancy, we estimate that on the average the number of offspring in an adult female's lifetime is eight. Since one out of eight infants therefore will be the offspring of a primiparous female, only seven out of eight individuals have any older sibling, living or dead. The probability that such a sibling will survive from conception to age 22 months (the mean interbirth interval with a surviving infant) is 0.46. Thus the probability that a liveborn infant will have a living next-older sibling is L (0.46) M 0.40. In such a stationary primate population, most adult females will not survive long enough to be grandmothers. Thus, an infant's available playmates usually will not include either siblings, nieces, or nephews. Cousins (probably offspring of half-sibs) are more likely. However, even this likelihood will be reduced because two reproductively mature sibs may not produce offspring sufficiently close in time.

Grandparents, especially of first- or second-born infants, and older siblings, especially of later-born infants, may play an important role in an infant's life if they survive. Moreover, the occasional cases of many surviving close kin may be quite dramatic in the impact on the social group as a whole, as well as on the individuals involved. Our aim is to point out that numerous surviving close kin will be uncommon, not that they will be unimportant.

The preceding examples were based on mean values, but in small social groups chance deviations from mean values are likely to be very large. To illustrate this, we consider gender of playmates rather than kinship. What is the chance that an infant will have in its group a potential playmate of the opposite sex that is within three months of its age, that is, another infant born during the six-month period centered on a given infant's birth date? Suppose that the infant lives in a group of 50 baboons, a size that would seem to be large enough to be buffered against small-sample effects and that is about average size for stationary baboon populations. Alto's Group is a group of that size.

During 1975 the 15 adult females of this group gave birth to 5 infants per 6-month period, none stillborn. Assuming equal sex ratios at birth, the probability that 5 out of 5 infants would be of the same sex is about .06, so that even if all infants survived, about 6 infants out of every 100 in social groups of this size would not have any available playmate of the opposite sex within 3 months of their own age. Furthermore, the probability that exactly 4 out of 5 infants will be of the same sex is about .30, which means that in groups of this size, almost a third of all half-year cohorts will include an individual with no same-sex associate. If some of these infants do not survive the first year of life, the chance that, at the time these individuals enter the juvenile play groups, some will have no choice in the sex of their playmates becomes even greater; mortality during the first year of life among liveborn baboons in Ambelosi has been 29% (J. Altmann *et al.*, 1977). Beyond that, lack of a sharply defined breeding season would further increase the chance that some infants will be born at a time of year in which few (or many) others are born, thereby exaggerating variability due to small-sample effects.

In our discussion of the BIDE model we pointed out that even if life history parameters are uniform throughout a population, some individuals will, by chance, find themselves in a much smaller group than will others. They will therefore be more susceptible to effects of small-sample fluctuations in number, gender, and kin relatedness of available playmates.

Group Size and Composition Affect Behavior

That brings us to the missing link in the system: the influence of group size and composition on behavior and social organization. We shall describe some of the few published examples. Perhaps the paucity of literature on this topic stems primarily from the fact that people have been largely unaware of the likelihood that demographic characteristics of groups will influence behavior, and therefore have not looked for such effects until recently.

EFFECTS OF HIGH DENSITY AND LARGE GROUPS The most extensive studies of demographic constraints on behavior and other biological processes are the experimental rodent population studies, pioneered by Calhoun (1963, 1973) and Christian (1961, 1971). These studies have dealt primarily with effects of crowding, and have demonstrated alterations in a wide variety of physiological, behavioral, and social processes when population density is high:

> In general, raising the population density increases the rate of individual interactions, and this effect triggers a complex sequence of physiological changes: increased adrenocortical activity, depression of reproductive function, inhibition of growth, inhibition of sexual maturation, decreased resistance to disease, and inhibition of growth of nursing young apparently caused by deficient lactation. (Wilson: 1975:84)

There are, in addition, numerous natural history descriptions of the effects of crowding on many aspects of behavior, including aggression, territoriality, competition, mating behavior, and so forth. Many species of vertebrates switch over from territoriality to dominance hierarchies when population density gets above a certain point (Wilson, 1975). If two large groups of blue monkeys meet at a fig tree, a fight ensues, whereas small groups coalesce peacefully (Aldrich-Blake, 1970). Numerous other density-dependent responses have been reviewed by Wilson (1975). The single most widespread response to increased population density throughout the animal kindgom is emigration.

Those interested in potentially adverse affects of crowding on people living in cities have referred to the effects described in the literature. Yet Draper (1973) has pointed out that !Kung bushmen living in their traditional, crowded bush camps show fewer signs of stress than do those !Kung who live in sedentary villages with more personal space and privacy. She suggests that crowding per se may not be stressful for humans.

Although many anthropological publications treat population growth as an effect of cultural practices, there is growing recognition that the causality may be reversed, that is, that cultural practices may be a result of population size and growth (Polgar, 1975). For example, it has been widely accepted that the development of agriculture during the Neolithic Age resulted from technological advances and may have been one cause of the subsequent population explosion, but Boserup (1965) proposed that the technology for some agricultural intensification is readily available to most primitive, nonagricultural peoples, and that the primitive farmer is inhibited from employing this technology by the fact that more intensive land use systems are more labor-demanding in terms of output per man-hour: "gathering" requires only, or primarily, the harvest component of agriculture. Increased population density resulted in an increased demand for food and a decreased availability of land per person, creating the pressure for agriculture (compare Bronson, 1975). Important though population size or density may be, in what follows we emphasize effects on behavior of the particular composition of the local population or group.

EFFECTS OF INCREASED LIFE EXPECTANCY The vast majority of animal species do not live for any appreciable period after they produce zygotes, thereby precluding parental care or any other family-specific relationships that extend beyond one generation. Furthermore, many species that do outlive their own zygote production are spatially isolated from their offspring because the offspring are put into a special habitat (for example, marine turtle eggs) or because the zygotes are spewed out into the environment and dispersed (for instance, most aquatic invertebrates). Parental care is precluded unless there is generation overlap in time and space. Long life expectancy and low dispersal rates are prerequisites to kin-based social systems.

EFFECTS OF OVERLAPPING REPRODUCTIVE PERIODS If overlap between generations extends beyond the onset of reproduction of the filial generation, parent-offspring incest becomes possible. However, Slater (1959) pointed out that for purely demographic reasons, the likelihood of incestuous mating in humans under primitive conditions is very small: at sexual maturity, an individual is unlikely to have a surviving parent, or to have a surviving sibling that is sexually mature and of the opposite sex.

Consider, again, Alto's Group of baboons. A female will, on the average, give birth to her first infant at the age of six. In the 10–11 years before her death she will produce about 8 infants at intervals of 19 months. Even if her first offspring is a male who survives to maturity (an improbable event: $p = .10$, if male mortality rates were as low as females', which is unlikely), a mother would be unlikely to live past the age at which even her firstborn son first reaches full adult status, at about 8–10 years of age, and is breeding. Thus, mother-son incest will be rare, even with no special mechanisms (migration, taboos, and so on) to prevent such mating.

Since daughters breed at a younger age than sons do, it would at first seem that father-daughter incest would be more likely. However, by the same token, fathers are older than are mothers when they first produce offspring. Moreover, changes in adult male dominance rank, correlation of rank with reproductive success, and mortality rates that probably are higher than those of adult females may make such an event at least as unlikely as mother-son incest. The available data indicate that a father is unlikely to be in high reproductive rank positions when his daughter matures five to six years later (Hausfater, 1975; Saunders and Hausfater, 1978), thus making father-daughter incest even less likely.

As a result, the only form of incest that could occur with an appreciable frequency is between brother and sister, particularly between older brothers and younger sisters. In a moderately promiscuous society, such as that of savannah baboons, the siblings are usually half-sibs. In monogamous primate species, such as gibbons, *Hylobates lar* (Carpenter, 1940) and titi monkeys, *Callicebus moloch* (Mason, 1966, 1978), in which siblings are usually full sibs, dispersal of the young before sexual maturity may be an important mechanism for reducing such potentially detrimental inbreeding.

EFFECTS OF MALE-FEMALE COMPOSITION Increased levels of aggression among adult males coincide with the onset of estrus in adult females in many group-living mammals, including several primate species, for example, rhesus, *Macaca mulatta* (Chance, 1956; Wilson and Boelkins, 1970), baboons, *Papio sp.* (Hall and DeVore, 1965) and ring-tailed lemurs, *Lemur catta* (Jolly, 1967). Dunbar (this volume, Chapter 4) has explored some of the consequences of variations in adult sex ratios for socio-sexual interactions in gelada baboon groups. Hausfater (1975) showed that in yellow baboons, the rate of agonistic behavior in adult males was significantly lower on days when at least one female was in estrus than when none were, possibly as a result of greater intermale spacing. However, when at least one female was in estrus, males were more frequently involved in inconsistent dominance interactions, a prerequisite to rank change, and were wounded more often; that is, when at least one female was in estrus, fights involving males were less common but more dangerous. Since the expected number of females simultaneously in estrus is a probabilistic function of the number of females in the group (Altmann, 1962), the rate of aggression in a group will depend on its adult composition.

EFFECTS OF SMALL-SAMPLE FLUCTUATIONS Effects on behavior of small-sample demographic variations are poorly understood, but the potential is enormous. Anyone who has read the recent literature on the rhesus monkeys of Cayo Santiago cannot but be impressed by the pervasive influence of kin relationships on almost every aspect of social behavior and group organization in these animals (Sade, 1972, 1977). We pointed out above that an abundance of close kin is characteristic of expanding primate populations.

Consider social play. Among several species of cercopithecine primates, it has been shown that when given a choice, monkeys tend to play with other monkeys of the same sex and same age class. Conversely, Chivers (1974) believes that the low frequency and intensity of play in small, monogamous groups of siamangs (*Hylobates syndactylus*) may be attributable to the large age gap, at least two or three years, between infants. As we indicated above, the range of available social partners such as playmates may be sharply curtailed in a small population by the recent pattern of births and deaths. The resulting differences in social environment may have major effects on behavioral ontogeny. For example, Green (1978) has claimed that in humans the incidence of homosexuality depends on the composition of peer play groups at age 8–10. In a small !Kung band the play group usually consists of children of both sexes and a wide range of ages (Draper, 1976). Draper has suggested that this play group composition determines the type of play that is feasible and excludes competitive games.

Here is another example. By chance, six out of seven of the surviving 1973 infants in Alto's Group of baboons in Amboseli are female. Furthermore, there are no surviving females who were born in 1971 or 1972. For several years matters remained relatively calm and the female dominance hierarchy quite stable (Hausfater, 1975 and in preparation). This continued until mid-1976, when this entire cohort of juvenile females reached three years of age and began challenging adult females to whom they had previously been subordinate. Since then, many changes in dominance rank have occurred. The agonistic behavior has involved other individuals as well, including younger siblings of these adolescent females. It seems likely that these younger siblings will effect their own dominance changes at a younger age than did their sisters, partially as a result of their involvement in their older sisters' interactions. Neither the remarkable peacefulness of the group before this time, nor the chaos since then make much sense without knowing the demographic history of the group.

This phenomenon of composition-specific social relationships should temper our comparisons of species or even populations that have been studied at different demographic stages. For black and white colobus monkeys (*Colobus guereza*) Dunbar and Dunbar (1976) have described between-group differences in behavior that appear to be related to differences in group size and composition. These demographic characteristics of colobus groups may change in a systematic way over time as one group type develops into the next. Altmann (1968) and Altmann and Altmann (1977) have described techniques for using class-specific rates of behavior and interactions obtained from samples on one population to generate expected values for another population with a different composition.

EFFECTS OF DISPERSAL We have mentioned the dispersal of offspring from their natal group. Among animals in which there is considerable variability between species or groups in such dispersal, it may be a major cause of differences in social structure. For example, the phenomenon of "helpers at the nest," when it involves sibling assistance in parental care, is well developed in some species of birds (Skutch, 1976), but is largely precluded in many others because of post-fledgling dispersal of the young.

Dispersal of primates from their natal group has been observed in a number of field studies. We have already mentioned the role such migration might play in preventing incest between brothers and sisters. In almost all primate species most migrants are adult or subadult males. In a few cases, we now have good descriptions of the direct effects that such intergroup movements have on the behavior of the migrants and on those individuals whose groups the migrants join. Perhaps the most extreme case is that of langurs (*Presbytis entellus*), in which adult males, after immigrating into a group, kill the infants in it (Mohnot, 1971; Hrdy, 1974, 1977; Sugiyama, 1967). In addition, intergroup migration may have considerable indirect effects resulting from alterations in group composition.

It would be especially informative to know whether such intergroup migration tends to stabilize group composition, that is, whether there exists, for any age-sex class, a level of representation (number or proportion of individuals) in social groups above which emigration by members of that class tends to exceed immigration, and conversely when that class is under-represented (Altmann and Altmann, 1970, Figure 11). If so, such demographic processes will tend to stabilize corresponding composition-dependent social relations.

Inherited Social Environment

We have presented a brief survey of demography-dependent behaviors. Clearly, we are dealing primarily with terra incognita, but no less important for that.

What are the consequences for behavior of the feedback loop that results from the influence of behavior on life history processes, and thus on group composition? Surely it would be premature to attempt to answer that question, but we would like to point out one tantalizing implication of these relations. The variance in some behaviors has a heritable component; that is, some differences in behavior can be attributed in part to differences in genotype. The remainder of the variance is attributable to environmental differences, broadly speaking, and we have tried to make a case for the importance for behavior of one aspect of the environment, namely, group composition—or if you will, the social environment. But if group composition is the result of life history processes, and if these life history processes, in turn, have a heritable component, then some part of the variance in the social environment is heritable!

The concept of a heritable component to the environment may explain why the behavior of animals often exhibits a degree of species-typical stereotype that is much narrower than one would expect from their experimentally demonstrable capacity for learning and for other environmentally induced variability. If, as seems often to be the case, the members of a species develop certain forms of behavior and social responses through interactions with their parents, their siblings, and other members of their social group, such group processes are made possible by life history patterns that have a heritable component. Imprinting is practical only when there is a reasonable guarantee that the appropriate social partners will be available at the right time in an individual's life history.

Acknowledgments

Research support by grant MH 19617-07 from the U. S. Public Health Service. Helpful comments on the manuscript were provided by Joel Cohen, by Montgomery Slatkin, and by several of the Burg Wartenstein conferees. Our research on baboons in Amboseli, Kenya, was carried out under permit number ADM 13/001/C351/5,13,38,53 from the Office of the President, and was greatly facilitated by Joseph Kioko, warden of Amboseli National Park, and his predecessors.

References

Aldrich-Blake, F. P. G. 1970 Problems of social structure in forest monkeys. Pages 79-101 in *Social behavior in birds and mammals.* E. J. H. Crook. New York: Academic Press.

Altmann, J., Altmann, S. A., Hausfater, G., and McCuskey, S. A. 1977. Life history of yellow baboons: Physical development reproductive parameters, and infant mortality. *Primates* 18:315–330.

Altmann, S. A. 1962. Social behavior of anthropoid primates: Analysis of recent concepts. Pages 277–285 in *Roots of behavior: Genetics, instincts, and socialization.* Ed. E. L. Bliss. New York: Hoeber-Harper.

Altmann, S. A. 1968. Sociobiology of rhesus monkeys III: The basic communication network. *Behaviour* 32:17–32.

Altmann, S. A. 1972. [Review of] Casual groups of monkeys and men: Stochastic models of elementary social systems, by Joel E. Cohen. *Am. J. Phys. Anthrop.* 36:447–449.

Altmann, S. A., and Altmann, J. 1970. Baboon ecology. *Bibliotheca Primatol.* 12:1–220.

Altmann, S. A., and Altmann, J. 1977. On the analysis of rates of behavior. *Anim. Behav.* 25:364–372.

Boserup, E. 1965. *The conditions of agricultural growth.* Chicago: Aldine.

Bronson, B. 1975. The earliest farming: Demography as cause and consequence. Pages 53–78 in *Population, ecology and social evolution.* Ed. S. Polgar. The Hague: Mouton.

Calhoun, J. B. 1963. The ecology and sociology of the Norway rat. Paper No. 1008, U .S. Dept. of Health, Education and Welfare, Public Health Service.

Calhoun, J. B. 1973. Death squared: The explosive growth and demise of a mouse population. *Proc. Roy. Soc. Med.* 66:80–88.

Carpenter, C. R. 1940. A field study in Siam of the behavior and social relations of the gibbon (*Hylobates lar*). *Comp. Psychol. Mon.* 16: 1–212.

Chance, M. R. A. 1956. Social structure of a colony of *Macaca mulatta. Br. J. Anim. Behav.* 4:1–13.

Charlesworth, B. 1973. Selection in populations with overlapping generations. V. Natural selection and life histories. *Amer. Nat.* 107: 303–311.

Chivers, D. J. 1974. The siamang in Malaya: A field study of a primate in tropical rain forest. *Contrib. Primatol.* No. 4:1–335.

Christian, J . J. 1961. Phenomena associated with population density. *Proc. Nat. Acad. Sci.* 47:428–449.

Christian, J. J. 1971. Population density and reproductive efficiency. *Biol. Reprod.* 4:248–294.

Cohen, J. E. 1969. Natural primate troops and a stochastic population model. *Amer. Nat.* 103:455–477.

Cohen, J. E. 1972. Markov population processes as models of primate social and population dynamics. *Theo. Pop. Biol.* 3:119–134.

Cole, L. C. 1954. The population consequences of life history phenomena. *Quart. Rev. Biol.* 29:103–137.

Draper, P. 1973. Crowding among hunter-gatherers: The !Kung bushmen. *Science* 182:301–303.

Draper, P. 1976. Social and economic constraints on child life among the !Kung. Pages 199–217 in *Kalahari hunter-gatherers*. Eds. R. B. Lee and I. DeVore. Cambridge, Mass.: Harvard University Press.

Dunbar, R. I. M., and Dunbar, E. P. 1976. Contrasts in social structure among black-and-white colobus monkey groups. *Anim. Behav.* 24: 84–92.

Emlen, J. M. 1970. Age specificity and ecological theory. *Ecology* 51: 588–601.

Fisher, R. A. 1930. *The genetical theory of natural selection*. 272 pp. Oxford: Clarendon Press.

Gadgil, M., and Bossert, W. 1970. Life history consequences of natural selection. *Amer. Nat.* 104:1–24.

Gauthreaux, S. A., Jr. 1978. The ecological significance of behavioral dominance. Pages 17–34 in *Perspectives in ethology*, vol. 3. Eds. P. P. G. Bateson and P. H . Klopfer. New York: Plenum Press.

Green, R., 1978. Sexuality and aggressivity: Development in the human primate. Pages 515–528 in *Recent advances in primatology*, vol. 1. Eds. D. J . Chivers and J . Herbert. London: Academic Press.

Hall, K. R. L., and DeVore, I. 1965. Baboon social behavior. Pages 53–110 in *Primate behavior: Field studies of monkeys and apes*. Ed. I. DeVore. New York: Holt, Rinehart and Winston.

Hausfater, G. 1975. Dominance and reproduction in baboons (*Papio cynocephalus*). A quantitative analysis. *Contrib. Primatol.* no. 7: 1–150.

Hrdy, S. B. 1974. Male-male competition and infanticide among the langurs (*Presbytis entellus*) of Abu, Rajasthan. *Folia Primatol.* 22: 19–58.

Hrdy, S. B. 1977. *The langurs of Abu (Female and male strategies of reproduction)*. 336 pp. Cambridge, Mass.: Harvard University Press.

Jolly, A. 1967. Breeding synchrony in wild *Lemur catta*. Pages 3–14 in *Social communication among primates*. Ed. S. A. Altmann. Chicago: University of Chicago Press.

Keiding, N. 1977. Statistical comments on Cohen's application of a simple stochastic population model to natural primate troops. *Amer. Nat.* 111:1211–1219.

Kendall, D. G. 1949. Stochastic processes and population growth. *J. Roy. Stat. Soc.*, Ser. B 11:230–264.

Keyfitz, N. 1977. *Applied mathematical demography*. 388 pp. New York: Wiley.

King, J. A. 1973. The ecology of aggressive behavior. *Ann. Rev. Ecol. Syst.* 4:117–138.

Lewontin, R. C. 1965. Selection for colonizing ability. Pages 79–94 in *The genetics of colonizing species.* Eds. H. G. Baker and G. L. Stebbins. New York: Academic Press.

Mason, W. A. 1966. Social organization of the South American monkey, *Callicebus moloch:* A preliminary report. *Tulane Stud. Zool.* 13: 23–28.

Mason, W. A. 1978. Ontogeny of social systems. Pages 5–14 in *Recent advances in primatology,* vol. 1. Eds. D. J. Chivers and J. Herbert. London: Academic Press.

Medawar, P. B. 1952. *An unsolved problem in biology.* 52 pp. London: Lewis.

Mohnot, S. M. 1971. Some aspects of social changes and infant-killing in the hanuman langur, *Presbytis entellus* (Primates: Cercopithecidae) in western India. *Mammalia* 35:175–198.

Polgar, S. 1975. *Population, ecology, and social evolution.* 354 pp. The Hague and Paris: Mouton.

Sade, D. S. 1972. A longitudinal study of social behavior of rhesus monkeys. Pages 378–398 in *The functional and evolutionary biology of primates.* Ed. R. Tuttle. Chicago: Aldine-Atherton.

Sade, D. S., Cushing, K., Cushing, P., Dunaif, J., Figueroa, A., Kaplan, J. R., Lauer, C., Rhodes, D., and Schneider, Jr. 1977. Population dynamics in relation to social structure on Cayo Santiago. *Yrbk. Phys. Anthropol.* 20:253–262.

Saunders, C. D., and Hausfater, G. 1978. Sexual selection in baboons (*Papio cynocephalus*): A computer simulation of differential reproduction with respect to dominance rank in males. Pages 567–561 in *Recent advances in primatology,* vol. 1. Eds. D. J. Chivers and J. Herbert. London: Academic Press.

Sheps, M. C., and Menken, J. A. 1973. *Mathematical models of conception and birth.* 428 pp. Chicago: University of Chicago Press.

Skutch, A. F. 1976. *Parent birds and their young.* 503 pp. Austin: University of Texas Press.

Slater, M. K. 1959. Ecological factors in the origin of incest. *Amer. Anthrop.* 61:1042–1059.

Stearns, S. C. 1976. Life-history tactics: A reiveiw of ideas. *Quart. Rev. Biol.* 51:3–47.

Sugiyama, Y. 1967. Social organization of hanuman langurs. Pages 221–236 in *Social communication among primates.* Ed. S. A. Altmann. Chicago: University of Chicago Press.

Wilson, A. P., and Boelkins, R. C. 1970. Evidence for seasonal variation in aggressive behaviour by *Macaca mulatta. Anim. Behav.* 18: 719–724.

Wilson, E. O. 1975. *Sociobiology: The new synthesis.* 697 pp. Cambridge, Mass.: Harvard University Press.

Chapter 4

Population Demography, Social Organization, and Mating Strategies

R. I. M. Dunbar

In this chapter, Dunbar continues the line of argument begun in Chapter 3. He notes that if one simply examines the local ecology, one may fail to find observable correlations between habitat and social organization. As the Altmanns have pointed out, life history processes affect social organization, but life history processes may not directly reflect local ecology. Dunbar illustrates quite nicely that among primates it is unlikely that any natural population ever achieves a state of population equilibrium, for there are so many random processes (both ecological and nonecological), that constant perturbations will not allow a constant age structure to be reached.

Random processes introduce perturbations to idealized stationary life tables which may not dampen out even over a full generation. In small populations, such random processes may be frequent, such that life history processes may be insulated from the local ecology by the constant noise introduced by random events. As a consequence, social organization should not be expected to be finely tuned to each detail of the ecology, but only to the broadest, most pervasive pressures.

Dunbar also alerts us to the possibility that mating strategies may not be fixed in a particular species. Using the gelada as an example, he points out that males employ different strategies to obtain females, depending on particular demographic factors as well as their ontogenetic stage. Statements concerning social strategies or patterns in a species may, therefore, be time dependent and we might expect that different investigators would describe very different patterns of social organization and social behavior in the same population solely as a consequence of having studied the group under different demographic states.

Introduction

The relationship between social organization and ecology in mammals has been extensively investigated, both at the interspecific and intraspecific levels, for two decades. While it is clear that local ecological conditions will place constraints on the behavior of the animals concerned, and may in this way influence their form of social organization, other factors can influence these variables. That these other factors have tended to be overlooked doubtless reflects the fact that form of social organization and gross ecology are obvious, easily abstracted features of a population, whereas anything more complex requires considerable analytical expertise and the expenditure of a great deal of time for their extraction.

Even so, as early as the mid-sixties, Rowell (1967) drew attention to the fact that a primate population (or group) is not stable over time, but rather is in a constant state of flux as a result of the various life processes (births, deaths, migrations, and maturations). These processes can radically alter such simple features as the socionomic sex ratio, group size and composition, and so on, within a relatively short period of time and, importantly, they may do so independently of local ecological conditions. Likewise, in discussing the structure of *Colobus guereza* groups, Dunbar and Dunbar (1976) observed that the distribution of group types in a population can change with time as a result of maturational processes within individual group.

Hitherto, a strong ecological bias in both theory and field research (itself largely a consequence of a lack of data of sufficient depth) has tended to obscure the fact that mating strategies are an important part of an animal's life strategy. Insofar as mating is a prerequisite for reproduction (the less direct benefits of kin selection aside), it will play an important role in determining an animal's behavior, and through this the species' form of social structure (Goss-Custard *et al.*, 1972).

My intention in this paper is to explore some of the ways in which social structure can be influenced by factors which are not directly ecological in origin. I should, however, make it clear at the outset that I do not wish to appear to dismiss ecological factors as unimportant: on the contrary, they are fundamental determinants of what animals can and cannot do, and, indeed, we shall have cause to discuss their influence briefly below. Rather, my aim is to draw attention to some aspects of a complex, integrated system that tend to get overlooked.

I shall be developing two main themes: (1) that the age–sex structure of a population can be influenced by social and demographic processes independently of any effects due to local ecological conditions, and (2) that changes in the age–sex structure of a group (or population) can influence the mating strategies open to an individual.

Population Processes and the Sex Ratio

Variations in Age–Sex Structure

Of necessity, animals are born, mature, and die; in addition, they may migrate, either from one group to another, or from population to population. The cumulative effect of these processes over time determines both the age–sex structure of the group (or population) and its size. These processes, however, are far from stable: the sex ratio at birth, although on average close to 50:50, can show marked variation from year to year. In addition, the number of animals born each year can fluctuate widely (even if the number of breeding females remains constant). This fluctuation in the birth rate will be translated, in due course, into variations in the adult sex ratio.

Table 1 gives the age–sex structure of our main gelada baboon study band at Sankaber (Simien Mountains National Park, Ethiopia) in 1972 and 1975. Since mortality among immatures is low (survivorship[1] up to the age of 1.5 years of age is estimated to be 92%, and up to 4 years of age 88%), and appears to fall equally on both sexes, these figures give reasonable estimates of the sex ratio at birth during successive years. It can be seen that, although on average the sex ratio for all immature (less than 4 years old) classes is very close to 50:50 (and no individual year departs significantly from such a distribution), the variation in the proportion of males is surprisingly large (41% to 61%). Data given by Dittus (1975) for a *Macaca sinica* population in Ceylon shows fluctuations of similar magnitude in the neonatal sex ratio.

Table 1. Age–Sex Structure of the Sankaber Gelada Band in 1972 and 1975.

Age–class (yrs)	1972		1975	
	Males	Females	Males	Females
0–1[a]	22.5	26.5	15.5	14.5
1–2	10.5	11.5	11	13
2–3	12	13	27.5	17.5
3–4	24	22	14	20
4–5[b]	10		13	
5–6	16		11	
Adult	34	89	42	86

[a]Unsexed immatures were assigned as a half to each sex; there was usually only one unsexed immature, and never more than three, per age class.

[b]Females are counted as adults from the age of four years, whereas males are not considered to be adult until the age of six years.

Not only may the neonatal sex ratio vary from year to year within a given population, but so also may the birth rate. The mean birth rate per female per year among our gelada baboons varied from as low as 0.171 to as high as 0.571. The data for *Macaca fuscata* show that the percentage of females giving birth each year varied from under 40% to over 90% during the period 1955 to 1971 (Koyama *et al.*, 1975). Data for the *Macaca mulatta* colony at La Paguera given by Drickamer (1974) shows a range from 68% to 79% of mature females giving birth in a year during the period 1964–1972, with as few as 41% doing so in 1963. Koford (1966) gives a range of 77% to 86% for the Cayo Santiago rhesus colony during the period 1960–1964.

Causal Factors

There are three possible classes of causal factor that could contribute to fluctuations in the birth rate, namely (1) environmental conditions, (2) demographic processes, and (3) social processes. Acting through fluctuations in the birth rate, each of these may influence subsequent adult sex ratios. In addition, chance variations in the neonatal sex ratio due to small sample bias may directly affect adult sex ratios in later years. It is perhaps as well to bear in mind the possibility that individuals may be able to influence the sex of their offspring for a variety of functional reasons (Trivers and Willard, 1973).

ENVIRONMENTAL FACTORS Although there are innumerable factors that could be implicated at this level, they can act only by affecting (1) the likelihood of conception or (2) the likelihood of postconception mortality (either pre- or postnatally). Other than predation (which will generally act either by reducing the number of breeding females or by affecting the survival of maturing offspring), food availability is probably the most common cause of a reduced birth rate. Food availability, of course, covers a multitude of possibilities, ranging from gross energy availability through the availability of particular nutrients to an excess of particular toxins. By way of an illustrative example, I shall consider only gross energy availability.

Food shortage can act to prevent conception either by suppressing reproductive activity altogether, by increasing the frequency of anovulatory cycles, or by delaying maturation. It has been shown that some species at least require a minimum nutritional intake in order to be able to initiate öogenesis (for example, marmots, Andersen *et al.*, 1976; weaver birds, Jones and Ward, 1976; and possibly even patas monkeys, Rowell, 1977). There is circumstantial evidence to suggest that severe food shortage may suppress reproduction in baboons (Hall, 1963) while Zimmerman *et al.* (1975) found that a low protein diet depressed sexual activity in captive rhesus macaques. Postconception food shortage is most likely to act either by causing preterm abortion of the fetus or by causing increased neonatal mortality (due, for example, to inadequate lactation).

DEMOGRAPHIC FACTORS The existence of age-specific death rates is well known; less familiar perhaps is the fact that there are also age-specific fertility rates. These have been extensively documented in a variety of vertebrate species, including humans, and there is increasing evidence for their existence in other primates, too. Drickamer (1974), for example, found that the proportion of rhesus females giving birth each year ranged from 69% for 4-year-olds to 89% for 7-year-olds. Likewise, Dittus (1975) found that the fertility of his *Macaca sinica* females was highest in middle-aged individuals and lowest in young and old females.

Drickamer (1974) was also able to show for rhesus that the neonatal mortality rate varied according to the mother's age: younger mothers were more likely to lose their infants than were older (more experienced?) mothers. Moreover, among humans at least, the risk of fetal abnormalities increases with the age of the mother (Milham and Gittelsohn, 1965). Since this is usually attributed to the deterioration of the ova with age, there is no reason why a similar phenomenon should not occur in other species.

Given age-specific fertility rates, it is not difficult to see that the number of births in a given year will depend not only on the number of females in the population, but also on the age structure of the female complement. Populations weighted in favor of high fertility age classes will obviously yield more births than populations of the same size which are weighted in favor of the low fertility classes. Consider the simple example of a species that has only two age classes of females, young and old, whose respective fertilities are 50% and 25%. A population of 100 females made up of 25 young and 75 old individuals would be expected to produce 31 births in a year, whereas one composed of 75 young and 25 old females would produce around 44 births—no less than 40% more.

SOCIAL FACTORS In this section I shall consider just two possible factors, namely reproductive synchrony and effects due to interference from other individuals. These will act in quite different ways, the first by affecting the reproductive rate per unit time (but without affecting the overall output over an extended period), the second by actually influencing the instantaneous reproductive rate itself.

Females of nonseasonally reproducing species may nonetheless be in reproductive synchrony (for example, *Papio hamadryas,* Kummer, 1968; gelada baboons, Dunbar and Dunbar, 1975). In some cases at least, this synchrony is the result of social processes. Takeover fights, for instance, are known to precipitate sexual cycling among the females of the group in a number of species (langurs, Sugiyama, 1965; lions, Bertram, 1975; gelada, unpublished data). Even humans show some tendency towards a seasonal distribution of births (Cowgill, 1966), and, although the mechanism remains unclear, it is known that group-living human females do tend to synchronize their menstrual cycles (McClintock, 1971).

In this context, it is perhaps significant that those studies that have reported seasonal peaks in births among *Papio cynocephalus* supersp. baboons have tended to disagree about the timing of the peaks (compare data given by Lancaster and Lee, 1965; Rowell, 1969; Altmann and Altmann, 1970). Traditionally, authors have looked for environmental differences between study areas in order to account for these differences in timing. However, variability of this kind is precisely what we would expect to find if females synchronized their reproductive cycles for essentially social rather than environmental reasons.

In the Simien Mountains of Ethiopia, for example, neighboring bands of gelada baboons may be completely out of phase in the timing of their birth peaks even though it is quite clear that the females within each band are in close synchrony (Dunbar and Dunbar, 1975).

Where females are in reproductive synchrony and do not normally give birth every year, the number of births per female will obviously vary from year to year. Among the gelada, for example, each female gives birth about once every two years (Dunbar and Dunbar, 1975). With the degree of synchrony that exists among the females, this means that about two-thirds of the females in the population give birth one year, while the remaining third do so the following year. This situation is then compounded by the fact that males and females enter the breeding cohort at different ages: females become reproductively mature at puberty (3–3½ years of age), whereas males do not have the opportunity to contribute to the species' gene pool until they are about six or seven years old. It is not difficult to see that the ratio of males to females entering the breeding cohort in any given year may vary widely if the females of high output years enter at the same time as males from low output years, and vice versa. This will result in oscillations in the breeding sex ratio that may be further emphasized or damped by random fluctuations in the primary sex ratio.

The second possible factor is the presence of other individuals. A female's reproductive output may be enhanced by the presence of additional animals (if these individuals confer an advantage through, say, greater protection from predators) or it may be decreased (as when, for example, the presence of additional females results in higher infant mortality due to increased levels of aggression). The basis for the first of these claims is widely recognized; the second is perhaps less often appreciated. I shall briefly discuss two possibilities, (1) mutual interference lowering each individual female's reproductive output by the same amount, and (2) the situation in which some females are adversely affected while others are not affected at all.

The existence of an inverse relationship between group size and mean number of offspring produced per female is known for several species (for example, marmots, Downhower and Armitage, 1971). Rowell (1969) found an inverse relationship between group size and the relative frequence of perinatal mortality in her Ugandan baboon troops which resulted in a decelerating growth rate over time. These appear to be largely density-dependent effects in what we can consider the density of animals per group (social density as it were), in the same light as we do the more conventional density per area. Perhaps the most familiar examples of these density-dependent effects come from the rodent work of the fifties and sixties (Southwick, 1955a, b; Christian *et al.*, 1965). Typically in these cases, the increase in density led to a gradual breakdown in social order, this in turn resulting in a declining birth rate and ultimately the extinction of the population (Calhoun, 1973).

In some cases, however, what at first sight seems to be a density-dependent phenomenon actually turns out to be a demographic one in which the socionomic sex ratio is the critical factor. In such cases, an increase in the number of females relative to the number of males may generate levels of competition and aggression among the females that are detrimental to successful reproduction. In the extreme case reproductive activity may cease altogether. In one of our Bole Valley (Ethiopia) *Colobus guereza* groups, the aggression and "tension" generated in the group by the presence of a number of additional peripheral males appeared to have been so disruptive that no infants were born into the group or successfully reared over a period of four years (Dunbar and Dunbar, 1976).

The second possibility is that some females are able to actively reduce the fertility of other females in their group without suffering any (or at least much) loss in fertility themselves. Obviously, the most likely candidate is social dominance. Dominant females could, for example, simply practice infanticide, although I know of only one apparent case of this in the literature (African hunting dogs, van Lawick, 1974). An alternative tactic might be for the dominant female to prevent subordinate females conceiving either by denying them access to the male(s) or by causing an increase in the frequency of anovulatory cycles and/or premature abortion through harassment and stress. Such a relationship between dominance rank and reproductive output has been found for gelada females: in this case, dominant females appear to achieve their advantage by harassing subordinates, with the result that the latter take longer to become fertilized than do dominants (Dunbar and Dunbar, 1977). In the extreme case, of course, the dominant female may be able to prevent *all* other females in her group from breeding, as seems to be the case in wolves (Zimen, 1976) and African hunting dogs (Frame and Frame, 1976).

Birth Rates and Future Adult Sex Ratios

In this section I want to demonstrate by means of some elementary simulations that variations in the birth rate and/or primary sex ratio can have dramatic (and surprisingly persistent) effects on the adult sex ratio some years later. It has been known ever since Leslie (1959) first pointed it out that the existence of a time lag between birth and maturity can set up cyclic oscillations in population size that take considerable time to dampen. The present simulations show that this result also applies to the population's socionomic sex ratio.

The simulations are based on the following assumptions:

1. Each female gives birth to one infant every 2 years
2. Half of all females breed each year
3. Females give birth for the first time at 4 years of age and die at aged 20 years
4. The fertility rate is constant across all ages
5. The sex ratio at birth is exactly 50:50
6. Mortality between birth and age 20 is negligible.

By simple iteration it is possible to determine the age structure of the population at any time in the future, given knowledge of its initial composition. In the present case we began with a population of 100 4-year-old females at year 0 and the iterations were carried out over a period of 75 "years."

The first point to be noted is that, from an admittedly extreme starting point, it takes no less than 65 years (more than three times the expected lifespan of an individual) for the variation in the rate of increase of each age class from year to year to be reduced to zero (that is, for the population to achieve a stable age distribution, see Figure 1). However, Leslie (1959) has shown that the time taken to achieve a stable age distribution is more or less constant, irrespective of the initial age distribution, so that the value obtained in the present case is unlikely to be seriously in error if a less biased age structure is taken. The point is simply that, in a long-lived species which as a relatively long interbirth interval and a long maturity lag, and disruption of the equilibrium age-structure takes so long to dampen that it is unlikely that any natural population ever achieves a state of equilibrium. Indeed, potentially disruptive environmental factors, such as droughts, often occur more frequently than the time taken to dampen their effects on the population's age structure. In East Africa, for instance, years of severe drought appear to

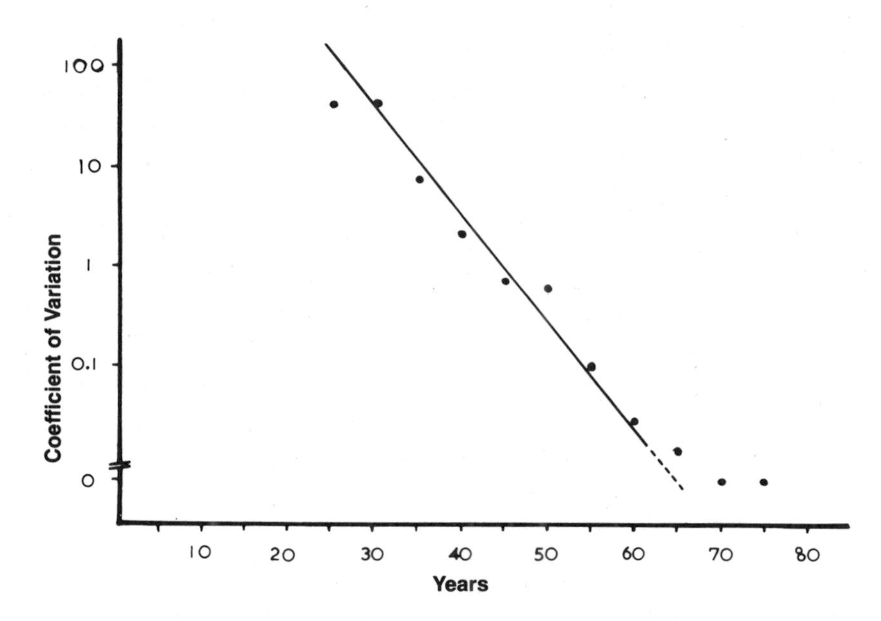

Figure 1. Coefficient of Variation in the percentage increase in each age class from year to year plotted against time from a simulation of population growth for a species with a constant birth rate (see text for details).

occur on a fairly regular cyclical basis with a periodicity of around 10–15 years (Wood and Lovett, 1974). It is more than likely, in other words, that at least minor perturbations will always be present.

The second point to be made is that disruptions of the birth rate can affect subsequent adult sex ratios. Figure 2 shows what happens to the sex ratio of the breeding cohort if no infants are born in year 0. In this case, we begin with a population with a stable age distribution (year 75 in Figure 2). We add the additional assumption that, while females are considered to be reproductively mature at 4 years of age, males do not reach maturity until the age of 6 years. However, like females, males die at age 20 years. The effects of a sudden short-fall in births were then followed through over the subsequent 30 years. It can be seen that, since the zero-birth year enters the female cohort first, there is an initial fall in the sex ratio, followed by a rise as the zero-year enters the adult male cohort two years later. It takes three more years for the sex ratio to return to approximately normal levels, and even then it continues to oscillate slightly

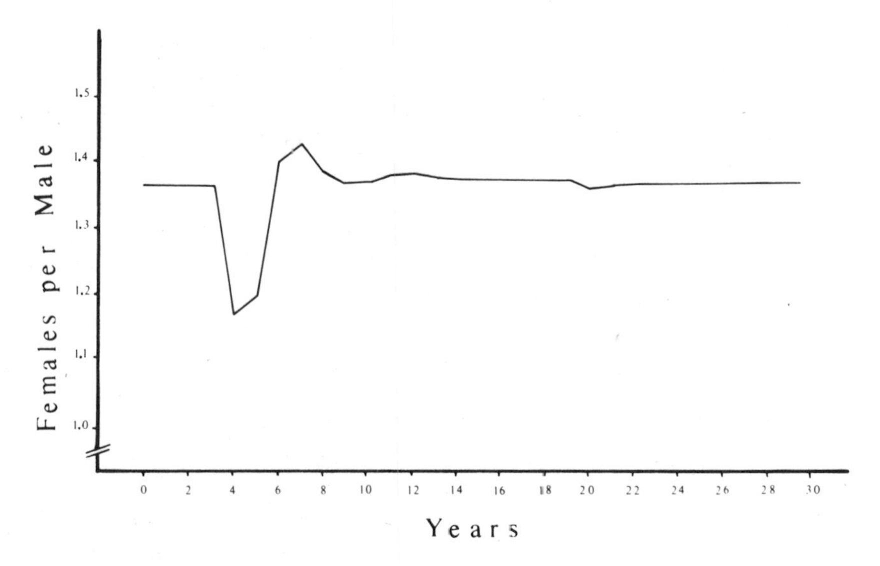

Figure 2. Changes in the adult male: female ratio in a population at equilibrium (year 75 from Figure 1) following a year (year 0) in which no infants are born.

above this level until there is a further minor perturbation in year 20 as the zero-year leaves the population at death, this time simultaneously for both sexes. Had we assumed differential mortality for the two sexes, this perturbation would have been rather more prominent. These final oscillations do not become completely damped even by year 30, although their magnitude is obviously very small.

The magnitude of these perturbations will be affected by a difference between the sexes in any age-related parameter. It will, for example, be inversely related to the animals' mean lifespan, whereas it will be positively related to the discrepancy in the ages at which the two sexes enter the adult cohort. With the assumptions of the present simulation, the magnitude of the oscillation is 18.5% of the value at equilibrium; however, it rises to 21% if the age at which males mature is raised to 7 years (while all other parameters remain unchanged).

Of course there will probably be few occasions in nature when the disruptive event is quite so extreme. Nonetheless, any tendency for the pattern of births to deviate from a uniform distribution (as previously assumed), or for the sex ratio at birth to vary, will tend to exacerbate fluctuations due to other causes. The main point here is simply that isolated events can cause surprisingly large disruptions that persist for many years. In long-lived, slowly reproducing species like primates, the cumulative effect of a variety of small perturbations (due both to chance and to specific causes) are likely to result in populations achieving stable age distributions only rarely.

The influence of birth rates and neonatal sex ratios on future adult sex ratios can be more realistically illustrated using data from the compositions of our main study band of gelada (Table 1). The age-sex structure of the band in each of the two years shown can be used to predict future adult sex ratios by subtracting some function representing deaths during the year from the current total of adults, and then adding the current number of 3-year-old females and 5-year-old males to the remainder. For simplicity, I have here assumed (1) a negligible death rate for nonadults, (2) a constant death rate for both sexes of adults (4 or more years old) derived from the observed death rate for all adults (0.1172 deaths/animal/year), and (3) no migration. Proceeding on this basis, we can determine what the breeding sex ratio will be in future years, given the composition of the population for each year in Table 1. (The breeding cohorts here include all females older than 4 years and all males older than 6 years.)

The results of the computations are given in Table 2. It can be seen that (1) the baseline ratios for the two years differ (1:2.6 as against 1:2.0), and (2) both ratios undergo considerable change over time. In general, these changes tend towards a levelling of the sex ratio, but this is not always so. Finally, (3) the growth rate for each sex from year to year is far from constant: indeed, in some years, the growth rate is zero or even negative. These irregularities mean that the growth rates for the two sexes are to all intents and purposes independent, given the time lags that are involved. Consequently, their ratios change continuously.

The Adult Sex Ratio and Mating Strategies

I shall now consider some consequences that variations in the adult sex ratio can have on an individual's mating strategies. I shall begin by briefly reviewing some of the evidence showing that adult sex ratios do vary over time in natural populations, and then go on to discuss the effects of changes in the socionomic sex ratio on intrasexual competition. Finally, I shall use some of

Table 2. Predicted Adult Sex Ratios for the Sankaber Gelada Band Based on Its Compositions in 1972 and 1975 (see Table 1).

Years from	*1972*			*1975*		
Start	*Males*	*Females*	*Ratio (F:M)*	*Males*	*Females*	*Ratio (F:M)*
0	34	89	2.618	42	86	2.048
1	44.1	100.6	2.281	46.8	95.9	2.049
2	46.7	101.8	2.180	51.4	102.2	1.988
3	59.9	101.4	1.693	56.3	103.2	1.833
4	62.2	116.0	1.865	71.1	105.6	1.485

Numbers of animals of each sex in any given year were obtained using the age-sex composition of the band as given in Table 1, and assuming a constant mortality rate for all adults (namely the observed average adult mortality rate of 0.1172 deaths/animal/year) and negligible mortality for immatures.

our gelada data to illustrate the way in which changes in the structure of a population can affect the choice of mating tactics open to an individual.

Group Size, Composition, and the Sex Ratio

Table 3 gives data on the structure of our main Simien gelada band at two different periods of time. It will be noted that between 1971 and 1974 there were marked changes in the structure of the population, despite the fact that population size and composition remained virtually unchanged. These changes involved (1) an increase in the mean number of mature females per reproductive unit (and hence in overall mean unit size), (2) an increase in the proportion of multimale reproductive units (with some units containing up to 6 mature males), and (3) an increase in the proportion of adult males in all–male groups. The latter was due in part to a decrease in the number of reproductive units and in part to a slight increase in the relative number of adult males in the population. That this is not an isolated case is indicated by the fact that changes also occurred in the nearby Geech population of gelada between 1971 and 1973 (compare Dunbar and Dunbar, 1975, with Ohsawa and Kawai, 1975, and Mori and Kawai, 1975). In this case, the Khadadit band, for example, increased in size from 14 reproductive units (with 2 multimale groups) to 17 units (with no multimale groups).

Changes in the adult sex ratio over time have been documented for a number of other species. Rowell (1969), for instance, found that the adult sex ratio of her study groups of *Papio anubis* varied from approximately 1:1 to 1:1.5 over a five-

Table 3. Structure of the Sankaber Gelada Band in 1971 and 1974.

	1971	*1974*
Total band size	262	272
Adult male:female ratio	1:2.78	1:2.26
Number of reproductive units	25	17
Mean size of reproductive units	9.32	13.53
Mean number of females per reproductive unit	3.52	5.12
Mean number of males per reproductive unit	1.20	1.53
Percent of reproductive units which were multimale	16.0	41.2
Number of all-male groups (AMGs)	3	4
Mean size of AMGs	7.67	10.50
Percent of adult males in AMGs	6.3	31.6
Percent of adult males "owning" reproductive units	78.1	44.7

year period, due, in part at least, to the migration of males. In addition, however, differential maturation combined with changing birth and death rates almost certainly played a part. In the Arashiyama A troop of *Macaca fuscata*, Koyama *et al.* (1975) found that the percentage of males in the adult section varied from as low as 20% to as high as 40% over the period 1955–1972. Likewise, Southwick and Siddiqi (1977) showed that the percentage of males varied from around 22% to 35% in a semiprotected troop of *Macaca mulatta* at Chhatari in India.

Migration and Intrasexual Competition

The most obvious (and perhaps general) effect of an imbalance in the socionomic sex ratio is that it will increase the level of competition among members of the sparse sex for mates. In most cases the effect will probably be greatest among the males, but in some cases, at least, there may be similar consequences for females. In species that live in one-male reproductive units, the males usually compete for control over the whole unit irrespective of the reproductive condition of the individual females at the time. Once "ownership" has been established, however, the females may compete amongst themselves to some considerable extent for access to the group's breeding male.

Among males in multimale groups, competition for access to reproductive females will tend to increase levels of intermale aggression as the ratio of females to males declines. The presence of cycling females is known to cause an increase in the level of aggression among males in a number of species (Wilson and

Boelkins, 1970; Packer, 1977; Jolly, 1967): it is particularly interesting to note, however, Hausfater's (1975) finding that as the number of cycling females increases beyond one (that is, as the ratio of cycling females to males in the group increases further), so the level of aggression (as indicated by the frequency of wounding) tends to decline again.

Two consequences of this are likely to be the instability of relationships among the animals concerned (for example, Hausfater, 1975; Packer, 1977) and increased migration (for example, Boelkins and Wilson, 1972). It seems intuitively likely that a male whose reproductive expectations are low due to a high male-male competition would do better to move to another group which had fewer (or weaker) males. It appears to be the case that most (if not all) males in baboon and macaque troops migrate at some point during their lives, very often as subadult or young adult males (Rowell, 1969; Packer, 1975; Boelkins and Wilson, 1972; Drickamer and Vessey, 1973; Sugiyama and Ohsawa, 1975). This is to be expected, since a change of strategy will be more beneficial for a younger animal than it will be for an older one (Dunbar, in press).

In certain cases females may be affected in like manner. Among the gelada, for instance, there is a linear relationship between a female's dominance rank in her unit and her reproductive output (Dunbar and Dunbar, 1977). Consequently, as the number of females in a group increases, so the reproductive expectations of the lowest ranking females decrease. At some point their expectations will be so low that it will pay them to move to other groups where their *effective* ranks will be higher. That this can be an important determinant of female behavior is evidenced by the fact that the females of both gorilla and the African hunting dog may migrate when their breeding success becomes too low (Harcourt *et al.*, 1976; Frame and Frame, 1976). In the case of the gelada it may place an upper limit on group size by making females in large groups more willing to desert their males, thereby causing groups above a certain size to be inherently unstable (see the following section). It is important to appreciate that it is *relative* rank that is important here, not absolute fighting ability (or whatever social skills and subterfuges may confer a reproductive advantage). The last-ranking female in a group of four will have a higher reproductive success than the last-ranking female in a group of ten, whatever their abilities relative to each other might be.

Gelada Baboon Mating Strategies

Gelada males have two main strategies whereby they can acquire females with whom to form their own harems, namely (1) by fighting an incumbent unit leader and taking over the entire unit intact, or (2) by joining a unit as a follower and building up a nuclear unit with one or two of the unit's juvenile or peripheral females (Dunbar and Dunbar, 1975). (There are, in addition, a number of alternative strategies, but these are by and large less important in terms of frequency of occurrence.) I do not wish to digress into a discussion of the complex story of these social processes, but certain points are important to an understanding of how population structure can affect a male's options. Three points should be noted.

1. In order to take over an entire unit, the young male has to "persuade" the females of the unit to interact with him, irrespective of whether or not he can physically defeat the incumbent unit leader. This, in turn, is partly dependent on the number of females in the unit. Kummer (1975) has suggested that a female's "loyalty" to her male may depend on the extent to which she interacts with him, and we have previously noted that females whose dominance rank is low will be most willing to seek a new male. Consequently, it is hardly surprising to find that gelada males rarely (if ever) attempt to take over small units with only a few females. Our field data indicate that a minimum of 5 postpuberty females are required for an attempted takeover to be successful, and males show little interest in units with fewer than 5 females.

2. The average reproductive output that a male can expect over his lifetime is much the same for a takeover strategist as it is for a follower, even though the former starts his reproductive career with more females than does the latter. This is due to the fact that, while reproductive output per unit time is an approximately linear function of the number of females that a male "owns," tenure as a unit leader is much lower for a takeover strategist than it is for a follower. Over a lifetime, these differences more or less cancel each other out. Thus, there is no intrinsic reason why a male should prefer one strategy over the other, since the expected net gain is the same either way.

3. The risks incurred from a takeover fight are much greater than those incurred by a follower, since, by fighting, the former necessarily risks both defeat and serious injury. Consequently, males are only likely to attempt a takeover if the

expected gain is high. By and large, older males (with a shorter reproductive life remaining to them) should opt for the takeover strategy, while younger males will prefer the less risky follower strategy: there is evidence from the field to suggest that this is in fact so.

Given condition (1) above, a male's options are likely to be influenced and constrained by the structure of the population, since a successful takeover depends on the availability of large units. Hence, the lower the proportion of units "at risk," the smaller will be the number of males that can acquire harems by takeover. The remainder will have either to become followers (a possibly suboptimal strategy for older males) or wait for the structure of the population to change through the normal maturational processes (and perhaps risk the possibility of failing altogether). Conversely, when the proportion of units at risk is high, more males will be able to take over units. Figure 3 shows that the probability of a unit being taken over during the course of a year increases as a function of the mean number of females per unit in the band as a whole. Two points are worth noting. First, the regression fitted to the data points indicates that the limit is reached at a mean unit size of 11 females, at which point the probability of all units being taken over during the course of a year is unity. No unit in any population in the Simien Mountains had more than 9 adult females. Second, with an average of 3 females per unit, the likelihood of a unit being taken over is zero. A unit with 3 adult females can be expected to have approximately 1 postpuberty juvenile, so that, as noted above, the probability that a unit will be taken over (that is, the risk that it runs) is negligible for units with 4 or fewer postpuberty females.

In general, for a constant population size and composition the proportion of males "looking for harems" will be related to mean-unit size. The larger the mean size of units, the fewer units there will be for a given number of females, and hence the fewer the number of males who will be able to "own" units: consequently, for a constant number of males, the more males there will be in all-male groups waiting to acquire females. As a result, the pressure on the reproductive units (that is, the risk of being attacked) will be higher. An example of how the relative deployment of males and females in reproductive and all-male groups can affect population structure and the animals' behavior comes from a natural experiment involving the Bole Valley gelada population (see Dunbar, 1977). Our present understanding of the situation is, briefly, as follows.

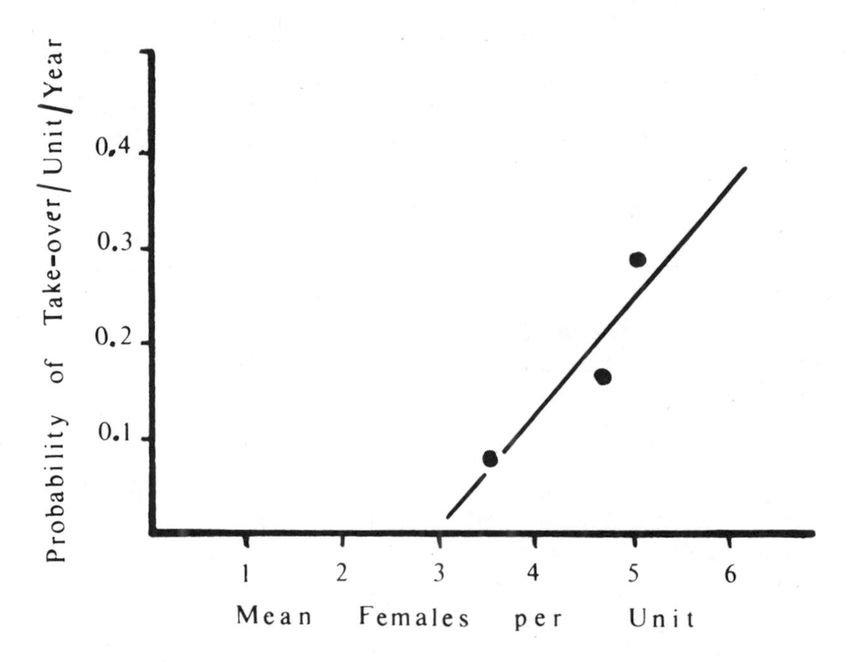

Figure 3. Probability of a gelada reproductive unit being taken over during the course of a year plotted against mean number of adult females per unit for three different bands.

Every eight years most of the adult males of the population are shot by the local tribesmen. This results in a large number of leaderless groups of females which apparently fuse to form large units under the control of the few remaining adult males. As the immature males mature into the population over the succeeding years, there is intense competition for access to the harems and the young males are able to break up these large units by stealing females from them. Thus, at the start of the cycle, the adult sex ratio is heavily biased in favor of females, and the population consists of a few very large units. As time passes the sex ratio tends towards that typical of undisturbed populations, and at the same time the number of reproductive units increases while their size decreases as they are broken up by the young males. During the early part of the cycle young males are able to steal females from the large, unstable units and the frequency of attacks on reproductive units by all-male groups is high. In contrast, later in the cycle, the males cannot do this so easily since small units are more stable, and the frequency of attacks appears to decline in consequence.

In practice, of course, the ratio of takeovers to follower-entries will depend on the adult sex ratio at the time and on whether the males have any alternative strategies, such as migration, open to them. High emigration of surplus males may result in a sudden reduction in the competition for units, and hence in the pressure on those units that are at risk. Migration, as we have noted, can be expected to be a general response to a disadvantageous situation, and it is doubtless for this reason that gelada all-male groups are highly mobile, often spending considerable periods of time with different bands. Mobility of this kind very probably permits young males to locate the most favorable conditions in the area. There is some evidence to suggest that many (but by no means all) gelada males acquire harems in bands other than their known or presumed natal bands.

Among the gelada there is a tendency for unit size to increase steadily with time as a result of the maturation of female offspring (contrary to our former supposition—see Dunbar and Dunbar, 1975). Unit size can only decrease as a result of fission, death, or (rarely) emigration. Consequently, unless the combined effect of these last three processes is equal to or greater than the natural rate of maturation, mean-unit size will tend to increase. The likelihood that these two rates will be more or less in balance is probably negligible for most populations, so that mean-unit size will tend to fluctuate over time. As a result, what appears to be the typical male mating strategy will vary with time. Some evidence that this may be so in the gelada population at Sankaber in the Simien Mountains is given in Figure 4, which shows that the proportion of males entering units who did so by way of a takeover (as opposed to becoming a follower) increased as a linear function of the proportion of units in the population that were at risk.

Conclusions

This paper has attempted to suggest that some features in the behavior and social organization of a population of animals may respond to factors other than environmental conditions. There are several implications of this, which I shall discuss only briefly.

First, in almost all cases, our study periods are far too short. As a result we are rarely able to distinguish clearly between

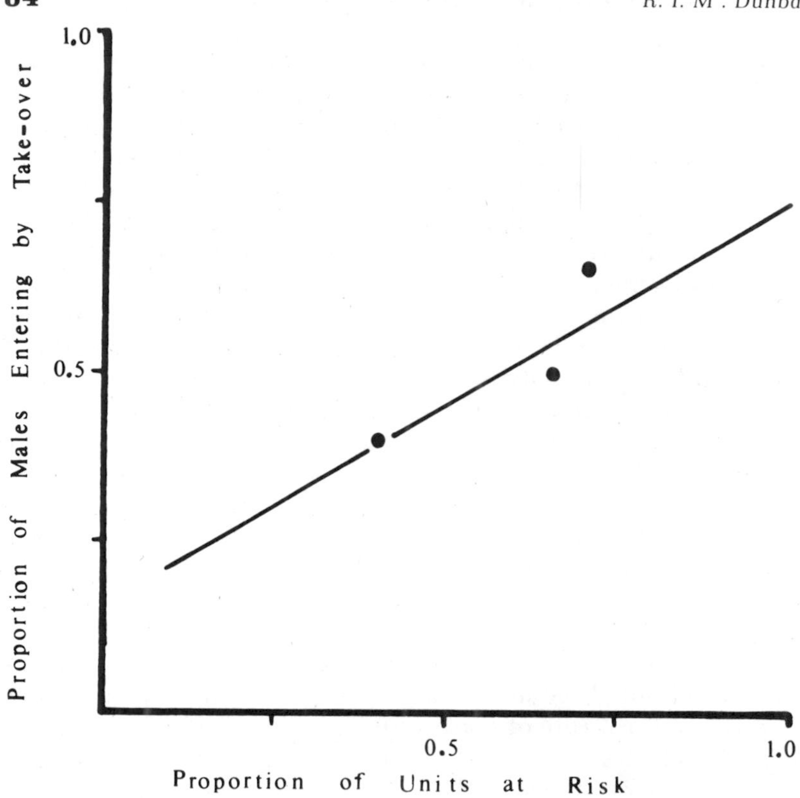

Figure 4. Proportion of males entering gelada reproductive units who did so by means of a takeover (as opposed to entering as a follower) plotted against the proportion of units in the band that were "at risk of being takeover" (that is, had 4 or more adult females): data for three different bands.

those features of social organization that are due to (1) gross ecological conditions, (2) local or short-term (for example, seasonal) changes in ecology, and (3) the kinds of demographic and social processes discussed above.

Second, to argue for species or population specific characteristics on the basis of such short studies is to oversimplify an extremely complex situation. Uncritical faith in the efficacy of explanations based on differences in local ecological conditions may be both misleading and may overlook important areas for research.

Third, much greater attention has to be paid to the dynamic aspects of social systems than has hitherto been the case. We need to know not only why a species has the mean value it does, but also why individual populations or groups differ from the species' mean in the way they do. Not all of these differences can be dismissed as random variation or statistical error of no biological significance.

Finally, it should, perhaps, be added that animals face life-historical problems other than the purely ecological. For evolution to operate, individuals must reproduce, and this inevitably means competition for mates. Individuals have, therefore, to apportion their available time and energy between the sometimes conflicting needs of maintenance and reproduction (Gadgil and Bossert, 1970). Reproductive strategies are not always determined completely by ecological conditions, although there is no doubt that they may be often influenced by them, directly or indirectly. It is perhaps pertinent to note in conclusion Ghiselin's (1966) caution against interpreting "adaptive" solely in terms of environmental referents: to do so, he observes, causes one both to overlook the animal's reproductive features and to fail to understand why Darwin became so interested in sexual selection.

Notes

1. Survivorship was determined indirectly from the observed mortality rates within age classes during the same time period and is thus based on a "synthetic" life table (*sensu* Keyfitz, 1968). It is not permissible to use the actual population age structure given in Table 1 to construct a life table: doing so is valid *only* if it can be assumed that the population has a *stationary* age distribution (Caughley, 1966, 1977; Caughley and Birch, 1971). This restriction has commonly been overlooked (for an example, *see* Dittus, 1975, whose proofs of stationarity *assume* that the population is in fact stationary).

References

Altmann, S. A., and Altmann, J. 1970. *Baboon ecology: African field research.* Chicago: University of Chicago Press.

Andersen, D. C., Armitage, K. B., and Hoffman, R. S. 1976. Socioecology of marmots: Female reproductive strategies. *Ecology* 57:552–560.

Bertram, B. C. R. 1975. Social factors influencing reproduction in wild lions. *J. Zool. Lond.* 177:463–482.

Boelkins, R. C., and Wilson, A. P. 1972. Intergroup social dynamics of the Cayo Santiago rhesus (*Macaca mulatta*) with special reference to changes in group membership by males. *Primates* 13:125–139.

Calhoun, J. B. 1973. Death squared: The explosive growth and demise of a mouse population. *Proc. Roy. Soc. Med.* 66:80–88.

Caughley, G. 1966. Mortality patterns in mammals. *Ecology* 47:906–918.

Caughley, G. 1977. *Analysis of vertebrate populations.* Chichester: Wiley.

Caughley, G., and Birch, L. C. 1971. Rate of increase. *J. Wildl. Manage.* 35:658–663.

Christian, J. J., Lloyd, J . A., and Davis, D. E. 1965. The role of endocrines in the self-regulation of mammalian populations. *Recent Prog. Horm. Res.* 21:507–578.

Cowgill, U. M . 1966. Season of birth in man. Contemporary situation with special reference to Europe and the southern hemisphere. *Ecology* 47:614–623.

Dittus, W. P. J. 1975. Population dynamics of the toque monkey, *Macaca sinica.* Pages 125–151 in *Socioecology and psychology of primates.* Ed. R. Tuttle. The Hague: Mouton.

Downhower, J . F., and Armitage, K. B. 1971. The yellow-bellied marmot and the evolution of polygamy. *Amer. Nat.* 105:355–370.

Drickamer, L. C. 1974. A ten-year summary of reproductive data for free-ranging *Macaca mulatta. Folia Primat.* 21:61–80.

Drickamer, L. C., and Vessey, S. H. 1973. Group changing in free-ranging male rhesus monkeys. *Primates* 14:359–368.

Dunbar, R. I. M. 1977. The gelada baboon: Status and conservation. Pages 363–383 in *Primate conservation.* Eds. Prince Rainier of Monaco and G. H. Bourne. New York: Academic Press.

Dunbar, R. I. M. In press. Social systems as optimal strategy sets. In *Perspectives on adaptation, population and environment.* Ed. J. B. Calhoun.

Dunbar, R. I. M., and Dunbar, P. 1975. *Social dynamics of gelada baboons. Contrib. Primatol.* vol. 6. Basel: Karger.

Dunbar, R. I. M., and Dunbar, P. 1976. Contrasts in social structure among black-and-white colobus monkey groups. *Anim. Behav.* 24:84–92.

Dunbar, R. I. M., and Dunbar, P. 1977. Dominance and reproductive success among female gelada baboons. *Nature, Lond.* 266:351–352.

Frame, L. H ., and Frame, G. W. 1976. Female African wild dogs migrate. *Nature, Lond.* 263:227–229.

Gadgil, M., and Bossert, W. H . 1970. Life historical consequences of natural selection. *Amer. Nat.* 104:1–24.

Ghiselin, M. T. 1966. On semantic pitfalls of biological adaptation. *Philos. Science* 33:147–153.

Goss-Custard, J . D., Dunbar, R. I. M., and Aldrich-Blake, F. P. G. 1972. Survival, mating and rearing strategies in the evolution of primate social structure. *Folia Primat.* 17:1–19.

Hall, K. R. L. 1963. Variations in the ecology of the chacma baboon, *Papio ursinus. Symp. Zool. Soc. Lond.* 10:1–28.

Harcourt, A. H ., Stewart, K. S., and Fossey, D. 1976. Male emigration and female transfer in wild mountain gorillas. *Nature, Lond.* 263:226–227.

Hausfater, G. 1975. *Dominance and reproduction in baboons* (Papio cynocephalus): *A quantitative analysis. Contrib. Primatol.* vol. 7. Basel: Karger.

Jolly, A. 1967. Breeding synchrony in wild *Lemur catta.* Pages 3–14 in *Social communication among primates.* Ed. S. Altmann. Chicago: University of Chicago Press.

Jones, P. J., and Ward, P. 1976. The level of reserve protein as the proximate factor controlling the timing of breeding and clutch-size in the red-billed quelea *Quelea quelea. Ibis* 118:547–574.

Keyfitz, N . 1968. *Introduction to the mathematics of population.* Reading, Mass.: Addison-Wesley.

Koford, C. B. 1966. Population changes in rhesus monkeys: Cayo Santiago 1960–1964. *Tulane Stud. Zool.* 13:1–7.

Koyama, N., Norikoshi, K., and Mano, T. 1975. Population dynamics of Japanese monkeys at Arashiyama. Pages 411–417 in *Contemporary primatology.* Eds. M. Kawai, S. Kondo and A. Ehara. Basel: Karger.

Kummer, H. 1968. *Social organization of hamadryas baboons.* Basel: Karger.

Kummer, H . 1975. Rules of dyad and group formation among captive gelada baboons (*Theropithecus gelada*). Pages 129–159 in *Proc. Symp. 5th Cong. Int. Primat. Soc. (1974).* Eds. S. Kondo, M. Kawai, A. Ehara, and S. Kawamura. Tokyo: Japan Science Press.

Lancaster, J. B., and Lee, R. B. 1965. The annual reproductive cycle in monkeys and apes. Pages 486–513 in *Primate behavior: Field studies of monkeys and apes.* Ed. I. DeVore. New York: Holt, Rinehart and Winston.

Leslie, P. H . 1959. The properties of a certain lag type of population growth and the influence of an external random factor on a number of such populations. *Physiol. Zool.* 32:151–159.

McClintock, M. K. 1971. Menstrual synchrony and suppression. *Nature, Lond.* 229:244–245.

Milham, S., and Gittelsohn, A. M. 1965. Parental age and malformations. *Human Biol.* 3:13–22.

Mori, U., and Kawai, M. 1975. Social relations and behaviour of gelada baboons: Studies of gelada society (II). Pages 470–474 in *Contemporary primatology.* Eds. M. Kawai, S. Kondo and A. Ehara. Basel: Karger.

Ohsawa, H., and Kawai, M. 1975. Social structure of gelada baboons: Studies of gelada society (I). Pages 464–469 in *Contemporary primatology.* Eds. M. Kawai, S. Kondo and A. Ehara. Basel: Karger.

Packer, C. 1975. Male transfer in olive baboon. *Nature, Lond.* 255:219–220.

Packer, C. 1977. Reciprocal altruism in *Papio anubis. Nature, Lond.* 265:441–443.

Rowell, T. E. 1967. Variability in the social organization of primates. Pages 219–235 in *Primate ethology.* Ed. D. Morris. London: Weidenfeld and Nicholson.

Rowell, T. E. 1969. Long-term changes in a population of Ugandan baboons. *Folia Primat.* 11:241–254.

Rowell, T. E. 1977. Variation in age at puberty in monkeys. *Folia Primat.* 27:284–296.

Southwick, C. H. 1955a. The population dynamics of confined house mice supplied with unlimited food. *Ecology* 36:212–225.

Southwick, C. H. 1955b. Regulatory mechanisms of house mouse populations: Social behaviour affecting litter survival. *Ecology* 36: 627–634.

Southwick, C. H., and Siddiqi, M. F. 1977. Population dynamics of rhesus monkeys in northern India. Pages 339–362 in *Primate conservation.* Eds. Prince Rainier of Monaco and G. H. Bourne. New York: Academic Press.

Sugiyama, Y. 1965. On the social change of hanuman langurs (*Presbytis entellus*) in their natural condition. *Primates* 6:381–418.

Sugiyama, Y., and Ohsawa, H. 1975. Life history of male Japanese macaques at Ryozenyama. Pages 407–410 in *Contemporary primatology.* Eds. M. Kawai, S. Kondo and A. Ehara. Basel: Karger.

Trivers, R. L., and Willard, D. E. 1973. Natural selection of parental ability to vary the sex ratio of offspring. *Science* 179:90–92.

van Lawick, H. 1974. *Solo: The story of an african wild dog.* Boston: Houghton Mifflin.

Wilson, A. P., and Boelkins, R. C. 1970. Evidence for seasonal variation in aggressive behaviour by *Macaca mulatta. Anim. Behav.* 18: 719–724.

Zimen, E. 1976. On the regulation of pack size in wolves. *Z. Tierpsychol.* 40:300–341.

Zimmerman, R. R., Strobel, D. A., Steere, P., and Geist, C. R. 1975. Behavior and malnutrition in the rhesus monkey. Pages 241–306 in *Primate behaviour: Developments in field and laboratory research,* vol. 4. Ed. L. A. Rosenblum. New York: Academic Press.

Chapter 5

The Phylogenetic and Ontogenetic Variables that Shape Behavior and Social Organization

John D. Baldwin
Janice I. Baldwin

The Baldwins, in this chapter, point to yet one more source of variability in observed social organizations. Not only will genetic factors fail to produce one to one correlations between social behavior, social organization and ecology, but one must remember that many nongenetic factors influence behavior. It is especially true that among the Primates, the mammalian order which we regard as having the highest levels of behavioral plasticity, learning may profoundly influence expressed social behavior. The ability to learn may in itself be a result of past selection, but what modifications an individual makes in its behavior, as a consequence of experience, may not be readily predicted from an evolutionary framework.

The Baldwins take the position that social organization may reflect direct responses to the ecology as individuals modify their behavior in response to the laws of conditioning and learning. Fixed action patterns may account for only a small fraction of primate behavior and ontogenetic studies will be absolutely essential complements to phylogenetic studies. Moreover, in social animals such as the primates, ontogenetic processes will include the influence of the social environment on behavior. Socialization processes and the establishment of traditions may therefore account for much of the social organization we see. Thus, learning contingencies may both permit more direct response to the immediate environment and simultaneously introduce other intervening variables between ecology and behavior, inasmuch as some responses will be determined not by direct solution of environmental problems, but rather by an interaction with past conditions mediated through learned traditions acquired in the socialization process. The variance of behavior and the variance of social organization then become important subjects for study in their own right and not just necessary considerations in the study of normative patterns for a species.

Introduction

During the past decades there have been an increasing number of field studies demonstrating the prevalence and importance of within-species variability in primate behavior and social organization. For example, different troops of Japanese macaques show variations in traditions, styles of troop leadership, levels of aggression, rates of peripheralization, socionomic sex ratio, troop size, and so forth (Itani, 1959; Itani *et al.*, 1963; Kawai, 1964, 1965; Miyadi, 1964; Mizuhara, 1964; Frisch, 1968; Yamada, 1971). Within-species variance in behavior and social organization has also been discovered in field studies on chimpanzees (Reynolds and Reynolds, 1965; van Lawick-Goodall, 1968; Sugiyama, 1968; Suzuki, 1969), rhesus monkeys (Neville, 1968, 1969; Singh, 1968; Southwick, 1972), Gibraltar macaques (Burton, 1972, 1977), langurs (Sugiyama, 1964, 1965, 1967; Jay, 1965; Yoshiba, 1968), vervets (Gartlan, 1966; Struhsaker, 1967a, b; Gartlan and Brain, 1968), baboons (DeVore and Hall, 1965; Rowell, 1966; Crook and Aldrich-Blake, 1968; Paterson, 1973), spider monkeys (Durham, 1971), squirrel monkeys (Baldwin and Baldwin, 1971, 1973), and several others.

Spuhler and Jorde (1975) conducted a multivariate quantitative analysis on 19 primate variables across 29 population samples and found that phylogenetic and/or genetic factors accounted for less than one half of the variance. They conclude: "In general, local ecological settings, and probably local social traditions transmitted by learning, are equally, if not more, important determinants of primate social behavior ... " compared with phylogenetic determinants.

The thesis of this paper is that these data on behavioral variability in primates can play a key role in the growth and further development of ethology and sociobiology.

Early ethology was primarily concerned with instincts and genetically determined behavior that emerges with little influence from the local environment (for example, Heinroth, 1910; Lorenz, 1935, 1937; Tinbergen, 1950, 1951). The classic deprivation experiments were designed to demonstrate that species-specific behavior patterns would develop even when an individual was (presumed to be) totally isolated from opportunities to learn those patterns. However, this approach to the study of behavior drew criticisms (for instance, Lehrman, 1953; Schneirla, 1956; Hebb, 1953) because it ran the risk of neglecting ontogenetic determinants, such as nutrition, exercise, nonspecific sensory stimulation, and learning.

By the 1960s it was clear that ethologists were concerned that both phylogenetic and ontogenetic causes of behavior should be interwoven in their theories of behavior. Tinbergen (1963) stressed that there are four problems that must be dealt with for the complete analysis of behavior: three that Huxley had emphasized—causation, survival value, and evolution—and one that Tinbergen added—ontogeny. Lorenz (1965) responded to the criticisms of ethology by describing his perception of modern ethology as encompassing both phylogenetic and ontogenetic causes.

However, theory often precedes data by a considerable period, and the traditional concerns for phylogeny and adaptation continued to predominate in ethological research. In 1970 Kummer (1971) stated that research on primate behavior was still biased toward functional interpretations of adaptations and that more attention should be focused on proximal causal factors. In part the imbalance was due to ethology's historical orientation toward the phylogenetic causes of behavior; but in addition, the right kinds of data were not sufficiently abundant to stimulate interest in ontogeny and proximal causal mechanisms.

With the advent of sociobiology, there has been renewed interest in the phylogenetic causes of behavior. The work of Hamilton (1964a, b, 1970), Trivers (1971, 1972), Wilson (1975), and others stresses population genetics, evolutionary biology, natural selection, and the genetic control of behavior. In Wilson's (1975) encyclopedic text on sociobiology, ontogeny receives only scant attention compared with evolutionary factors. Socialization is reduced to a "multiplier effect" (pp. 11–13) which purportedly allows small genetic influences to create large effects on behavior and social organization. The emphasis is that "natural selection is the agent that molds virtually all of the characteristics of species" (p. 67).

As long as the behavior observed by ethologists and sociobiologists appeared to be relatively invariable, there seemed to be little need to study ontogeny or proximal causal mechanisms. If an organ, anatomic structure, or behavior develops with little or no variance in all members of a given species (even in deprivation experiments) it is easy to conclude that phylogenetic and genetic causes adequately explain the phenomenon. Wilson (1975:22) espouses the ethological notion that "behavior and social structure, like all other biological phenomena, can be studied as organs, extensions of the genes that exist because of their superior adaptive value."

Because the behavior and social organization of many species are often reported to be quite stereotyped and "organlike," phylogenetic explanations continue to dominate in much animal research. As a consequence, ontogenetic mechanisms have received inadequate attention and hence are not as well understood as phylogenetic causes. The field data on primate variability are changing our perception of the organlike fixedness of behavior. Research on the origins of behavioral variability promises to entwine both phylogenetic and ontogenetic causal factors and hence, to advance the ultimate theoretical synthesis outlined by Lorenz and Tinbergen in the 1960s.

Proximal Causes of Behavior

Although the proximal, ontogenetic causes of behavior have been underemphasized in ethology and sociobiology, they impinge on the developing organism from the moment of fertilization to the end of life. Proximal mechanisms allow behavior to be shaped by current environmental conditions and certain randomizing processes.

Nutrition plays a central role, since food, vitamins, and minerals provide the building blocks, and calories provide the energy, from which all the organs of the developing individual are constructed. Because those organs are the behavioral machinery that mediates the eventual behavior and social patterns shown by the individual, the environment can have a substantial effect on behavior through nutritional inputs (Zimmerman *et al.*, 1972; Elias and Samonds, 1974). Poisons, disease, thermal extremes, desiccation, and physical injury can damage the behavioral machinery of the developing organism and hence also exert environmental influences on behavior. Physical exercise has been shown to be crucial in developing many of the muscle sets involved in the individual's full behavioral capacity (Fagen, 1976). Sensory stimulation to the perceptual and central nervous systems serves in a similar manner to "exercise" those components of the behavioral machinery, and is crucial for developing the individual's full behavioral capacity (Riesen, 1961, 1965; Volkmar and Greenough, 1972; Rosenzweig, 1976). Finally, learning experience (via classical and operant conditioning, along with observational learning) shapes the response patterns of the individual throughout its life. All these proximal environmental factors can induce variance into the behavioral patterns shown within a species.

Not only is individual behavior shaped by proximal environmental factors, but troop structure and social organization are also profoundly influenced by proximal causes. For example, environmental variance in food supply can affect physiological development sufficiently that animals in one environment may breed later and/or less frequently than in another (Rowell, this volume, Chapter 1). A brief change in breeding patterns can, in turn, affect the population structure of troops, the age composition of sociometric networks, and other aspects of the troop social organization for generations (Dunbar, this volume, Chapter 4). Differences in social organization guarantee that the animals will have different learning environments, different life histories, and different socialization experiences which can introduce significant variance in the behavioral repertoire, style, and individuality of the animals. Individual variability can, in turn, have dramatic effects on social organization and the socialization of latter generations. These reciprocal influences can be seen nicely in the Japanese macaques where the temperament of the leader male and the adult females strongly influences troop social organization and peripheralization (Mizuhara, 1964; Yamada, 1971). Because the animals living in troops with nonaggressive leaders and tolerant females are exposed to very different adult models and reward structures for aggression and competition than are the young raised in troops with aggressive leaders and rejective females, it is not surprising that the temperament of many of the troop members of both troops is influenced by the adults' individuality (Yamada, 1971).

Of all the ontogenetic mechanisms that allow within-species

variability in primate behavior and social organization, learning is probably the most important. Learning mechanisms help adjust each animal's behavior to current environmental conditions, overriding or displacing fixed, innate patterns established under strong genetic control. With the evolution of increased cortical size and complexity within the primate order, there is a general trend for increased learning ability and discriminative (or "cognitive") capacity (Rumbaugh and McCormack, 1967; Rumbaugh, 1970). Increased learning capacity reduces the impact of direct genetic control over behavior for two reasons. First, increased learning potential makes possible a more extensive repertoire of learned behavior which becomes predominant over unlearned behavioral units. Second, increased learning capacity allows animals to survive with less and less preprogrammed behavior, which in turn allows innate patterns to drop out of their repertoire.

As a consequence, in primates it is necessary to turn to the study of learning mechanisms in order to account for the particular behavioral variants observed in a given environment. In addition to increased general learning capacity, the increased role of observational learning (or imitation) seen in the monkeys and apes greatly facilitates the acquisition of traditional behavior. As traditional patterns accumulate over generations, they increase the complexity of learned responses each generation can acquire from its social environment. Finally, the evolution of increased learning capacity and linguistic ability in hominids allowed the social environment to become several magnitudes more complex and learned responses to deviate ever further from the strict control of genetic preprogramming.

As a species' current environment takes greater control over the shaping of each individual's behavior, phylogenetic mechanisms account for less and less of the behavior observed. In an interview with the *Harvard Crimson,* Wilson estimated that perhaps only 10% of human behavior could be traced to biological causes (Sahlins, 1976:65). In his primer on sociobiology, Barash (1977:41) presents a figure that approximates the genetic component of behavior in humans at 13% and in nonhuman mammals as between 16% and 42%. Although such numbers are only crude estimates, it is clear that when dealing with primates, especially anthropoid apes, early hominids, and humans, proximal causes become crucial in explaining most of the specifics of behavior.

Ecology Shapes Behavior

Both phylogenetic and ontogenetic mechanisms tend to produce adaptive (but can produce maladaptive) patterns, reflecting the influences of ecological conditions.

In the evolution of any genetically controlled trait, it is the environment that selects which individuals survive and reproduce and which do not. An animal breeder, a conspecific, a predator, an infection or some other environmental factor determines whether a given individual will live, die, or reproduce at a given point in time. Figure 1 shows the manner by which selection can alter the frequency of a genetically determined trait in a population by favoring the propagation of one subset of the variance and selecting against other subsets. Because the environment is the agent of selection, ecology has long played a key role in explaining the phylogenetic history of genetically controlled traits.

During behavioral ontogeny, the environment also has a powerful effect on behavior, though on a different time scale and operating through different mechanisms. An animal's genetic inheritance establishes a potential range of behavior that may develop, with many degrees of freedom open for environmental influence. As is shown in Figure 2, environmental constraints determine which subset of a species' potential will develop, with some parallelism to the selection shown in Figure 1. Abundant food and nutrients enhance the likelihood that the genetic potential will be fulfilled (upward arrows) whereas deprivation conditions thwart development (downward arrows). Abundant environmental sensory stimulation facilitates nerve development and behavioral potential, whereas sensory deprivation has adverse effects. Injury, poisoning, disease, and lack of exercise all add further constraints on the degrees of freedom allowed in a species' genetic potential. Thus, each individual is clearly a product of its current environment.

Learning provides a powerful mechanism for molding an animal's behavior to current environmental conditions. As a result of classical conditioning, one primate may learn a food aversion elicited by the sight of small red berries, whereas a conspecific in a different environment may learn a food aversion elicited by large green fruits. Sickness resulting from bad food is an unconditioned stimulus which elicits the aversive reflex of regurgitation. The local environment determines which condi-

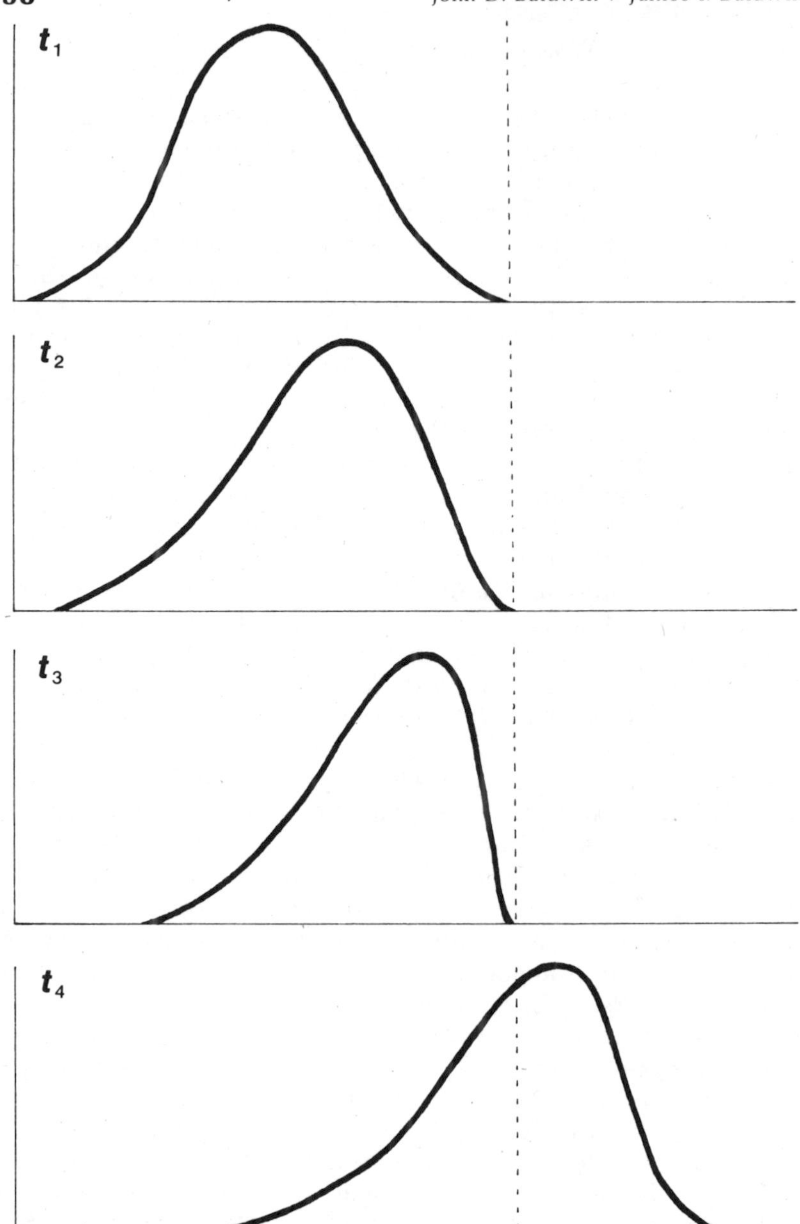

Figure 1. The natural variation of a given trait in an isolated population at an initial time (t_1) becomes modified if natural selection favors one extreme of the variance and not the other. Without mutations, the trait could not evolve (at t_2 and t_3) further than the limit of the initial variation (dotted line). With mutations to augment the initial variation, the trait can evolve beyond the initial limit (t_4).

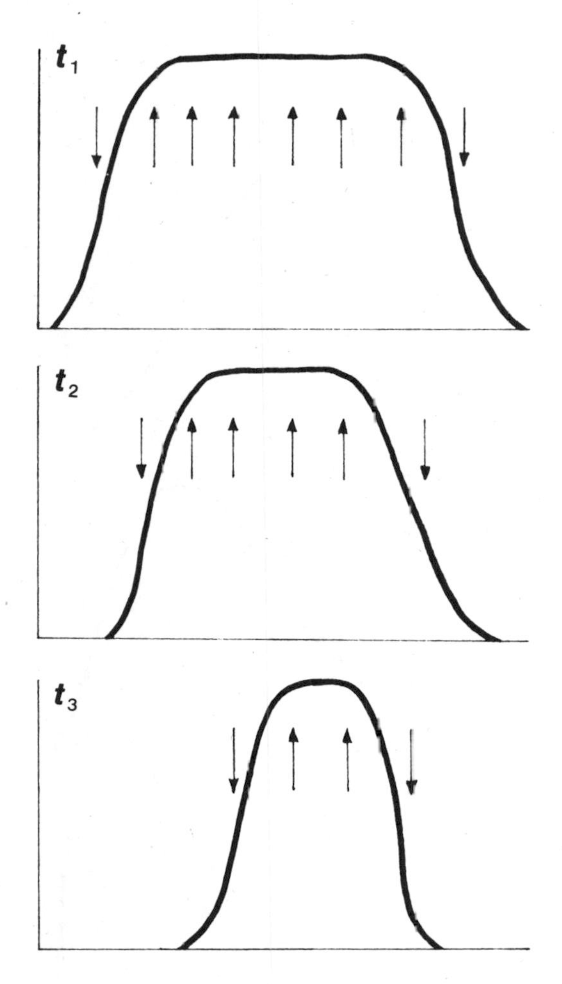

Figure 2. Early in an individual's developmental history (t_1), a broad range of the species potential is open. Good nutrition, exercise, sensory stimulation, and other advantageous conditions (upward arrows) keep part of the potential open. Injury, disease, thermal extremes, and other damaging conditions (downward arrows) progressively restrict the range of traits that can be realized from the species potential.

tional stimulus—for example, red berries or green fruits—will become associated with this biologically established reflex. Operant conditioning modifies behavior via differential reinforcement, where one subset of behavioral variance is reinforced while another is extinguished or punished. As is shown in Figure 3, frequencies of behavior are shaped by the environment in

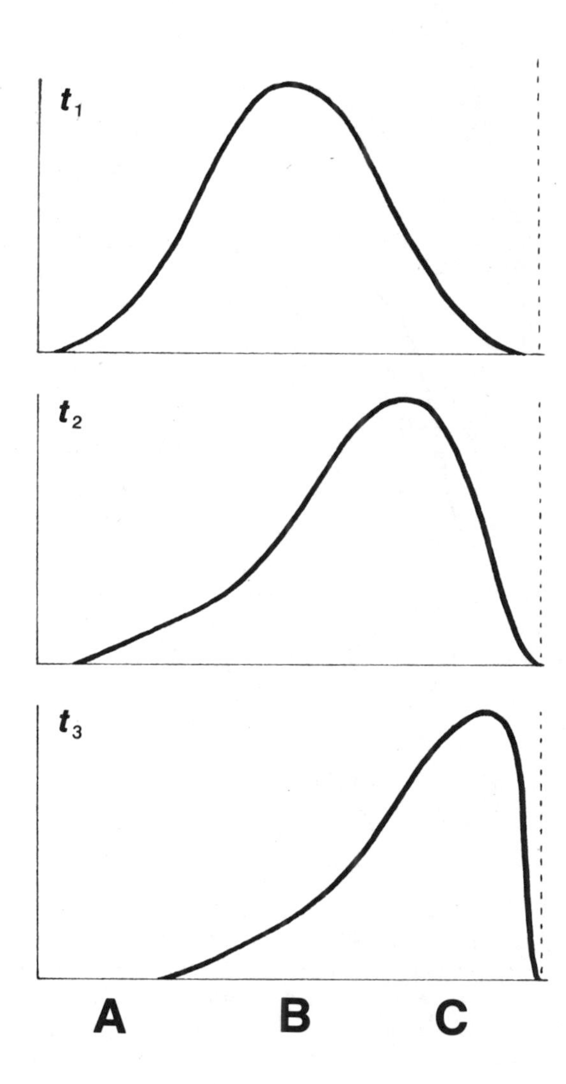

Figure 3. The natural variation in a particular operant performance at an initial time (t_1) becomes modified as differential reinforcement causes a reduction in the frequency of the nonreinforced variants (A and B) and an increase in the frequency of the reinforced variant (C). Without the possibility of extra response variation, the final operant performance could not be shaped further than the limit of the initial variation (dotted line).

ways which strongly parallel the environmental selection of genetically determined traits (Figure 1). For example, a low muscle tension behavior (A) may be inadequate to open a palm nut, while medium (B) and high (C) muscle tension behavior are progressively more successful. The primate who began with variable muscle tension when pulling to open the nut (at t_1) would be shaped toward high muscle tension patterns as the low muscle tension patterns were extinguished and the high muscle tension patterns were reinforced.

Observational learning is affected by the environment in two ways: first, the observer is influenced by the behavior of the model it sees or hears, such that the observer may learn and replicate the behavior; second, the reinforcement patterns that follow the imitative performance determine whether the behavior will become a high or low probability response in the observer's repertoire. For example, if a juvenile howler monkey attempts to imitate an adult male's jump across a certain wide arboreal gap, the juvenile may take a painful fall which will partially inhibit (1) the acquisition of the tendency to jump at the gap in questions, and (2) future imitation of similar adult male behavior. Traditional behavior is also shaped by environmental reinforcers. If sweet potato washing or termiting produces more reinforcement (in the quantity or quality of food) than other feeding behavior, the traditional pattern will be established in the behavior repertoire of any new animals who imitate the response and maintained in each animal's behavior repertoire after initial learning. If the environment changes such that an old tradition no longer produces reinforcement, the behavior will be extinguished and eventually drop out of the troop's repertoire of traditions. Likewise, specific items of linguistically encoded knowledge also rise and fall in frequency depending on the reinforcement patterns in a given environment. For example, many modern farmers are learning complex verbal information about the proper use of machinery, pesticides, fertilizers, and new hybrids because these practices are associated with high levels of reinforcement. Simultaneously, much of our culture is losing the verbal and nonverbal traditions relating to horse-powered farm technology because of the lower levels of food and profit reinforcement associated with horses, smithies, cart-wrights, and so forth.

It has long been recognized that ecological factors play a key role in explaining phylogenetically determined traits. It is equally important to realize that—in species where behavior is highly influenced by ontogeny—current ecological factors will shape much of behavior and account for much of the variance seen within a species. However, the time scale and intervening mechanisms involved in the phylogenetic selection and ontogenetic shaping of behavior are extremely different. It may take hundreds of years to change the genetic inheritance of a primate species; yet radical behavioral changes can be induced within seconds (for example, injury or learning), hours (for example, disease or poisoning), or longer (for example, nutrition or sensory stimulation). As ecological pressures modify gene frequencies, the effects of environmental selection are stored in the genetic code, a very conservative storage mechanism that remains unaffected by most short-term environmental perturbations. The ecological factors that shape ontogeny are stored extragenetically and allow for greater flexibility and plasticity. The brain is an especially designed organ for rapid processing and storing of environmental influences. Transmission routes for genetically encoded information follow kin lines, whereas learned behavior can be acquired directly from the nonsocial environment or from any subset of the social environment, kin and nonkin alike. Because primate behavior *is* variable and *is* shaped significantly by ontogenetic factors, the failure to disentangle the various causal determinants of behavior can result in a variety of errors when considering the means of behavioral acquisition, rate of behavior change, role of kin, questions of altruism,[1] and so forth.

An Analysis of Variance

Because behavioral variance within species is a key data source for expanding on genetic theories and for advancing the integration of phylogenetic and ontogenetic causes of behavior, it is instructive to examine the nature of that variance in closer detail. There are multiple sources of variance which can be superimposed upon each other, accumulating to account for a substantial component of any given behavior.

Genetic variation is the least difficult for biological theories of behavior to assimilate. Mutations serve as the primary source of novel genetic variations in a population. Sexual reproduction spreads mutations and enhances the mixing of genes within the population. The gene pool of any given population contains enough variance that no two individuals are quite the same, unless monozygous twins; and this variance among individuals is the raw material on which natural selection can operate. In the long term, populations with moderate variance can usually survive and adapt to greater environmental change than populations with limited variance.

Although genetic variance has been dealt with in biological theories of behavior, there are many degrees of freedom in behavioral machanisms not accounted for by genetic factors alone. Variations not explained by genetic variability are induced by either (1) environmental influences, or (2) randomizing mechanisms.

The first source of environmentally induced variance was discussed in the previous section. Local differences in nutrition, toxins, disease, exercise demands, sensory stimulation, and learning conditions create within species variations that can be traced to current environmental conditions.

Second, the behavioral mechanisms that mediate learned behavior induce randomness into learned behavior, creating behavioral variants which cannot be predicted from either genetic or environmental causes. Variability in the performance of a given operant behavior provides the raw material from which new operant performances can be conditioned (much as mutations provide the raw material for natural selection). Hence, the machinery of behavior has been selected to introduce a moderate level of variance into operant behavior (or was not selected to produce perfect replication of operant performances). Figure 3 portrays a hypothetical case of operant conditioning without randomization mechanisms. Starting from an original variable response (perhaps a reflex), differential reinforcement shapes the high muscle tension subset of the variance to high frequency: eventually the animal performs at muscle tension C, and lower muscle tension performances A and B have been extinguished. Without randomness superimposed on performance C, there would be no chance for the animal to learn muscle tension levels higher than level C.

Figure 4 depicts the more realistic situation: the original variability of the operant behavior is augmented by response induction, modeling effects, generalization from other high muscle tension acts, faulty replication of prior patterns, influences from exogenous variables (such as differences in hormone titer, general arousal level, fatigue), and so forth. If high levels of muscle tension (D) produce more reinforcement than does level C, the random variance above C will provide the animal with chances to be shaped by the environmental reinforcers. Thus, randomness in operant performances can be beneficial in facilitating operant conditioning. As discussed elsewhere, learning principles help explain why the behavior of young primates has

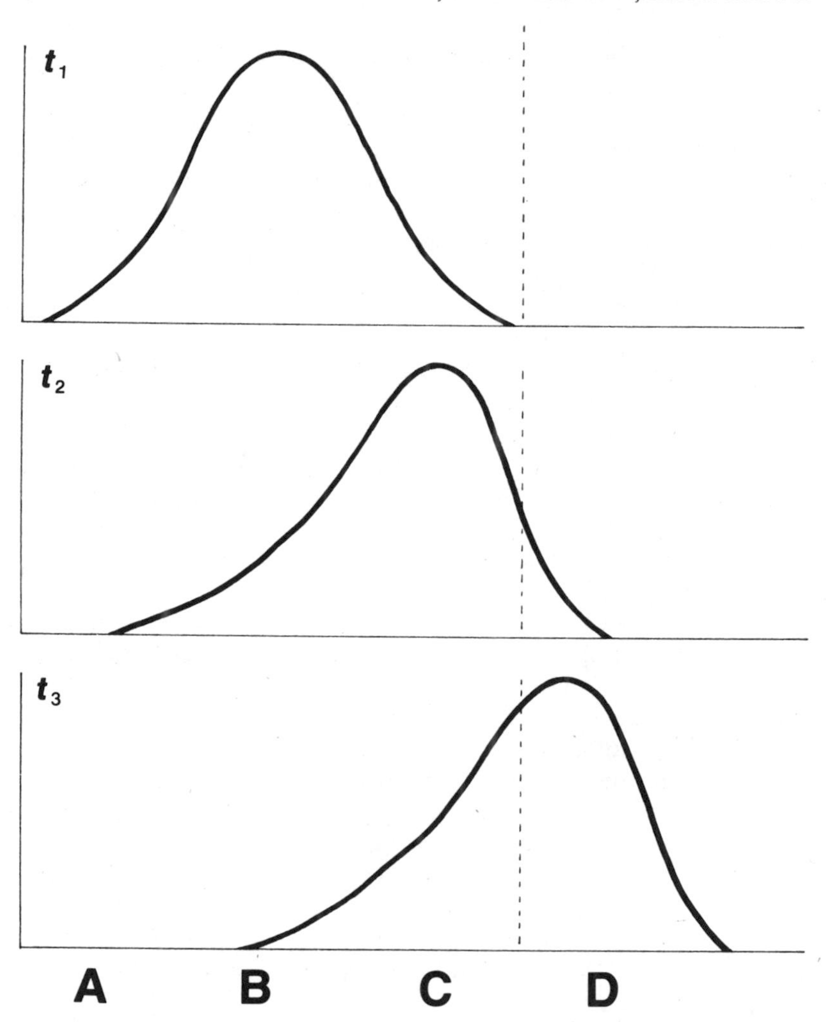

Figure 4. In contrast with Figure 3, this figure shows how randomized response variation is superimposed on the initial variance to allow operant conditioning to shape operant patterns beyond the limits of the initial variation (dotted line).

more randomness than that of older, more practiced animals (Baldwin and Baldwin, 1977:375–382).

Since observationally learned and traditional behavior are maintained or suppressed by the patterns of environmental reinforcement and punishment, these types of learned behavior also benefit by the extra level of adaptive shaping afforded by behavioral variance. For example, when a young chimpanzee imitates tool construction for termiting, the imitation is seldom a perfect copy of the model's performance (van Lawick-Goodall, 1968) and hence needs to be shaped by the reinforcement or punishment resulting from the success or failure of the tool. If there were no response variance, an imperfect imitation might fail to produce food reinforcement and each unvaried repetition yield the same nonreinforcing results. But response variance gives the animal more opportunities to learn, because some imitations will succeed better than others and these become reinforced while the poor imitations are extinguished. Behavioral variance is the raw material from which the environment shapes increased operant skill via differential reinforcement.

There is at least one other feature of primate learned behavior that introduces extra levels of variance into operant performances. Novel sensory input has reinforcement properties for primates. Although extremely novel stimuli tend to be aversive, moderately novel stimuli have the properties of positive reinforcers and can condition higher frequences of those operant behaviors which commonly precede the novel inputs (Baldwin and Baldwin, 1977, 1978). Exploration, play, and innovative behavior all tend to produce novel experiences and hence are reinforced by the novelty effects. If a primate repeats activity X without variance, it produces less novelty and hence will receive fewer novelty reinforcers than if it repeats the activity with variance. Thus, novelty reinforces animals for varying old patterns, for innovating, for creating new patterns. Young primates tend to be influenced by novelty reinforcers more frequently than older animals because older individuals have already explored most of the possibilities in their environments and (due to familiarization) ceased to find most things novel. It is the young who still have the greatest amount of untapped novelty to explore and who are more frequently reinforced by novelty reinforcers for further exploration, play, and creativity. Hence, younger primates tend to generate a higher frequency of the explorative and playful behavior which may lead to useful innovations. The habits of exploration, play and creativity tend to be extinguished and replaced by less variable routines by middle or late adulthood (Baldwin and Baldwin, 1977:375–382). Again, the behavioral variance generated by creative and innovative activity is the raw material from which new operants can be conditioned by success and failure: and it is the young who are most likely to be the innovators and perfecters of effective new ways of coping with the environment. The protocultural innovations made by Japanese macaques are most often made by juveniles who are old enough to have learned a repertoire of skills but young enough to be frequently reinforced for novel, playful, and innovative behavior (Tsumori, 1967). In addition, some animals (for example, Imo) appear to be more successful innovators and to have contributed more to their troop traditions than others (Kawai, 1965). As Kitahara-Frisch (1977a) points out, play is closely involved in the invention and transmission of traditional behavior. Tool use also appears to be facilitated by play experience (Schiller, 1957; van Lawick-Goodall, 1970; Kitahara-Frisch, 1977b). Given the importance of traditions and tool use in hominid evolution, the randomness introduced by playfulness and novelty reinforcers might be expected to be greater in these species than in other primates.

Increased Ontogenetic Control Over Behavior

Using a cost-benefit analysis, it is clear that in many species, individuals can benefit in increased fitness when natural selection produces behavior mechanisms that facilitate behavioral plasticity. Phylogenetic controls are extremely conservative and slow to respond to environmental changes. In addition, innate responses are usually tied to simple cue stimuli, which makes them error prone when the environmental stimuli are complex, variable, or unusual. Learning permits rapid behavioral change and subtle discriminative capacity (Moynihan, 1976:217). Opportunistic species and species living in variable environments would be especially handicapped if the species' behavior repertoire were strongly fixed by genetic factors. In these species, the evolutionary advantage would go to those animals with behavioral mechanisms that allowed them to be shaped rapidly by current environmental conditions, to learn by observing their troopmates, and to accumulate traditional behaviors associated with reinforcement. Thus, in many mammalian species, the genes have relinquished their hegemony over behavior, allowing increased control by the current environment. As Gartlan (1973) observed, the genes determine how much variance a species may exhibit, but not the variant shown. The current environment and randomizing factors explain which variants will appear in any given individual. Thus, an adequate analysis of the relationship of behavior to ecology will require data on a species' phylogenetic history and individual ontogenetic development.

Although it has been recognized that biological factors put constraints on learning, it is important to realize that increased specialization for learning also puts constraints on natural selection, closing certain options while opening others. Species that have been selected to have a high learning capacity can no longer have as much of their behavior directly controlled by genetic "design." Also, learning potential creates a "preparedness" for certain evolutionary directions to pay off well and for others to be counterproductive. For example, as less behavior becomes genetically determined and more degrees of freedom enter a species' repertoire, it becomes advantageous for the young to have a longer dependency period, protected by others, in which they can learn from their troopmates and from the nonsocial environment (Washburn and Hamburg, 1965:613). Thus, increased dependence on learning constrains evolution to favor certain possibilities, but not others. Because of the advantages of minimizing interference between fixed action patterns and learned responses, evolution would favor dropping fixed patterns from the species' repertoire. Since environments in which the young must learn to survive and be social can be variable, there would be a selective pressure to overbuild diverse and redundant mechanisms so many behavioral patterns could be learned via different routes.

Since learning produces behavioral outputs via different mechanisms than those used by natural selection (for example, transmission of behaviors among nonkin rather than by kinship alone), the optimizing functions predicted by genetic theory, selfish genes, and evolutionary biology are compromised, too. For primates to gain the increased fitness resulting from behavioral plasticity, many of the optimal genetic strategies attributed to insects or birds had to be compromised.

Genetic Idealism

Variance—whether induced by genetic or learning mechanisms—plays an important role in allowing the environment to select or shape future behavioral patterns which tune the species to changing ecological conditions.

Prior to Darwin many biologists had difficulties seeing within species variance (Mayr, 1972). One of the major intellectual obstacles that hindered the development of evolutionary thought was that of idealism, or "essentialism." Plato was the earliest philosopher to articulate idealism, and his impact has not ceased resonating through modern thought. Thomas Aquinas further entrenched idealism into the modern Western thought: "The observed vast variability of the world has no more reality, according to this philosophy, than the shadows of an object on a cave wall, as Plato expressed it in his analogy" (p. 984). Only the unchanging "ideas" behind the flux are important. Idealism made it easy to believe that each species was perfectly adapted to its environment, as if a benevolent Designer had optimized each species' fit to its ecological niche. Operating within this paradigm, the scientist could observe a real lion but be thinking of an idealized lion, the invariant form located in a Platonic heaven. This type of idealism hinders the perception of variance, and was a major obstacle in the discovery and acceptance of species variability and evolutionary theory.

Modern ethology and sociobiology often describe maximizing conditions and optimally adaptive traits, drawing attention to idealized, formal models to the point that variance becomes invisible. Variance is often described as merely "noise" that is superimposed on the system, a nuisance which clutters or confounds otherwise lovely mathematical equations. In his valedictory as president of the International Congress of Genetics, Curt Stern captured the leitmotif of genetic idealism nicely when he said: "The eggs or sperm of a lion are the lion themselves, stripped of all ephemeral attributes" (Douglas, 1973). This form of genetic idealism allows one to forget all the "slop" and imperfections introduced when the developing lion has suboptimal nutrition, a serious infection that retards development, an injured paw, or unique learning experiences. The strategy of genetic idealism is pleasing to those who wish to believe that genetic causes will explain "everything important" about behavior. The slop and noise are unimportant, ephemeral imperfections; the Platonic essence is explained by the genes. The "selfish genes" (to use Dawkins' terms) can be seen as the designer, director and choreographer of the great pageant of life. However, when the variance which is not accounted for by genetic factors is substantial, disregarding it may be unwise. The Platonic lion, even if it were to be "realized" under conditions of optimal nutrition and exercise along with minimal injury and disease, would not be able to survive on our fair, but imperfect, planet. The genetic code would not provide it with the ability to stalk prey, kill, and coordinate with conspecifics, for these skills are learned as part of the ontogenetic contribution to behavior (Adamson, 1960; Schaller, 1972:165). The data on primate behavioral variability can serve as a healthy stimulant for furthering the growth of ethology and sociobiology. The variance is too large to be neglected. A synthesis of proximal and distal causes of behavior is needed to cope with the data.

Problems with "Proof by Adaptiveness"

One common practice which reinforces the belief in genetic idealism is that of taking evidence of adaptiveness as proof of selection (or genetic causes). Since biologists often cannot reconstruct the evolutionary history of a trait or behavior, data on adaptiveness are taken as indirect evidence that the trait or behavior was selected to fulfill certain adaptive functions. Functional analyses of play, aggression, territoriality, and other behavior often assume when the adaptive functions of the behavior have been located that an adequate demonstration has been presented to show that the trait evolved to serve those functions and that no additional explanation is needed.

The ability of observers to invent numerous "adaptive" interpretations for almost any behavior (even maladaptive ones)[2] creates a grave risk of erroneously inferring selective and genetic causes for any behavior that looks adaptive. Adaptiveness and function are easy to postulate but difficult to demonstrate. When considering functional "explanations" of primate behavior, it is often difficult to muster enough concrete data to reject the null hypothesis that there may be no function. However, the seeming elegance and easy applicability of adaptive "explanations" remains popular.

In addition, the presence of adaptive-looking behavior does not prove the existence of a strong genetic base for the behavior. Holding species genetics constant, animals can learn either adaptive or maladaptive behavior depending on current environmental conditions (for example, Kelleher and Morse, 1968; McKearney, 1969; Byrd, 1972).

The discovery that behavior is highly variable introduces another major challenge to the venerable tradition of explaining traits and behaviors by locating plausible adaptive functions. In the Analysis of Variance section (preceding), several randomization mechanisms were discussed that make it possible for a sizeable number of traits to be nonadaptive or even maladaptive and still remain common in a population (for genetic *and* ontogenetic reasons). The more randomized a species' behavioral repertoire is, the more likely it is that a given behavior will not be suited to analysis by adaptiveness. Because the observer has no a priori methods for discriminating which behaviors to exclude from an analysis by adaptiveness, there is a heightened risk of using this method when dealing with highly variable species. In addition, the existence of several studies showing that species X does behavior Y in all environments studied is not evidence that the behavior is under strong genetic control. The behavior may be very responsive to ontogenetic control, but common qualities of all the study environments may have produced the same variant at each study site.

Finally, behavioral variance is not merely "noise" that sullies an otherwise ideal system. Variance is the raw material for future behavior. It is bought, however, at the price of current adaptiveness: a certain percent of ongoing practices must be freed from the constraints of being adaptive in order to become the variance for future selection and shaping. Therefore, the more innovative, creative, and adaptable a species or individual is, the greater will be the proportion of its current behavior which is not constrained by the necessities of being directly adaptive.

Biology Is Not Destiny

The primate literature is full of descriptions that tie behavior to anatomy. Ethologists often compare behavior with organs, claiming that behavior evolves much as organs do and that it can be treated as another functional organ. In addition, anatomical structures are seen to dovetail with behavior, much as two other functional organs might have evolved to work together. For example, adult male baboons have large canines which, according to functional logic, are designed for fighting, aggression, and dominance; hence, male anatomy has predisposed some writers to focus on the "appropriate" behavioral correlates and stress male aggression and dominance. Only several years afterward did Rowell (1966) enlarge our understanding of male baboon behavior, describing forest dwelling baboons where the males were not dominance oriented, aggressive, or tyrannical. In species with behavioral flexibility, biology is not destiny. Functional logic that analyzes behavior as an organ ceases to explain as much as it might have in invertebrates and other species with relatively invariable behavior.

Two examples will suffice. First, Sugiyama (1976) described the langurs of the Himalayan mountains and compared their behavior and social organization with that of other langurs. The

Himalayan study revealed even broader behavioral variance in this species than was already documented. In his conclusion Sugiyama points out that langurs often are much more terrestrial than one would predict from an analysis of their anatomy alone—which is clearly adapted for an arboreal, leaf-eating way of living. Because of the considerable variance in their terrestrial behavior, Sugiyama stresses that "the so-called anatomical characteristics of this species do not always work as the limiting factors in its daily life." Rather, current ecological conditions, especially the distribution of food and shelter, determine the animals' allocation of time on the ground.

Second, Burton (1977) compared sex-role behavior in 21 free-ranging cercopithecoid societies and found much more variance in sexual behavior and roles than one would presume from most biological theories of behavior. The variance between species is certainly greater than would be predicted from the idealized male and female reproductive "strategies" for maximizing inclusive fitness. Especially important is the within-species variance, which demonstrates the absence of direct genetic control over biologically important behavior. In addition, it is likely that the studies cited by Burton underestimate the variability. We have no idea how much more variance will be reported in the primate literature once researchers (1) give up the habit of describing behavior in terms of fixed action patterns, rituals, stereotyped sequences, species-typical responses, and ethograms, and (2) focus on the more challenging task of describing the variability rather than the idealized norm.

In their enthusiasm for genetic explanations, ethologists and sociobiologists often write as if they were unaware of the influence of ontogenetic factors on behavior. Thus, the tradition has been established to see biology as destiny. Now the times are changing.

Conclusion

Because many observers of animal behavior have attempted to describe behavior in terms of fixed action patterns, ethograms, stereotyped displays, or species-typical responses, and to explain these "fixed" patterns in terms of idealized reproductive strategies, genetic idealism has flourished. For many years, the Weltanschauung of genetic idealism and the descriptive vocabulary stressing fixed patterns hid the existence and importance of variance. Laboratory and field observations and experiments were biased toward searching for and describing invariance rather than the rich variability which actually exists. Thus, biological theories have run the risk of a Type I error, accepting as true many hypotheses which should not have been accepted about strong genetic causes of behavior.

The data on primate behavioral variance are providing a catalyst for breaking out of the circular trap of (1) assuming genetically fixed patterns, (2) recording data in terms of stereotyped categories, then (3) concluding that assumption (1) was correct. At present, it is difficult to predict how much *more* variance would be found in primate behavior if observers were to focus on variance, ontogeny, learning, and the means by which current environments shape behavior instead of emphasizing fixed action patterns, ethograms, and genetic determinism. We suspect that there is a wealth of data yet to be tapped that will produce a rapid growth in and reevaluation of animal behavior theory. In a paper stressing the importance of learning in determining primate behavior, Washburn and Hamburg (1965: 613) point out that, "in field studies observations are primarily of the results of learning rather than of the process itself." As observers are trained to focus on variance, ontogenetic factors, operant and classical conditioning, observational learning and tradition modification, they will be able to describe processes that are presently unnoticed and thus vastly expand our understanding of how behavior is shaped by current ecological conditions.

Ethology and sociobiology must grow if they are to cope with the complex problems of relating behavior to ecology. Ontogeny and learning are given lip service in current biological theories, but the genes and selection get the limelight. Behavioral flexibility takes a back seat while fixed patterns and genetic determinism dominate. We need new theoretical models that give due weighting to both phylogenetic and ontogenetic causes of behavior. This will be the ultimate synthesis.

Notes

1. Learned altruism is far different from the genetically determined altruism postulated by Hamilton (1964a, b, 1970) and subsequent sociobiologists. Learned altruism need have no kin favoritism or nepotism involved, can be either widely generalized or very discriminative, and can be gained or lost in a short-time interval (Scott, 1971; Mussen and Eisenberg-Berg, 1977).

2. Have your students write theoretical papers on the evolution and possible functions of acne, migraine headaches, ejaculatory incompetence, vaginismus, and so on. The number of imaginiative explanations is amazing. Proof by adaptiveness can be used to prove anything, even things that should not have been proven.
3. Prior generations of biologists, psychologists, and social scientists maintained interdisciplinary hostilities and barricades. Learning theorists criticized biologists and vice versa. Social scientists feared reductionism and shunned biology and psychology. By maintaining traditional allegiances, scientists only retard the development of a unified theory of behavior. Foresightful scientists and graduate students will break down the barriers and seek interdisciplinary training. Biophysics and biochemistry are clear examples of flourishing fields where interdisciplinary research has paid off in spectacular breakthroughs in scientific understanding. A unified behavioral science may expect similar rewards.

References

Adamson, J. 1960. *Born free.* New York: Bantam Books.

Baldwin, J. D., and Baldwin, J. I. 1971. Squirrel monkeys (*Saimiri*) in natural habitats in Panama, Colombia, Brazil and Peru. *Primates* 12:45–61.

Baldwin, J. D., and Baldwin, J. I. 1973. The role of play in social organization: Comparative observations on squirrel monkeys (*Saimiri*). *Primates* 14:369–381.

Baldwin, J. D., and Baldwin, J. I. 1977. The role of learning phenomena in the ontogeny of exploration and play. Pages 343–406 in *Primate bio-social development: Biological, social, and ecological determinants.* Eds. S. Chevalier-Skolnikoff and F. E. Poirier. New York: Garland Publishing.

Baldwin, J. D., and Baldwin, J. I. 1978. Reinforcement theories of exploration, play, creativity and psychosocial growth. Pages 231–257 in *Social play in primates.* Ed. E. O. Smith. New York: Academic Press.

Barash, D. P. 1977. *Sociobiology and behavior.* New York: Elsevier.

Burton, F. D. 1972. The integration of biology and behaviour in the socialization of *Macca sylvanus* of Gibraltar. Pages 29–62 in *Primate socialization.* Ed. F. E. Poirier. New York: Random House.

Burton, F. D. 1977. Ethology and the development of sex and gender identity in non-human primates. *Acta Biotheor.* 26:1–18.

Byrd, L. D. 1972. Responding in the squirrel monkey under second-order schedules of shock delivery. *J. Exp. Anal. Behav.* 18:155–167.

Crook, J. H , and Aldrich-Blake, P. 1968. Ecological and behavioural contrasts between sympatric ground dwelling primates in Ethiopia. *Folia Primat.* 8:192–227.

DeVore, I., and Hall, K. R. L. 1965. Baboon ecology. Pages 20–52 in *Primate behavior: Field studies of monkeys and apes.* Ed. I. DeVore. New York: Holt, Rinehart and Winston.

Douglas, J. H. 1973. Genetics: A science that is coming of age, *Sci. News.* 104:332–334.

Durham, N. M. 1971. Effects of altitude differences on group organization of wild black spider monkeys (*Ateles paniscus*). *Proc. 3rd Int. Congr. Primat.* vol. 3, pp. 32–40. Basel: Karger.

Elias, M. F., and Samonds, K. W. 1974. Exploratory behavior and activity of infant monkeys during nutritional and rearing restriction. *Amer. J. Clin. Nutr.* 27:458–463.

Fagen, R. M. 1976. Exercise, play, and physical training in animals. Pages 189–219 in *Perspectives in ethology*, vol. 2. Eds. P. P. G. Bateson and P. H. Klopfer. New York: Plenum.

Frisch, J. E. 1968. Individual behavior and intertroop variability in Japanese macaques. Pages 243–252 in *Primates: Studies in adaptation and variability.* Ed. P. C. Jay. New York: Holt, Rinehart and Winston.

Gartlan, J. S. 1966. Ecology and behaviour of the vervet monkey, *Cercopithecus aethiops pygerythus,* Lolui Island, Lake Victoria, Uganda. Ph.D. thesis, University of Bristol.

Gartlan, J. S. 1973. Influences of phylogeny and ecology on variations in the group organization of primates. *Symp. Fourth Int. Congr. Primat.* vol. 1, pp. 88–101. Ed. E. W. Menzel. Basel: Karger.

Gartlan, J. S., and Brain, C. K. 1968. Ecology and social variability in *Cercopithecus aethiops* and *C. mitis.* Pages 253–292 in *Primates: Studies in adaptation and variability.* Ed. P. C. Jay. New York: Holt, Rinehart and Winston.

Hamilton, W. D. 1964a. The genetical evolution of social behaviour, I. *J. Theoretical Biol.* 7:1–16.

Hamilton, W. D. 1964b. The genetical evolution of social behaviour, II. *J. Theoretical Biol.* 7:17–52.

Hamilton, W. D. 1970. Selfish and spiteful behaviour in an evolutionary model. *Nature, Lond.* 228 (5277):1218–1220.

Hebb, D. O. 1953. Heredity and environment in mammalian behavior. *Brit. J. Anim. Behav.* 1:43–47.

Heinroth, O. 1910. Beiträge zur Biologie, nämentlich Ethologie und Psychologie der Anatiden, *Verh. V. Int. Ornith. Kongr.*, Berlin, pp. 258–702.

Itani, J. 1959. Paternal care in the wild Japanese monkey, *Macaca fuscata fuscata. Primates* 2:61–93.

Itani, J., Tokuda, K., Furuya, Y., Kano, K., and Shin, Y. 1963. The social construction of natural troops of Japanese monkeys in Takasakiyama. *Primates* 4:1–42.

Jay, P. 1965. The common langur in North India. Pages 197–249 in *Primate behavior: Field studies of monkeys and apes.* Ed. I. DeVore. New York: Holt, Rinehart and Winston.

Kawai, M. 1964. *The ecology of Japanese monkeys.* Tokyo: Kawade-Shoko.

Kawai, M. 1965. Newly acquired pre-cultural behavior of the natural troop of Japanese monkeys on Koshima Islet. *Primates,* 6(1):1–30.

Kelleher, R. T., and Morse, W. H. 1968. Schedules using noxious stimuli. III. Responding maintained with response-produced electric shocks *J. Exp. Anal. Behav.* 11:819–838.

Kitahara-Frisch, J. 1977a. Tools or toys: What have we really learned from wild chimpanzees about tool use? *J. Anthrop. Soc. Nippon.* 85:57–64.

Kitahara-Frisch, J. 1977b. Considerations on the subcultural behavior of Japanese macaques (*Macaca fuscata*). Pages 371–386 in *Morphology, evolution, primates.* Ed. T. Umesao. Tokyo: Chuo-Koronsha.

Kummer, H. 1971. Immediate causes of primate social structures. *Proc. 3rd. Int. Congr. Primat.* vol. 3, pp. 1-11. Basel: Karger.

Lehrman, D. S. 1953. A critique of Konrad Lorenz's theory of instinctive behavior. *Quart. Rev. Biol.* 28:337–363.

Lorenz, K. 1935. Der Kumpan in der Umwelt des Vogels. *J. Ornithol.* 83: 137–213, 289–413.

Lorenz, K. 1937. Über die Bildung des Instinktbegriffes. *Naturwiss.* 25: 289–300, 307–318, 324–331.

Lorenz, K. 1965. *Evolution and modification of behavior.* Chicago: University of Chicago Press.

Mayr, E. 1972. The nature of the Darwinian revolution. *Science,* 176: 981–989.

McKearney, J. W. 1969. Fixed interval schedules of electric shock presentation: Extinction and recovery of performance under different shock intensities and fixed-interval durations. *J. Exp. Anal. Behav.* 12:301–313.

Miyadi, D. 1964. Social life of Japanese monkeys. *Science,* 143:783–786.

Mizuhara, H. 1964. Social changes of Japanese monkey troops in the Takasakiyama. *Primates.* 5:27–52.

Moynihan, M. 1976. *The new world primates: Adaptive radiation and the evolution of social behavior, languages and intelligence.* Princeton, N.J.: Princeton University Press.

Mussen, P., and Eisenberg-Berg, N. 1977. *Roots of caring, sharing and helping.* San Francisco: Freeman.

Neville, M. 1968. A free-ranging rhesus monkey troop lacking adult males. *J. Mammal.* 49:771–773.

Neville, M. 1969. Male leadership change in a free-ranging troop of Indian rhesus monkeys (*Macaca mulata*). *Primates.* 9:13–27.

Paterson, J. D. 1973. Ecologically differentiated patterns of aggressive and sexual behavior in two troops of Ugandan baboons, (*Papio anubis*). *Am. J. Phys. Anthrop.* 38:641–647.

Reynolds, V., and Reynolds, F. 1965. Chimpanzees in the Budongo

Forest. Pages 368–424 in *Primate behavior: Field studies of monkeys and apes.* Ed. I. DeVore. New York: Holt, Rinehart and Winston.

Riesen, A. H. 1961. Stimulation as a requirement for growth and function in behavioral development. Pages 57–80 in *Functions of varied experience.* Eds. D. W. Fiske and S. R. Maddi. Homewood, Ill.: Dorsey.

Riesen, A. H. 1965. Effects of early deprivation of photic stimulation. Pages 61–85 in *The biosocial basis of mental retardation.* Eds. S. F. Osler and R. E. Cooke. Baltimore: Johns Hopkins Press.

Rosenzweig, M. R. 1976. Effects of environment on brain and behavior in animals. Pages 33–49 in *Psychopathology and child development.* Eds. E. Schopler and R. J. Reichler. New York: Plenum.

Rowell, T. E. 1966. Forest living baboons in Uganda. *Zool. J. Lond.* 149:344–364.

Rumbaugh, D. M. 1970. Learning skills in anthropoids. Pages 1–70 in *Primate behavior: Developments in field and laboratory research,* vol. 1. Ed. L. A. Rosenblum. New York: Academic Press.

Rumbaugh, D. M., and McCormack, C. 1967. The learning skills of primates: A comparative study of the apes and monkeys. Pages 289–306 in *Progress in primatology.* Eds. D. Starck, R. Schneider, and H. J. Kuhn. Stuttgart: Gustav Fisher.

Sahlins, M. 1976. *The use and abuse of biology: An anthropological critique of sociobiology.* Ann Arbor: University of Michigan Press.

Schaller, G. B. 1972. *The Serengeti lion.* Chicago: University of Chicago Press.

Schiller, P. H. 1957. Innate motor action as a basis of learning: Manipulative patterns in the chimpanzee. Pages 264–287 in *Instinctive behavior.* Ed. C. Schiller. New York: International Universities Press.

Schneirla, T. C. 1956. The interrelationships of the "innate" and the "acquired" in instinctive behavior. Pages 387–452 in *L'Instinct dans le comportement des animaux et de l'homme.* Ed. P.-P. Grassé. Paris: Masson.

Scott, J. F. 1971. *Internalization of norms: A sociological theory of moral commitment.* Englewood Cliffs, N.J.: Prentice-Hall.

Singh, S. D. 1968. Social interactions between the rural and urban monkeys, (*Macaca mulatta*). *Primates.* 9:69–74.

Southwick, C. H. 1972. Aggression among nonhuman primates. *Addison-Wesley module in anthropology,* 23:1–23. Reading, Mass.: Addison-Wesley.

Spuhler, J. N., and Jorde, L. B. 1975. Primate phylogeny, ecology and social behavior. *J. Anthrop. Res.* 31:376–405.

Struhsaker, T. T. 1967a. Social structure among vervet monkeys (*Cercopithecus aethiops*). *Behaviour* 29:83–121.

Struhsaker, T. T. 1967b. Ecology of vervet monkeys (*Cercopithecus aethiops*) in the Masai-Amboseli Game Reserve, Kenya. *Ecology* 48:891–904.

Sugiyama, Y. 1964. Group composition, population density, and some sociological observations of Hanuman langurs (*Presbytis entellus*). *Primates* 5:7–38.

Sugiyama, Y. 1965. Behavioral development and social structure in two troops of Hanuman langurs (*Presbytis entellus*). *Primates* 6:213–247.

Sugiyama, Y. 1967. Social organization of Hanuman langurs. Pages 221–236 in *Social communication among primates*. Ed. S. A. Altmann. Chicago: University of Chicago Press.

Sugiyama, Y. 1968. Social organization of chimpanzees in the Budongo Forest, Uganda. *Primates* 9:225–258.

Sugiyama, Y. 1976. Characteristics of the ecology of the Himalayan langurs. *J. Hum. Evol.* 5:249–277.

Suzuki, A. 1969. An ecological study of chimpanzees in a savanna woodland. *Primates* 10:103–148.

Tinbergen, N. 1950. The hierarchical organization of nervous mechanisms underlying instinctive behaviour. *Symp. Soc. Exp. Biol.* 4:305–312.

Tinbergen, N. 1951. *The study of instinct*. New York: Oxford University Press.

Tinbergen, N. 1963. On aims and methods of ethology. *Z. Tierpsychol.* 20:410–433.

Trivers, R. L. 1971. The evolution of reciprocal altruism. *Quart. Rev. Biol.* 46:35–57.

Trivers, R. L. 1972. Parental investment and sexual selection. Pages 136–179 in *Sexual selection and the descent of man, 1871–1971*. Ed. B. Campbell. Chicago: Aldine.

Tsumori, A. 1967. Newly acquired behavior and social interactions of Japanese monkeys. Pages 207–219 in *Social communication among primates*. Ed. S. A. Altmann. Chicago: University of Chicago Press.

van Lawick-Goodall, J. 1968. The behavior of free living chimpanzees in the Gombe stream reserve. *Anim. Behav. Mon.* 1:161–311.

van Lawick-Goodall, J. 1970. Tool-using in primates and other vertebrates. Pages 195–249 in *Advances in the study of behavior*. Eds. D. S. Lehrman, R. A. Hinde, and E. Shaw. New York: Academic Press.

Volkmar, F. R., and Greenough, W. T. 1972. Rearing complexity affects branching of dendrites in the visual cortex of the rat. *Science* 176: 1445–1447.

Washburn, S. L., and Hamburg, D. A. 1965. The implications of primate research. Pages 607–622 in *Primate behavior: Field studies of monkeys and apes*. Ed. I. DeVore. New York: Holt, Rinehart and Winston.

Wilson, E. O. 1975. *Sociobiology: The new synthesis*. Cambridge, Mass.: Harvard University Press.

Yamada, M. 1971. Five natural troops of Japanese monkeys on Shodoshima Island: II. A comparison of social structure. *Primates* 12: 125–150.

Yoshiba, K. 1968. Local and intertroop variability in ecology and social behavior of common Indian langurs. Pages 217–242 in *Primates: Studies in adaptation and variability*. Ed. P. C. Jay. New York: Holt, Rinehart and Winston.

Zimmerman, R. R., Steere, P. L., Strobel, D. A., and Hom, H . L. 1972. Abnormal social development of protein malnourished rhesus monkeys. *J. Abnorm. Psych.* 80:125–131.

Chapter 6

Ecological Influences on Australian Aboriginal Social Organization

Joseph B. Birdsell

Turning now from the more theoretical orientations of the previous chapters, Dr. Birdsell presents the results of his rigorous analysis of the extensive data on aboriginal Australians which he has compiled. In relating ecology to social organization, Dr. Birdsell has used the same techniques which are familiar to students of nonhuman primate social organization, and this in itself now allows us to bridge the gap between studies of nonhuman primates and the vast knowledge that has accumulated concerning selected human populations and their adjustments to local ecology.

The Australian aborigines exhibit a number of remarkable constancies across a diversity of habitats. The numerical composition of social units shows little variation across the continent. Whereas range sizes and land-use practices may reflect the quality of the environment, group size appears to be more conservative. The size of social units thus appears to be a tolerable adaptation to the entire range of habitat types, and in fact may be constrained more by traditional or other social requirements rather than environmental parameters such as resource distribution.

Dr. Birdsell emphasizes, in his work, how the human primate may respond to environmental pressures in a direct fashion which anticipates environmental sources of mortality and thereby gains control of these influences on demography and social structure. Traditions become established which not only permit adaptation to a specific habitat, but which are sufficiently general so that they require little modification to be suitable under a variety of conditions. Preferential female infanticide in the Australian aborigines supercedes all other causes of mortality in determining demography and social structure. It is, moreover, a sufficiently general adaptation so that it functions to adequately regulate populations under a variety of circumstances and mitigates the selective pressures which might otherwise operate to change life history processes themselves.

The Australian aborigine thus achieves a measure of self-regulation which maintains the population at below the environmental carrying capacity. This self-regulation is a development that removes human populations one step from the ordinary sources of mortality which would set upper limits on population size in other primates. The established traditions may not be conscious choices for the individuals involved, but they are nonetheless effective means of self-regulation which adapts the population to its ecological constraints.

Birdsell brings us one step further in his analysis of cultural contributions to social organization when he introduces us to the profound change which language makes on human adaptation to ecology. With this cultural acquisition, the human primate can undertake a journey to a resource which he has never seen, which is beyond the experience of any of his ancestors or the ancestors of any member of his social unit, and which may have become known to him through a chain of people, each having heard that a resource existed in a distant place they had themselves not ever directly experienced.

Language then adds a new dimension to human cultural and traditional heritage, but this addition should in no way obscure the profound similarities of human and nonhuman primate social organization and its response to the ecology. New specific mechanisms may be acquired but the same broad principles governing life history processes, demography, social behavior, social organization, and ecology continue to apply.

Introduction

Ecology is one of the important variables which strongly influences social organization among both economically simple human populations and in a wide variety of nonhuman primates

(Spuhler and Jorde, 1976). But the phylogenetic distances between man and all other primates are too great in time to expect that living hunter-gatherers can directly illuminate the way in which ecology affects social organization among the latter. Therefore in this chapter the goal is to specify how social organization, used in its broadest senses, has evolved ecological accommodation among the Aborigines of Australia. Little attempt will be made here to seek homologies or even analogies with the social organizations of various human primates.

Perhaps the best reason for examining economically simple humans is that we know more about them than has yet been recorded for any natural population among the other primates. The Aborigines of Australia are chosen as models for man at the hunter-gatherer level since they have been more thoroughly researched than any other such human groups. A great deal of our knowledge of them stems diretly from Norman D. Tindale's 50 years of fieldwork among them. Much of my own professional work is directly indebted to his efforts, and I gladly acknowledge the fact. The cultural and economic posture of the Aborigines is such that they serve as perhaps the best models available for reconstituting human behavior in the Upper Pleistocene. Many of their attributes can be projected even further back in time.

The Australians lived as generalized hunters and collectors, utilizing all economically efficient energy sources. They adapted to a wide range of environments, from stringent desert to lush tropical rain forests, with annual rainfall varying tribally from 4 to 120 inches. Their occupancy was so adaptive that they left no empty spaces in continental Australia. The aboriginal basic tool kit was Upper Paleolithic in character, while for the Tasmanians, with their very similar life-style, it was Middle Paleolithic, closely resembling the Mousterian industry of Western Europe. On the continent the addition of the dingo, a feral dog, and such regional lithic advances as microliths, pressure chip points, and polished stone axes did not greatly change the overall aboriginal extractive efficiency. Finally, the Aborigines at the time of contact were alone in inhabiting an entire continent in which they suffered no ecological competition from either pastoral or horticultural peoples.

Today there are no tribal Aborigines. The nearly 600 tribal groups have long since disintegrated under the impact of White contact. The 30,000 or so surviving full-blooded Australians subsist on White support-bases, such as government stations, missions, or cattle ranches. In recent years many have drifted into the little towns in northern Australia, there to become welfare problems. A few families still return to the "bush" for parts of the year but this is to ecologically underexploited regions, no longer in balance with their human occupants. The last free-living tribal peoples, the Ildawonga, came in out of the interior of the Western Desert in 1965 and are now wards of the government.

In a paper to be given elsewhere documentation is presented that the Aborigines of Australia developed a homeostatic living system and managed it very well. Some of the details will be summarized here as they are pertinent to this topic. Further evidence from Vorkapich (1976) shows that the Great Basin Shoshoni were also characterized by a self-regulating system. There is even some scanty evidence analyzed by Martin and Read (1976) that hunter-gatherers in Africa also achieved stable living systems. Different groups of hunters and gatherers residing in three different continents and exploiting totally different biota, have each demonstrated the ability of man at simple economic levels to adjust to ecological pressures and regulate both his cultural and biological behavior so as to maintain stable equilibrium systems.

Ecological Relationships

As a basis for this discussion it is convenient to use an ecological triangle of relationships as shown in Figure 1. That tripolar diagram attempts to show the interdependence of the primary variables which contribute to the ecology of our species. The upper apex represents the *environment* in its totality, including all other forms of life within it. The lower right-hand apex stands for *population biology,* as it is especially concerned with reproductive factors, demographic characteristics, and the forces that influence population life and death. The lower left-hand apex points to *society and culture* and includes all social institutions and patterned forms of group behavior. Each of these components contributes to the other two, and these interactions must be included in the analysis of a species' ecology. For mankind obviously the sociocultural pole is more heavily weighted than for other kinds of animals. But among economically simple human groups the environment is still a pervasive influence. Their living systems must be manipulated to accommodate to its restraints. In this paper social organization necessarily includes the interactions, with the the environmental pole and the pole of population biology. Adaptiveness pervades aboriginal behavior in Australia, and examples of maladaptiveness are difficult to cite.

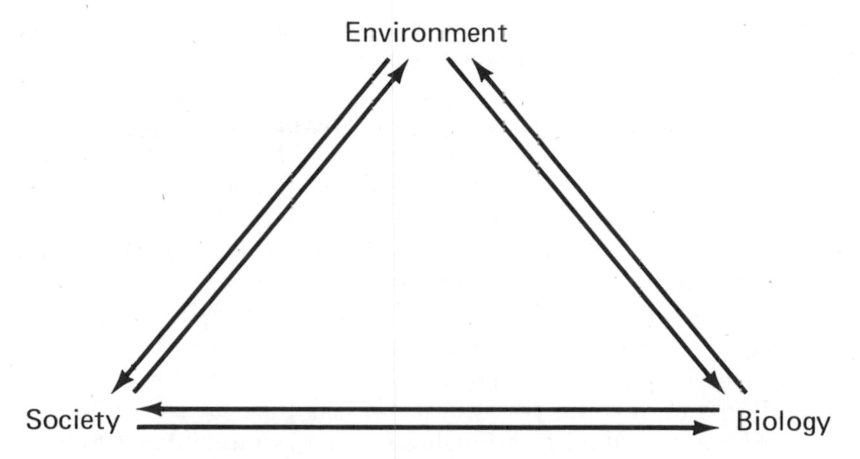

Figure 1. Basic triangle of ecological relationships.

Environmental Determinism in Aboriginal Australia

Ecological research originally undertaken on the Australian Aborigines (Birdsell, 1953) demonstrated a bivariate correlation of 0.81 between the area of the tribal domain and the mean annual rainfall it received. This value was sufficiently high to suggest that the environment importantly structured human distributions in the continent. More recent research, commenced at the Australian National University in 1973, and still ongoing, has proceeded to more complex levels investigating the impact of the Australian environment upon its original inhabitants. Using a larger basic series of 183 tribes derived from Tindale's magnificent monograph on tribal territoriality (1974), a simple correlation of 0.79 was obtained between the size of the tribal area and in this case median annual rainfall. The series represented those tribes in which all water resources were locally earned through rainfall. Accordingly, coastal tribes, tribes downriver from stream headwaters, as well as those involved with tribal fragmentation following the passage of initiation rites, have been eliminated. The next step was to further refine the series by finding the ecologically most homogeneous segment of it. It developed that tribes for whom summer rainfall totalled 45% or more of the annual amount yielded the highest area-rainfall correlations. In this series of 86 tribes the bivariate correlation rose to 0.87.

This value will no doubt be further enhanced with the completion of the multiple step regression analysis which involves a total of 66 environmental variables. There is a good measure of overlap between some of the variables in this analysis. No less than 17 of the 66 involved rainfall related factors. Other clusters included those relating to temperature, solar radiation, and botanical variables. A degree of multicollinearity is present. Further corrections will be made to rectify R^2 and to take into account Mallow's C_p. All of these reduce the value of the coefficient. With median annual rainfall accounting for so much of the explained variance in the system, it is not to be expected that the multiple coefficient for correlation will be greatly improved over the value 0.87 achieved for the bivariate analysis. Perhaps the final value will rise into the low 90s, yielding an unexplained variance of less than 20%.

There is another relationship in the system which serves to depress the coefficients of correlation. The method of analysis assumes that the size of the tribal populations is a constant, approaching 500 persons. In fact it is known to vary considerably. If it ultimately proves possible to correct for this factor which depresses the relationship between tribal population and the variables in the environment in which they live, the multiple step coefficient of correlation should certainly rise to the mid 90s. This natural system is highly structured and the author would be surprised if ultimately the unexplained variant is not reduced to the magnitude of about 10%. This is a surprisingly small value, for a number of errors in the data are still present and visible. These errors must be essentially random in nature, that is self-compensating, and so do not seriously harm the results. The tightness of this correlation, which is based upon the area occupied by a tribe, must be projected to cover the derived fact that tribal population numbers have a relatively constant central tendency throughout the range of Australian environments. This will be discussed more in detail later.

The Basic Units of Australian Social Organization

To provide materials for comparison with nonhuman primates Australian Aboriginal social organization is presented in several dimensions. The first involves a discussion of the nature of the basic units. These consist of two elementary units which are tightly structured, the family and the local group. A third, the dialect tribe, is an entity which results from the pattern of the density of communications. Each will be discussed in some detail, with central tendencies noted. Variants on the basic pattern will be given, and where possible an explanation offered. Later movements and dispersals in terms of these basic units will be discussed.

Perhaps the surprising thing is that among nearly 600 tribes in continental Australia, ranging from stringent desert with 4 inches of annual rainfall to tropical rain forests with in excess of 120 inches of rainfall annually, the consequences of ecological change are relatively slight. It would seem that human behavior is complex enough so that basic social adaptations carry through this extreme environmental range. Whether this is a greater constancy than is shown for the nonhuman primates would be worth some examination.

The Family in Australian Society

The procreative family is the basic unit in aboriginal society. Its members include an adult male, his wife or wives, and their children. If the offspring have all been sired by the man, it is a biological family. If some of the children are not his, it is still a social family. Polygyny is permissible throughout Australia but only a relatively few fortunate men attain this marital status. Among Western Desert tribes genealogies show that 90% of the marriages are monogamous. In the remaining 10% men may have multiple wives, but they only rarely number up to 5. In north coastal Australia "big men" have been reported having as many as 20 wives, but these are not present in his family all at one time. The wives cycle through his family as marriage and death affect them. The number of men who remain unmarried is consequently greater in North Australia than in the rest of the continent.

INFANTICIDE AS A PRIMARY STABILIZING DEVICE Population biology stringently structures the Australian aboriginal family. Infanticide is universally practiced and accounts for somewhere between 15 to 30% of the total number of births. It is preferentially female infanticide, since patrilocal band residence produces a bias in favor of the retention of male infants where possible. Figure 2 shows the frequency of families with different numbers of children. The configuration approximates a truncated Poisson distribution. It demonstrates that moderate-sized families are the commonest, as might be expected, and those both larger and smaller are less frequent. As these figures stand, the data do not quite approximate a fully stable population, for each parent produces 1.33 mean offspring. This discrepancy does not stand as evidence against a population well regulated by infanticide, but rather that the genealogical method offers no assurance that remembered children have indeed reached their mid-reproductive point in time. It is in fact surprising that the children remembered genealogically are no more than 33% more frequent than would be expected in a totally stable population.

Evidence for the widespread practice of preferential female infanticide comes from both early observers and from genealogical data which show biased sex ratios. The biological sex ratio is perfectly normal among Australian Aborigines and approximates 100% as it does with other populations. But in Figure 3 the sex ratios are badly biased in favor of males, the degree depending upon family size. In families in which one child is retained there are 186 males remembered for every 100 females. Thereafter in the more frequent families in which 2 or 3 children are retained, the sex ratio drops to about 130%, but this is still well above normal expectations. In the larger families, the sex ratio again rises sharply, and culminates in families of 5 surviving offspring in which it reaches 260%. When all families

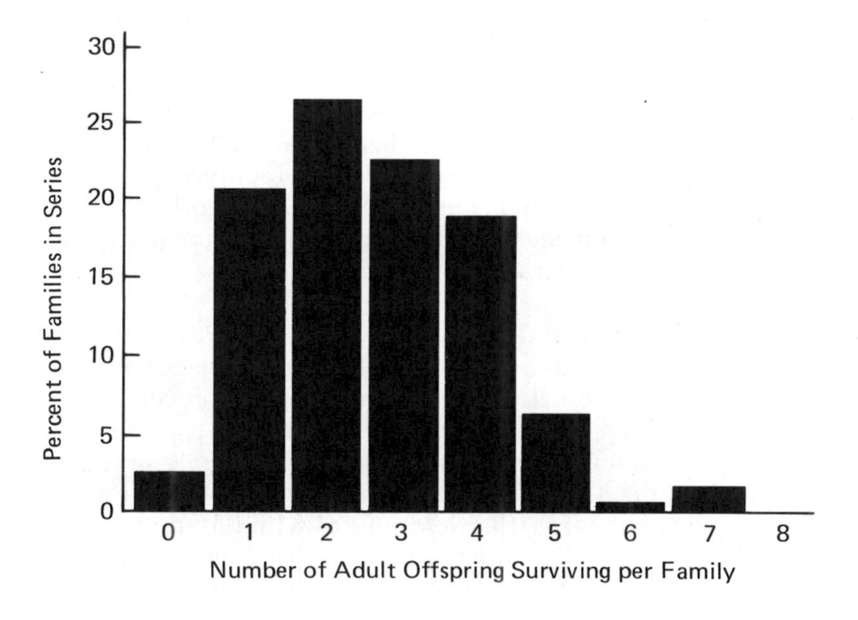

Figure 2. Frequency distribution showing range of reproductive success in 194 precontact Australian matings.

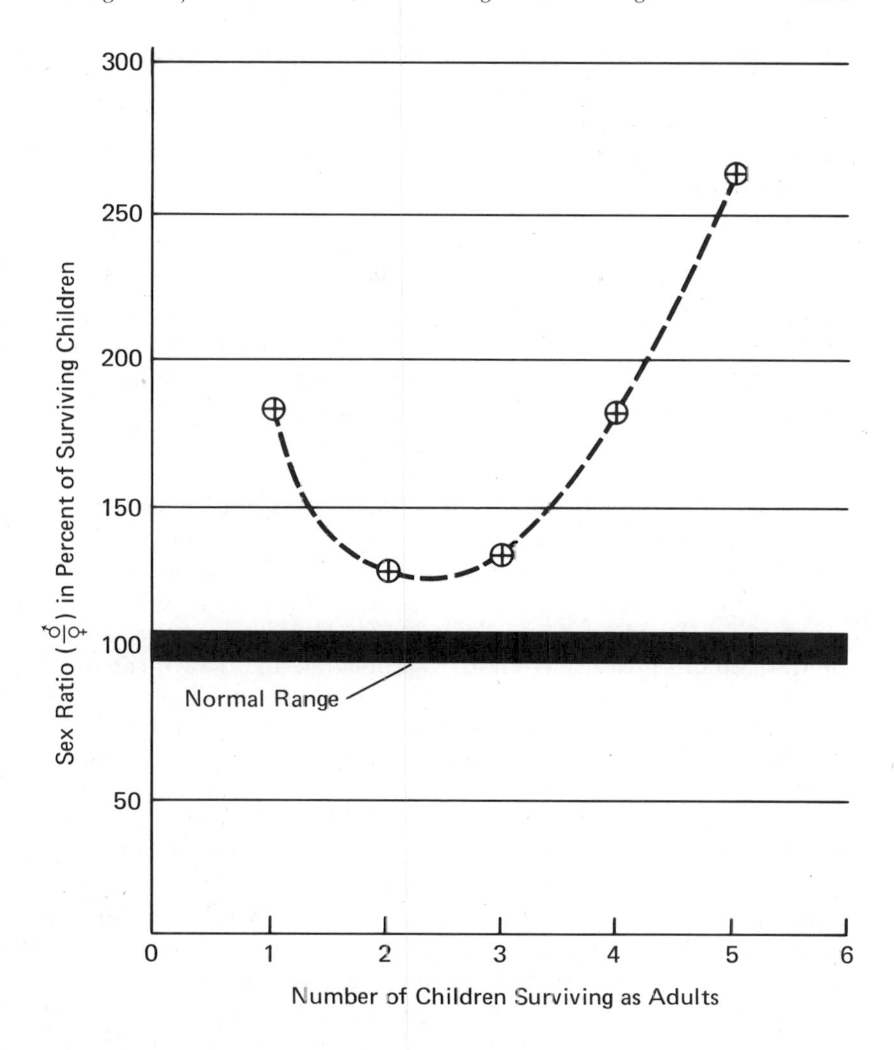

Figure 3. Sex ratios as indicators of preferential female infanticide in precontact Australian matings.

are taken together the data indicate 153 males per 100 females. These data are based upon precontact matings derived from genealogies collected in 1953 and 1954 by Norman B. Tindale and the writer while working around the fringes of the Western Desert of Western Australia. They independently attest to the tight control exercised over the recruitment in each generation of new members in the local groups.

Infanticide serves primarily and immediately to relieve

an aboriginal mother of the necessity of nursing two children simultaneously, since weaning occurs well after the third year, and pregnancy is not inhibited in all women by nursing. Further, infanticide allows the mother to continue her economic activities of food-gathering which she could not achieve if impeded by two young children due to poor spacing of births. As important, but less visible, is the effect of infanticide in throttling fertility by limiting the recruitment into the group. It is one of the most important behavioral devices operating to maintain their homeostatic living system. Among live births, that is children spared, infant mortality is high among the Aborigines as among other hunter groups. This seems to be true among the other primates, too. But there is no analogous behavior among the latter to correspond to intentional infanticide.

There are further indications, particularly in the fieldwork of Dr. Joan Greenway (personal communication), that indicate that the small society represented in the Australian local group makes its weight felt in the decision-making as to whether a given child shall be saved or killed. The intervenor at the time of birth is an old "granny woman," a respected member of the local group, who places her hand over the mouth and nose of the child as it emerges at birth and so prevents it from ever coming to life. It is implicit in her presence that she acts to express the opinion of the local group.

THE IMPACT OF INFANTICIDE UPON MARITAL PRACTICE Polygynous marriages would seem difficult to arrange in a society in which the sex ratio shows 153 adult men for 100 women. But the difficulty is solved by the fact that girls are married at puberty, while obtaining a wife is delayed among the men. The population pyramid in Australia is broad at the base, with younger members well represented. So by the simple device of culturally arranging that men receive their first, and possibly only wife, at about 25 to 35 years of age, while the girls become available for marriage at age 15, there are enough of the latter to go around. Indeed, a young man's first wife is very apt to be a hand-me-down widow from an older brother, real or classificatory.

Young wives are obtained through complex social bargaining which involves adjusting the debit and credit sides of female exchange between friendly local groups. Since it is primarily arranged when a young man has been initiated, the actual marriage is considerably delayed. In the first place the promise usually involves a child as yet unborn, Hence, the bridegroom must await the birth of a female baby who is in fact saved for recruitment into the group, and healthy enough to remain alive. Obviously with a further 15-year period involved in the bride growing up, the marriage of young girls is usually to men who have become middle-aged.

The whole of the living system is obviously adjusted to accommodate considerable neonatal mortality, which even in precontact times seems to have consisted primarily of pulmonary and gastric diseases. With the participation of society in the process, the variation in family size seems to have fluctuated within adaptive limits. It shows no trends in passing from the most stringent desert ecology to the best watered tropical rain forests. Individual family size varies as an expression of parental desires. The overall effect of cultural manipulation is such that two parental adults are in time replaced on the average by two offspring who reach their own mid-reproductive years. Numbers are carefully controlled in aboriginal society, a basic constraint to any homeostatic system. Population planning of a subtle sort is clearly indicated, although unnoticed by most sociocultural anthropologists.

Examples of Variant Behavior at the Family Level

In times of extreme stress, as with prolonged droughts in the desert, there are records of young children being sacrificed. These episodes again involve planning within individual families, for the youngest child is killed and fed to the next older child to insure its survival. This form of cannibalism is surely adaptive behavior. In very severe situations children may be nearly wiped out of some small populations through sacrifice, starvation, and consequent illness. There is one indication from Tindale's evidence (1965) of the Kaiadilt of Bentinck Island in the Gulf of Carpentaria, that serious water shortages can occur even in the monsoonal belt of the tropics. In this instance fresh water resources were drowned out by an unusually high tide so that the only potable fluid remaining was inside the water frogs excavated from deep in the mud. This required serious effort for survival, and it was noted that the men dug for themselves and survived better, and let the women and children take care of themselves. But in general the loss of children is readily compensated for in the years ahead by the very high rate of effective fertility shown by these people, as well as other humans.

MAXIMUM INTRINSIC RATE OF INCREASE AMONG THE AUSTRALIANS The most revealing example illustrating why aboriginal fertility must be tightly controlled is revealed in the natural experiment involving the so-called Nanja horde. This event served to define the maximum rate of fertility, and is the only known instance which involves a hunter-gatherer people. The best measure of potential fertility is the so-called maximum intrinsic rate of increase which has been defined as the rate of population increase in unlimited space and with boundless food resources. These conditions were approximated in the case of the Nanja horde.

In the late nineteenth century as White colonists encroached upon aboriginal living room on the Darling River in western New South Wales, a young man by the name of Nanja went "bush" with two young women. He effectively remained out of contact with the expanding settlers until he was rediscovered 30 years later when his descendant group had grown too large to remain invisible longer. When Nanja and his family were brought back to the river settlement they were found to number 29 persons.

Tindale (personal communication), who has reworked the available data on this crucial experiment, says that each of the two original women bore ten viable children who survived at least until the time of recapture. The production of six members of a third generation involved enforced incest. At its simple face level this indicates a maximum intrinsic rate of increase of fivefold per generation. Granted that it is a single and isolated case, nevertheless it is impressive since it occurred in a hostile environment. The rainfall away from the river was so low that no surface waters were available. Survival depended upon obtaining water from the lateral roots of mallee, a kind of scrub eucalypt. The little group was under the further restraint of building no fires in daytime, lest they give away their presence. While there are stochastic elements in this single experiment, it nonetheless dramatically provides an illustration of how great human fertility can be even at a hunter's level of economy. Small wonder that systematic infanticide was required to solve the problem.

A SERIES OF CASES IN WHICH INDIVIDUAL FAMILIES APPEAR TO HAVE FILLED EMPTY SPACE Families are the basic units in Australian social organization, and in a country with scant resources they are the primary task force. In arid country the search for food involves family units much of the year. Consequently they move with greater mobility than the local groups of which they are members. The following instances are suggestive of situations in which single families may have moved off into space empty of other human beings. That they produced aberrant small societies is suggestive.

In northwest Australia a large tribe, the Njangamarda, was discovered by Tindale to consist of two subtribes which used different and conflicting arrangements of the four-class social system. This prevented marriage between the two tribal subdivisions. Since dialect differences were small between the northern or so-called Kundal division of the tribe and the southerly Iparuka division, this schism apparently was a fairly recent one. It can be hypothesized as Tindale has (1974) that the smaller northern portion of the tribe originated through a man contracting a socially wrong marriage, and moving off into empty space to avoid the social and legal consequences. In his initial isolation he continued to use the four-class system, but one in which the marital arrangements were as incorrect as his own original one. If this hypothesis is correct, it is of further interest in showing that local populations can be wiped out by drought, and that recolonizing may be achieved by families who have broken tribal laws. For they are the ones tempted to undertake the venture.

In the northeast portion of South Australia there existed a tribe in which the language involved systematic lisping. Since there is no good reason in structural linguistics why this should occur, Tindale (personal communication) has suspected that an old man who himself lisped moved off into empty country. There he found a new local group which patterned their speech upon that of the original migrant male.

EXCEPTIONS TO NORMAL MARITAL CHOICES Marriages in aboriginal Australia almost universally involved band or local group exogamy. This practice reduced chances of biological inbreeding. The search for marital partners further sought to minimize biological relationship by seeking out brides who had no known relationship to the prospective groom. Both practices effectively served to prevent biological inbreeding among the Australians.

This matter is frequently misunderstood by anthropologists who follow the prescription of the kinship system model as it apparently regulates marriage. According to kinship reckoning, a four-class system preferentially promotes marriage between first cousins. The eight-subsection system does the same among second cousins. Vital facts have been ignored in the study of kinship since these are based upon classificatory terms, and bear little or no relationship to the biological relationship. Thus in the four-class system it is perfectly feasible to marry a "first cousin" who is your classificatory "mother's"-"brother's"-"daughter" but in fact is so biologically distant as to have no remembered relationship. The class systems relate people to each other in social space, but they do not involve any inbreeding in marriage. In fact, the thrust of the Aborigines is in the opposite direction, and they try to maximize distances biologically between spouses.

It is therefore of some interest that Tindale (personal communication) in recent fieldwork found a man of the Nakako tribe involved in a direct brother-sister incestuous marriage. The Nakako were a small tribe, lying on the border between South Australia and Western Australia, which has only recently come into contact and so has been little studied. But Tindale in extracting genealogies from a group he met in the desert did uncover this case of incest. Under repeated questioning the old man admitted the crime, but said that after many years his group "forgot about it." In short, a wrong marriage proved socially acceptable after time. In this case the incestuous marriage was a necessity, since they had no contact with any neighboring groups which had earlier gone into White settlements.

The Local Group or Band

The largest social unit which functions on a regular face-to-face basis in Australia is the local group or band. In the literature it is sometimes called the horde. Among the Aborigines it is universally patrilineal and patrilocal. With a single known insular exception, the local group is also everywhere exogamous. It is the territorial unit in aboriginal society and owns the hunting and gathering rights to its country. Members of other friendly local groups can enter and hunt over their land upon invitation. At appropriate seasons such invitations are freely given, and constitute a form of food reciprocity.

The adult members of a band consist of the men who are born there, and their wives who have been brought in from other local groups. All the grown women in a local group are strangers to that country. Daughters born there marry out at puberty, and so are lost to other bands. The ownership of band country continues through the male line from father to son, and so it is among the men that the social solidarity of the band is maintained. In a sense the local group can be viewed as an extended patrilineal family exercising its territorial rights over a land which has long been in their male lineage.

The men of a local group feel great spiritual attachment to their country. This is maintained through a system of local totems which involve ceremonial activities, as well as in the tracks of mythical ancestors which cross band country. Consequently the men are reluctant to leave their country for long, and if forced to do so, as by drought, return at the first opportunity. Their spiritual ties with their land are so strong as to compel them to return to die there if possible. This set of values effectively acts as a spacing mechanism to maintain adaptive densities throughout the continent.

MINIMAL BAND SIZE TENDS TOWARD A CONSTANT NUMBER Since the local group consists of a limited number of families related in the male line it might be expected that the processes of birth and death would introduce stochastic factors which would produce considerable variation in group size at a given moment in time. But the minimal number of persons in local groups shows an amazing constancy. In three different continents, under conditions lacking enriched resources such as permanent rivers and coastlines, the "magic number" for the average band size approximates 25 persons. In Australia this is based upon data provided by Tindale (personal communication) in which 11 bands seen by him under precontact conditions in arid country averaged 25 persons of both sexes and all ages. The range around this mean value was small. It can be inferred that the numerical size of the local group was under a rather tight system of controls.

It has been pointed out elsewhere (Birdsell, 1968) that local group size of this same magnitude characterizes the !Kung Bushmen of South Africa and the Bihor, a hunting and gathering group in north India. Marshall (1960) found 18 Bushman bands which again averaged 25 persons each. Among the Bihor, Williams (1969) censused 25 bands, and found them to average 26 persons each. It is to be stressed that this constant number, which approximates 25 persons, represents the minimal local group size. In other places environmental factors and the type of task forces needed may enlarge it.

It is not clear what system of forces tends to produce such constancy in local group size across these three continents. A band of 25 persons would only include a dozen or so males of all ages, and perhaps no more than 6 or 7 men capable of territorial defense. Local group territories are defended, but apparently only on rare occasions. Williams (personal communication) has suggested that the social functioning of the local group becomes ineffective if numbers fall much below 25 persons. With exogamous marriage practiced, the stochastic factors and the frequency and sex of births may dictate that a certain minimum size is needed in order to provide daughters in exchange to the men of other bands. The maintenance of these reciprocal exchange mechanisms over a reasonably short time span may be what dictates the basement number for local group size. It certainly is an important part of the situation. Unfortunately functioning local groups have not been intensively studied from this point of view, so that some mystery remains as to how these "magic numbers" are produced.

Peterson (1975) has pointed out that the Australians have developed a kind of greeting ceremony as a mechanism to facilitate the entry of strangers into the camp of the local group. Essentially the visitor or visitors formally sit down outside the camp at a considerable distance and in full sight. In time an influential older man from the local band will come out, approach them, and inquire as to their intentions. With these preliminary exchanges completed, the visitors are then invited into camp. This formalized arrangement prevents surprise meetings, and of course smooths social relationship.

TWO EXAMPLES OF AGONISTIC BEHAVIOR BETWEEN LOCAL GROUPS

Greeting ceremonies of the type described above serve to bring most visitors easily into the camp of the local group. But where visitors come from a great distance, have not been seen for perhaps a generation or more, and are not personally known, agonistic displays of aggression may take place. Tindale (1974) had the opportunity to document two such instances photographically in essentially precontact times. His plates 56–59 in Tindale (1974) indicate the sequence of events when strangers approached a Ngadadjara camp at Warupuju in August of 1935. As smoke fires to the west heralded the prospective arrival of the visitors, the young men prepared their weapons for possible action. The strangers, Nana tribesmen, approached the camp armed, and their womenfolk waited in the distance. Driven out of their own territory in the west by a lack of water the Nana tribesmen strode nervously about with their spears at ready as they identified themselves. A generation had lapsed since the last contact between these two groups and there remained some unfinished business involving a woman promised in marriage. Finally the old leader of the Ngadadjara, Katabulka, carried out a wooden dish of water as a sign of friendship. Spears were thrown, but intentionally allowed to fall short of targets. While the visitors had calmed down generally, one leading and tense Nana paced around stiff-legged and shouting loudly. The meeting was resolved without genuine fighting.

A similar incident recorded in Plates 60–67 in Tindale (1974) documented the arrival of a Pintubi local group from the west at the camp of Ngalia tribesmen in their own territory. The event constituted the first contact within the memory of the Pintubi of the Kintore Range with their eastern neighbors, the Ngalia, and occurred at Mount Liebig in August 1932. The Pintubi men, armed, with their bodies decorated with red ocher, appeared at the edge of the camp at Mount Liebig. They sat down with spears and spear-throwers ready to use and with additional spears behind them. In response a body of Ngalia men, fully armed, moved toward them and the visitors stood with spears in hand. An elder Ngalia, Ngoritjukurapa by name, approached the visitors with spear-thrower in hand. He talked to an elder Pintubi and in a friendly way inquired how the latter had become partially hamstrung, since he walked lamely. The Pintubi visitors, now more at ease, put their spears behind them and awaited a ceremonial welcome or reception. The resident Ngalia paraded about in a body at double time, and came together in a closed circle as they shouted in unison. They then moved slowly in a body toward the visitors, and by using totemic names and section terms established the generation level of each person, and hence their kinship status with respect to each other. Subsequently the Pintubi men were joined by their children and finally their women. Perhaps because of the tensions involved, a pregnant Pintubi had a miscarriage that night in camp. These two cases indicate that adaptive mechanisms existed among the Aborigines, which allowed strange groups to meet, to test intentions, and to establish individual positions within the network of kinship relations. Agonistic display here seems analogous with that reported for various nonhuman primate groups.

EXCEPTIONS TO THE BASEMENT NUMBER IN BANDS Favorable local resources resulted in larger local group sizes. In spite of the evidence that throughout most of Australia band size approximates 25 persons, there are indications that regional ecological factors allow this number to increase substantially. The data, and they are admittedly not altogether satisfactory, do suggest that in the more richly endowed environments local group numbers may increase to 50, or occasionally even 100 persons. Anthropological observers have not always been careful to distinguish between aggregations of several local groups and permanent territorial units. Larger local group populations seem to characterize the northern coastal stretchs of Arnhem Land in the Northern Territory. Here there is even some suggestion that owing to the richness of locally unearned marine food sources, terrestrial animals may not be as intensively exploited as elsewhere in Australia. Higher values again seem to reappear in the lower reaches of such permanently flowing rivers as the Murray and the Darling. These estimates are based upon the observations of early observers, whose very presence may have catalyzed local aggregations above the size of the real territorial groups. It is possible that the local groups of the smallish tribes of the Queensland rain forest area were also numerically inflated. This is based primarily on the evidence that tribes there seem to show a relatively limited number of bands, and consequently these may have been enlarged in numbers. Consequently, in ecologically favored areas, local group numbers considerably exceed the basement values. *This is the most dramatic instance of the impact of ecological variables on social organization in aboriginal Australia.*

The Kaiadilt Provide Further Exceptions

The Kaiadilt tribe of Bentinck Island in the southern portion of the Gulf of Carpentaria in north Australia provide notable exceptions to the general regularities which characterize local group organization in the rest of Australia (Tindale, 1962). Situated on a small island and isolated from all other Aborigines apparently for millennia, these people evolved an unusual reef form of hunting and gathering. Their subsistence came totally from the tidal zone of the tropical reefs surrounding the island. The total population maximum approximated 120 persons, and was divided into 8 dolnoro or local groups. The average size of these bands was only 15 persons, well below the continental mean. But among the Kaiadilt dolnoro range fluctuated markedly, for the smallest contained no more than 7 persons, whereas the largest numbered 29.

Unequal access to resources seems to have introduced an element of instability into this local scene. Suitable materials for building fish traps on this low sandy island were primarily concentrated in the country of the most northerly dolnoro. Not surprisingly they were both the most prosperous and the most numerous with 29 members. Perhaps because of limited total numbers in this island universe, combined with considerable variation in group size, dolnoro endogamy came to be practiced preferentially. It was not universal, but judging from Tindale's (1965) genealogical data endogamy seems to have been the preferred way of assuring that sons in the dominant local group could obtain wives without undue delay. Mean local group size here is well below the basement number, and seems to have caused dysfunctioning in the reciprocal system for exchanging women as wives. This seems to be a special instance in which a very local type of ecology modified band size, and the marital regulations which in all other parts of Australia decreed exogamous local group unions. Since these islanders showed a higher density than other groups of Aborigines, obviously it was dolnoro size, and not density proper, which was responsible for this series of exceptional traits.

The Nature of the Dialect Tribe in Australia

The anthropological concept of the tribe is both ambiguous and controversial. Therefore the nature of the dialect tribe in Australia must be closely specified. The Australian tribe is subject to a system of forces which do produce regularities, but this is through the molding of the local groups which comprise it. It has no structure beyond that provided by the local groups it contains. There is no political authority nor organization within it. Leadership in the tribe is nonexistent, except for the influence residing in the elder men who constitute an informal gerontocracy. These prestigious old men are important in all decision making within their local group, and so in broader gatherings. But it should be emphasized that they lead only through influence and have no authority as such.

Primarily the tribe in Australia is a language unit, and hence can be called a dialect tribe. It is a product of the dynamics of language differentiation, and so in a sense can be considered a "happening." Language is not completely homogeneous within these tribes, as recent fine-grained linguistic analyses have demonstrated, but it tends to be so.

The Australian tribe has certain recognizable attributes. It is a collection of geographically adjacent local groups. The tribal territory is the sum of the territories of these contiguous bands. The tribal boundary is the outer perimeter which encloses them. Tribes are known by distinct names, both to themselves, and to others. Proximity in space and in descent have produced a complex of customs and laws in which a tribe generally differs to a small degree from adjacent ones. But the best definition of a tribe turns upon the reproductive behavior of its members. The rate of intertribal marriage varies from approximately 15% (Tindale, 1953) to as low as 9% in some regions (Birdsell, no date). In all cases marriages involving partners across tribal boundaries are infrequent. This attribute allows *the tribe to be best defined as a deme, the minimal breeding population*. Social anthropologists may resent the intrusion of reproductive biology into their area, but it does produce a rigorous definition which has no exceptions in tribal Australia. It has the further great advantage that the Aborigines themselves recognize members of their own tribe as "of one blood."

Tribal Size is Self-Regulating

The bivariant coefficient of correlation of 0.87 for the optimum summer rainfall series of tribes is ample evidence that the numbers of persons per tribes do not vary greatly, nor at random. Tribal areas, and hence numbers of persons per tribe, are rigidly determined by a complex of environmental variables. A total of 45 population estimates (Birdsell, 1953) of which only 11 were considered to be essentially reliable, confirm the essential constancy of tribal numbers. But they leave the question as to whether the constant is approximately 500 persons per tribe, or more nearly closer to 450 persons per tribe, unanswered. The exact value of this constant central tendency may remain unresolved.

There are two further types of evidence which reflect the systems regulating tribal numbers as a constant virtually irrespective of ecological variation in the continent. The first of these involves a making and breaking tendency evident within dialect tribes. In a number of instances where tribal numbers have grown well beyond the "magic number" of 500 persons, dialectal differentiation is well underway. A split at some future date into two or three new tribal entities could be predicted. This type of dialect differentiation became evident as the numbers in the original tribe approached 1,000 persons.

However, those tribes which through circumstances become reduced in number tend to be unviable as entities. They show a marked tendency to coalesce with and be absorbed in neighboring larger tribes speaking closely related dialects. There are suggestions that when tribal populations fall below 200 persons, they no longer can maintain independent dialectal identity. A number of such actual instances are documented in Tindale's tribal monograph (1974).

A second approach to identifying the principles which regulate tribal numbers around a constant value involves an appropriate model (Birdsell, 1958). A model using hexagonal territorial shapes for the close packing of contiguous bands allowed the demonstration that the frequency of interactions across tribal boundaries varied with the number of bands constituting the dialect tribe. Where the bands are few in number, too many interactions occur across tribal boundaries, and so are intertribal in nature. Increasing the number of bands within the tribe reduces the relative number of intertribal contacts, and so produces an increase in internal cohesion culturally and linguistically. But the limitation of mobility to foot travel allows distance to become an isolating factor limiting the size of these larger entities. Empirical evidence indicates that dialect homogeneity cannot be maintained where the frequency of interactions becomes attenuated.

It is appropriate to view the creation of this cellular tribal structure in Australia as a *consequence of the density of communications between individuals.* Parameters in that type of model would include frequency, duration, and intensity of communication, as well as level of mutual comprehension. Unfortunately no empirical data have been collected to build into such a model, although its consequences are clear enough to stand by itself at this time.

REGIONAL VARIATIONS IN TRIBAL SIZE There are a few regions of Australia in which deviations from the normative number of 500 persons seems to be under ecological influences. They were located on coasts or on permanent rivers, and so are not included in the correlation analysis. Some other very large tribes were present in New South Wales, such as the Kamilaroi, the Wirdajuri, and the Wongaibon. Across the continent in the the central portion of Western Australia was the Wadjari, an equally large tribe. Early estimates of populations of these four tribes range from 1,000 to 2,000 persons. In both regions wild grasses grow profusely and their seeds provide the inhabitants with a major staple of their diet. Tindale (1974) considers that these abundant wild grains, which were stored in skins, may have altered the normal ecological relations between people and their land. The grass seeds were regarded as the staple food by the tribes utilizing them. Early explorers passing through New South Wales were astonished to see on alluvial flats prepared shocks of drying grass which gave the appearance of a European grain field. It is to be emphasized that no agriculture was involved here, although some primitive attempts to divert flood waters were utilized.

This intensive exploitation of grass seed could occur only in regions lying below the tropical belt, for in the tropics termites destroyed grasses faster than seeds could be collected. The eastern and western ends of the grass seed belt were connected by an arch of tribes also known to have depended heavily upon the grinding of grass seeds for staple fare. But except at the extremities, tribal sizes seem to have been quite normal in numbers so that the problem cannot be called solved. There is no particular explanation in terms of the model for the density of communications to explain why the four terminal tribes should so markedly exceed the norm of 500 persons per tribal unit.

Clusters of small tribes exist in several regions on the continent. These units each occupied a smaller area and were reduced in numbers compared to the average tribe. The west coast of Cape York Peninsula is characterized by a flat coastline in which tribal size is small. Tindale's own words (1974:112) describe the situation.

> Combination of a fertile area in the lowlands of the eastern side of the gulf with many estuaries, slow-moving fresh water streams and lagoons, and intervening higher inhospitable areas has encouraged the development of a series of settled and largely sedentary populations, each knowing, and caring little about other than immediate neighbors. With the relative ease of life has come enlarged populations—but not very large ones—moderate changes in dialect and variations in practices and beliefs, but on a common theme. What may at one time have been only a series of expanded hordes have acquired the manners and trappings and independence of small tribes. It seems noteworthy that such developments virtually are confined to areas where the special environmental characteristics outlined above are present.

A comparable situation seems to have existed on the northwest cost of the Northern Territory on the Daly River, where there are also clusters of small tribes of sedentary habit. Each seems to base its living on relatively permanent areas or slow-moving streams, lagoons and swamps, all of which yield fish. Those nearer the coast receive a constant renewal of their supplies of estuarine fish. This environment also provides ample vegetable foods.

A somewhat different category of small tribes is found in the dense tropical rain forests in the Cairns area of northeast Queensland. There existed a dozen small tribes in these tropical rain forests with populations ranging from 200 to 300 persons. A tropical rain forest is a difficult place for gathering food and worse for hunting it. So the tribal size here does not seem to be a consequence of the richness of food supplies, but rather as a result of the difficulties involved in moving through such jungles. These diminutive tribes do fit the density of communications model in that the environment tends to restrict the frequency of person-to-person contact. Elsewhere as in southern Queensland and even Victoria, tribes inhabiting dense rain forests, whether they be tropical or temperate in flora, again seem to be characteristically smaller in population numbers than in the more open country around them. These examples tend to reinforce the idea that frequency of communications are diminished in such situations.

Ecological Influences on Tribal Size

The size of tribal populations on most of the continent is surprisingly constant and immune to ecological pressures. Nevertheless there are a few situations in Australia which do seem to influence the numbers of people in a dialect tribe. Some of the grass-grinding tribal groups are unusually large in size, and while this may be due to the unusual availability of a staple food, others in the same situation are normal in size. Sedentary coastal tribes in both Cape York Peninsula and on the Daly River clearly reflect a local abundance of food and a tendency to settle down in smaller than usual tribal groups to sedentary ways of exploiting it. Rain forest environments, especially in the tropics, seem to impede personal mobility and so to reduce the size of the tribal population through a constriction of density of communications.

The Influence of Ecology Upon Patterns of Territoriality

Territoriality ranges all the way from the fine grained claims by families to certain assets in well-endowed country to the rare tribal claims which provide the only landowning unit. In general, however, territoriality resides on a broad basis with the local group. Obviously ecological variables play an important part in defining the types of territorial claims which will be adaptive in the long run.

TYPES OF FAMILY OWNERSHIP Assets claimed by the male head of a family, and occasionally by women in it, usually involve recurrent food resources of reliable character, but sometimes are attached to mineral rights. Data from early observers indicate that such family ownership characterized all of the better watered stretches of Australia, including all the coastal districts. Examples include the rights to black swan eggs on Raymond Island in King Lake in Victoria. These rights were held in the male line and at the time of contact were in the hands of an old man and several of his nephews. Elsewhere fish traps were individually owned and provided rather constant food resources. Even the great complex fish trap at Brewarrina on the Darling River in western New South Wales fell into this category. There each trap cell was owned by an individual. On the Bloomfield River in northeastern rain forest Queensland, certain *Zamia* trees were owned by women who bequeathed them to their daughters or other kinswomen. These cycads produced important nuts which were a staple food. This form of inheritance allowed women to retain special food assets in their father's band territory, and so gave greater flexibility to the distribution of food than would otherwise occur. The great bunya conifers in coastal Queensland, which produced an enormous harvest of pine nuts every third year, were individually owned and passed from father to son. Large numbers of invited people from neighboring tribes came at harvest time, but the cones were brought down from the tree by the owner and formally distributed to them. This type of control of food resources provided a base for reciprocal exchanges of food with others, and so was adaptive in nature.

The axe quarry at Lancefield, Victoria was owned by the head of the family in whose country the quarry occurred. He took care of it for his tribe which held overrights. The owner and his descendants split the stone into blanks which were traded through barter to other portions of the tribe and more distant local groups.

LOCAL GROUP TERRITORIAL RIGHTS Broader and more inclusive rights of ownership were exercised by local groups. Territorial rights of this nature were almost universally present throughout the continent in one form or another. Ownership in this sense refers to the right to utilize animal, plant, and mineral products of the land. It does not involve real estate in any Western sense of the word. Band boundaries were well known and tended to follow easily identifiable landmarks. In some portions of Victoria each local group owned the drainage system of a river tributary. This would insure a variety of land forms and food resources. Tindale reports (1974) that the Tanganekald tribe which inhabited a long strip of land separating the brackish lagoon waters of the Coorong from the sea off the southeastern corner of South Australia divided its land equitably among its bands. Each of the 22 local groups possessed a strip extending from the ocean front, through the sand hills and the inland strip into the lagoon proper. This insured equivalent ecological opportunity to each of the local groups which thus had access to four distinct types of terrain.

There are general indications that local group territories throughout Australia were adjusted by mutual accommodation to maximize the support base for each band. This insured that local group territories acted as an effective spacing mechanism for the human population, and further resulted in each being essentially self-supporting on its own land for perhaps 80 to 90% of the year, save in the more arid stretches. In desert regions band territoriality was manifest in the ownership of waterholes, for these determined what land could be reached for hunting and gathering. In general reciprocal visits involved surplus food, the trading of artifacts and materials, and the enjoying of ceremonial occasions when food was in abundance. Some visiting went on simply to get a change of diet from a seasonally overabundant food.

TRIBAL TERRITORIALITY The territory of the dialectal tribe is of course the sum of the ownership of land by its contained local groups. But there was probably no instance in Australia where local groups did not move about in a fairly flexible fashion within the overall pattern of band reciprocity. There was a good deal of visiting between local groups at the appropriate seasons. Bands on the boundaries of adjoining tribes tended to maintain friendly relations, and so occasionally camped together when visiting a common boundary or waterhole. A further escape valve was provided in desert country by the fact that it was crisscrossed by the tracks of mythical ancestors. Their travels were incorporated into totemic systems of beliefs. These tended to relate peoples across tribal boundaries at a considerable distance from each other. For example, men of the wildcat totem were bound together by their common ceremonial activities and their segmented ownership of the ancestor's track. Should drought affect wildcat men in one area they felt free to travel far down the ancestral track to take refuge among men of the same totem who owned a different section of the pathway.

The relative freedom of movement in crisis in the desert has led some (Berndt, 1959) to question whether tribes really existed in the arid reaches. This has been tested and answered (Birdsell, 1976). There can be no doubt that tribal space in cellular form characterizes the whole of Australia, including the desert, but in the latter regions adaptive behavior has obviously developed more flexible patterns of land use. For example, in the Wanman tribe in the Western Desert of Western Australia, the whole tribal territory belongs to all the Wanman people. All had equal rights to water at important water places. But within this overall tribal ownership, there still persisted local group ownership expressed in terms of an exploitable chain of waterholes (Tindale, 1974:19–20). So it may safely be concluded that the aboriginal adaptation to his country involved stages of increasingly flexible movement as decreasing rainfall, and the attendant occurrence of drought, made life more precarious there.

Geometrical Patterns of Tribal Spacings

Too little is known of the details of local group territoriality to analyze it in terms of its geometrical patterns. But the boundary of the tribes which the local groups comprise does lend itself to a search for geometrical regularity. Using the splendid maps available in Tindale's latest monograph (1974), it was found that 100 interior tribes dependent upon local rainfall for their water averaged 5.8 sides each. This is a numerical approach to a six-sided configuration, but of course is more irregular than the geometrical hexagon which provides for ideal packing in two-dimensional space. Nonetheless these values do indicate an accommodating effort to provide a minimum boundary for a contained space.

On the other hand, on the permanently flowing Murray River of southeastern Australia the waterfront is the primary focus for the food search. There 17 tribes averaged 4.3 sides per tribal domain. Their territories were arranged in a more or less linear fashion parallel to the river course.

Coastal tribes reflect a somewhat similar situation. For the west coast of Western Australia 16 tribes also averaged 4.3 sides for each territory. These three ecologically different tribal samples reflect optimizing principles in the distributions of people accommodating to their food resources.

Table 1 shows in more detail the fuller aspects of tribal geometry. The method of ascertaining the number of side for a given tribe is somewhat subjective, and depends upon how much inflection in a boundary is required to score a second side. In this survey a change of direction of 10 or 15 degrees was considered to produce a second side. Obviously sides of tribal boundaries determined this way do not necessarily correspond to the number of tribal neighbors involved in the example. Therefore Table 1 gives a count of the neighbors for each of the tribes in the three series. It will be noted that for coastal tribes an additional neighbor is arbitrarily added to compensate for oceanic space with no neighbors. Tribes which only touched apically upon the boundaries of the tallied tribe were not included in this count. To compensate for this treatment, a third column lists the number of neighbors including those with only apical contacts. In terms of social interaction this may well be the most reasonable count. Using these criteria the series of 100 inland tribes shows a rise to 6.2 neighbors adjacent to its boundaries.

These values are not to be taken as indicating more than trends, for these aboriginal tribes do arrange themselves in real ecological space according to rational principles of minimizing

Table 1. Variations in Tribal Geometry as Influenced by Ecology.

Region	*Number*	*Number of Sides*	*Number of Neighbors*	*Number of Neighbors with Apices*
Inland tribes	100	5.8	5.5	6.2
Murray River tribes	17	4.3	4.8	5.8
West Coast tribes	16	4.3	3.6	3.9
			(+1 = 4.6)	(+1 = 4.9)

travel distance within the tribal area, and optimizing food exploitation. Figure 4 gives a tribal landscape for 11 interior tribes in Western Australia. This is characteristic of the inland tribal pattern which numerically approaches a hexagon on account of tribal sides and neighbors. Figure 5 displays 11 contiguous tribes along the Murray River. This geometry is visibly simpler than that which characterizes interior tribes and is of course oriented along the waters of this permanently flowing river.

In one area in the northern portion of the Western Desert there were too few permanent waterholes to be incorporated centrally into each tribal territory. In this instance a single waterhole, Karbardi by name, was located at the apex of no less than five tribal territories. The waterhole was permanent in character and each tribal population had equal access to it. So that adaptive sharing seemed to develop where necessary for survival.

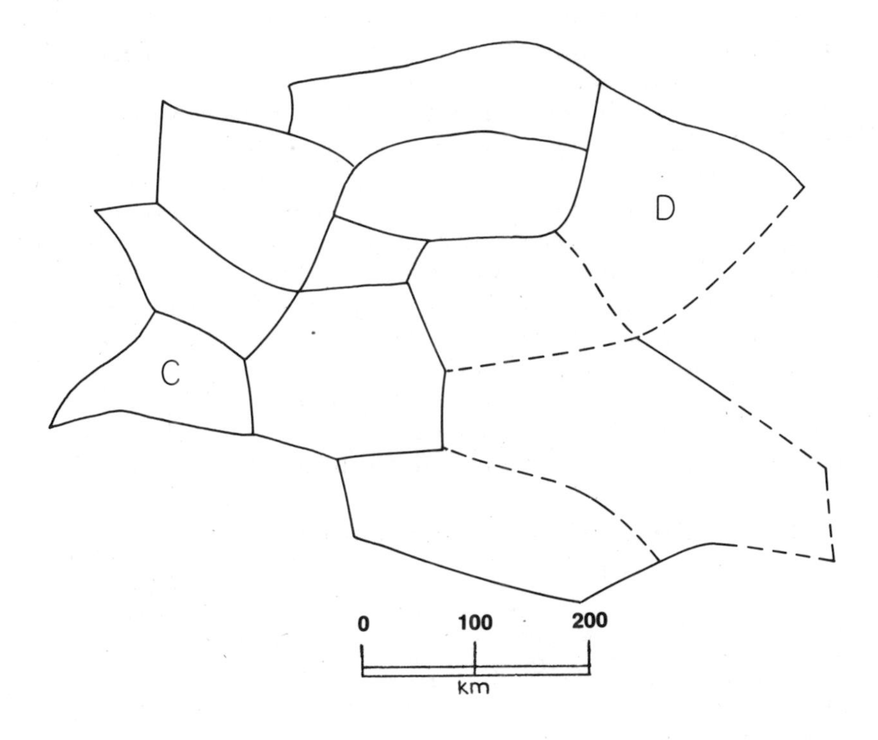

Figure 4. A sample of 11 inland tribes from Western Australia showing the tendency toward six-sided boundaries. They extend from C, the Ninanu tribe on the west to D, the Kartudjara tribe on the east.

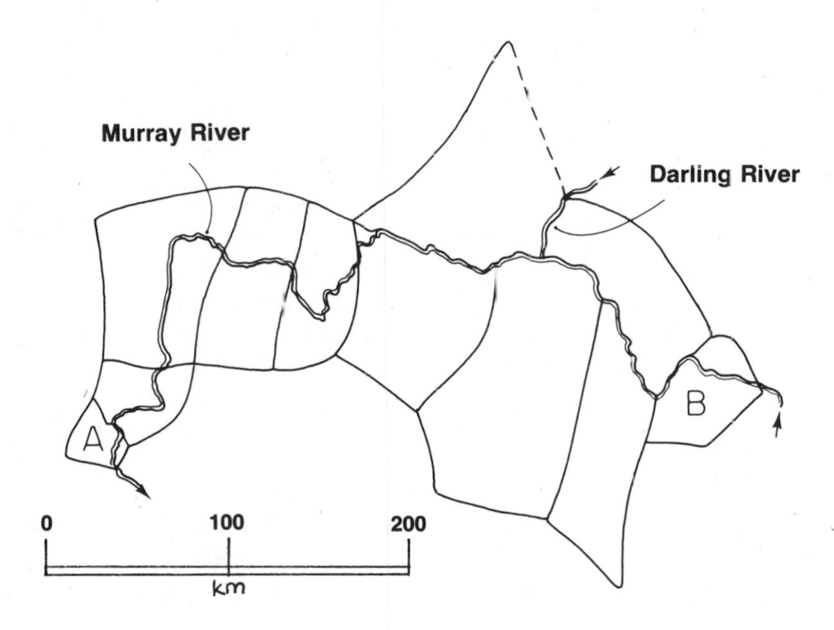

Figure 5. The distribution of 11 tribes along the lower stretches of the Murray River in South Australia indicating a tendency toward four-sided boundaries. This series ranges from A, the Ngaralta tribe on the west, to B, the Tatitati tribe on the east.

Patterns of Aboriginal Movements

Movements outside of the local group territory can be viewed from the perspective of the individual, the local group as a body, and as portions or the entirety of a tribe. As indicated earlier there is a good deal of seasonal visiting in aboriginal Australia which is both socially and ecologically adaptive. More to the point here is the question of permanent moves, who is involved, and what the motivating factors may be.

PERMANENT MOVES BY INDIVIDUALS Individual men may leave their local group permanently in order to escape punishment within them. Causes may involve a fight and death within the band. Or a man may elope with a woman whom he is ineligible to marry. Crimes against individuals are usually subject to trials by ordeal, and a man might flee from this test. These all constitute deep social affronts, and the offender best leave his local group for an extended period of time, and quite possibly permanently. Whether he will merely move to a distant local group where he is known, or cross a tribal boundary, perhaps depends upon the severity of the crime.

Women who have been induced to elope with a man of the wrong marital class of course are leaving their own local group under improper auspices. They too may be considered to have moved out of their band of birth and severed ties with it. Marriage by capture was reasonably rare in aboriginal Australia, but it did occur. Under these circumstances the woman victim of the practice would be carried so far away from her proper people that opportunities to return would be rare.

One of the common causes of flight among males involved the advent of new, severe initiation ceremonies. Both circumcision, and the more profound surgery involved in subincision, found many individuals unwilling to accept the rites. Among young men it was fairly common practice to escape ahead of the slowly advancing boundary of the rite and to find refuge in noninitiating groups. These might belong to bands of the individual's tribe, or if the rites had reached a tribal boundary, the next tribe over. Instances of this type were still occurring in early postcontact times.

THE DISPLACEMENT OF LOCAL GROUPS Temporary displacements of bands occurred whenever gatherings of one kind or another were held. These would include initiation ceremonies, trade fairs, invitations to utilize surplus food, and totemic rites along an ancestral pathway. It is very probable that the so-called intertribal gatherings of Australia never involved the movement of full tribes, but rather a series of local groups drawn from several tribes, and perhaps largely associated with the mythical tracks of ancestral figures. But these did not result in permanent displacement.

Reports of three instances involving the permanent displacement of local groups have been reported but in fragmentary nature. Two of these involve attempts to transfer land owned by one band to another. Among the Jualaroi tribe of central New South Wales (Fraser, 1892), it was reported that one local group had become so numerous that its own territory no longer served as a support base. It threatened to usurp the part of the territory of an adjacent band which was numerically smaller. The numerically dominant band threatened direct coercion and the matter was handed to a "tribal" council for settlement. This presumably consisted of a number of assembled local groups of the tribe. After considerable consultation among the unaffected elders of the tribe, a decision was reached that each band would be represented by its own single champion. The land in question would go to the winner in combat. The informant unhappily does not tell us how the decision fell. The case is of interest in indicating that formal mechanisms for settling land disputes were in existence at that time, and no doubt in precontact Australia.

Another rather similar type of conflict occurred among the Tanegekald tribe previously mentioned as living on the Coorong of South Australia. Norman Tindale (personal communication) reports that two neighboring local groups were constantly in dispute. Finally one, presumably the loser in most arguments, petitioned to the tribe to be allowed to move away to the eastern end of the Coorong and there occupy the territory of a band which had gone into the little colonial town of Kingston. By tribal consensus this request was granted. Although empty land was available for the move, the need to receive overt consent to change band residence is of judicial interest. Again, this case suggests that territorial disputes could be settled by aboriginal mechanisms.

Meggitt (1962) recounts a case in which a pitched battle over territorial assets occurred. One or more local groups of the Walpiri tribe apparently were driven by drought to the southeasterly waterholes of the Ngardi tribe. A prolonged spear fight resulted. The battle is reported to have terminated with the death of 20 men. This number of fatalities suggests that perhaps several local groups were involved on each side. The Walpiri were the winners and thereafter incorporated the disputed waterholes into their territory. The Ngardi accordingly withdrew. This event occurred in early postcontact times, but seems to have been an engagement in precontact spirit.

A NANGATARA HEGIRA Human speech vastly expanded the options for survival strategems. In desert Australia severe local droughts may cause heavy mortalities among the Aborigines. But as the following example shows, the ability to project past learning into future situations allows escape for some from this trap.

About 1943 at the end of a long continued local drought of unusual severity in Nangatara tribal territory, Paralji, even then an old man, undertook to save his local group. From an area generally north of salt Lake Tobin the little group began to work their way across their own tribal territory, traversing at least 25 waters to finally reach the refuge waterhole Karbardi lying in the northwest corner of their tribal domain. This important water was a fallback resource for no less than five tribal groups whose territories touched upon it in the form of apices.

Paralji previously had only visited Karbardi once in his life when his guardian took him on a journey as a part of his initiation into manhood more than half a century earlier. So that even this portion of the journey involved leading his band through much country with which he was not familiar. In time local food supplies began to fail at Karbardi so that Paralji, then leaving some of his horde members behind, but accompanied by several younger men with their wives and families, left this waterhole and proceeded west into unknown districts. He was guided chiefly by remembrance of lines of place names mentioned in Nangatara ceremonial song cycles. These are sung at totemic increase ceremonies, and they detail the wanderings of ancestral beings. To these traditional clues were added only such occasional marks which may have developed in the long term as a result of earlier movements of peoples. Their journey led them, in fact, along a line of meager waters between the tribal territories of the Njangamarda Iparuka and the Njangamarda Kundal tribes who had long lived apart. The old trails led from one spring to another. They eventually arrived in the native camp at Naldgi at Mandora Station, on the coast of Western Australia.

The minimal length of this hegira was 600 km, of which some 350 km traversed country known to Paralji only through tradition. The trip involved successfully locating 50 to 60 waterholes, and covered a 5 to 7 months' period of time. Since this was but one of many possible routes stored in the memory of the old Aboriginal, his total recall of the geographical locations and sequences of waterholes outside of his own radius of personal knowledge must have been enormous. This example was gathered by Tindale at Mandora Station in 1953. I am privileged to use it through his courtesy, in this abridged form, and a detailed account will be published by him in the future.

PERMANENT DISPLACEMENT OF TRIBES Evidence is growing that suggests that tribal stability in Australia may extend backwards some millennia in time. Tindale (1974) has discussed the antiquity of tribes from the point of view of the introduction of the dingo, and the words associated with the animal. Perhaps even more compelling is the case reported by Dixon (1977) of the Konkgonkji tribe on Cape Grafton, in rain forest northeast Queensland. There tribal memories recall in a very detailed way the time when certain offshore islands were once a part of the mainland. These recollections presumably refer to the time prior to the eustatic rise of sea level, which terminated some 10,000 years ago. A very similar story is recounted by Tindale (1974) on the other side of the continent by the Jaudjibaia tribe who in historic times occupied the Montgomery Islands lying well offshore of the Kimberley coast. They have a tradition that the islands were once a large country, but that a big flood came and drowned it so that it now consists only of small islands. This event too seems to refer to the postglacial rise of sea waters about ten millennia ago. Thus some tribal stability relating people to their present locales over long periods of time seems to be indicated. There are other instances of tribal memories going back thousands of years to such events as volcanic eruptions in South Australia, and the encroachment of rain forests on eucalyptus woodland on the Atherton Plateau in northeast Queensland.

On the other hand, Tindale (1974) has documented a good many cases relating to tribal movements in postcontact times. These usually involve the taking over of land from the members of another tribe who have moved into White settlements, or otherwise have been decimated by the consequences of permanent colonial contacts. But it is not difficult to project these backwards into purely aboriginal times. It may be that where a marked decline in numbers occurred for whatever reason, populous neighboring hordes, and perhaps even a large proportion of the tribe they represented, would move in with the return of good times and take over the territory previously held by another tribe. Certainly a good deal of give and take must have occurred throughout the occupancy of Australia for the peoples to become so regularly spaced as to approximate optimum geometry both in terms of local groups and the tribes themselves. At the time of White contact the distribution of both of these units was highly structured, regularized, and anything but haphazard.

Conclusion

The Aborigines of Australia lived in a homeostatic system which they regulated with considerable skill. Faced with rigorous environmental determinism of their densities, they managed their sociocultural behavior and their reproductive biology to maintain systemic stability. Even so, three basic social units, the family, the local group, and the tribe show amazing constancy numerically across a wide range of environments extending from the most arid deserts to tropical rain forests. Preferential and systematic female infanticide was a primary determinant of consequent population numbers and the maintenance of a suitable adjustment between the numbers of people and their environment. In spite of the preceding plasticity of behavior, the basic social units in aboriginal Australia remained essentially constant across a wide range of environments.

References

Berndt, T. M. 1959. The concept of "the tribe" in the Western Desert of Australia. *Oceania* 30:81–107.

Birdsell, J. B. 1953. Some environmental and cultural factors influencing the structuring of Australian Aboriginal populations. *Amer. Nat.* 87:171–207.

Birdsell, J. B. 1958. On population structure in generalized hunting and collecting in populations. *Evolution* 12:189–205.

Birdsell, J. B. 1968. Some predictions for the Pleistocene based on equilibrium systems among recent hunter-gatherers. Pages 229–240 in *Man the hunter.* Eds. R. B. Lee and I. DeVore. Chicago: Aldine.

Birdsell, J. B. 1976. Realities and transformations, the tribes of the Western Desert in Australia. Pages 95–120 in *Tribes and boandanos in Australia.* Ed. N. Paterson. Canberra: Australian Institute of Aboriginal Studies.

Birdsell, J. B. n.d. Intertribal marriages rates among aborigines of the Western Desert.

Dixon, R. A. 1977. *A grammar of yidij.* Berkeley: University of California Press.

Fraser, J. 1892. *The Aborigines of New South Wales.* Sydney: New South Wales Commission of the World's Columbian Exposition at Chicago.

Marshall, L. K. 1960. !Kung Bushman bands. *Africa* 30:221–252.

Martin, C. R., and Read, D. W. 1976. The relation of mean annual rainfall to population density for some African hunters and gatherers. In *The perception of evolution* . Ed. L. Mai. *Anthropology,* UCLA, 71: unpublished.

Meggitt, M. J. 1962. *Desert People.* Sydney: Angus and Robertson.

Peterson, N. 1975. Hunter-gatherer territoriality: The perspective from Australia. *Amer. Anthrop.* 77:53–68.

Spuhler, J. N., and Jorde, L. B. 1976. Primate phylogeny, ecology and social behavior. *J. Anthrop. Res.* 31:376–405.

Tindale, N. B. 1953. Tribal and intertribal marriage among the Australian Aborigines. *Hum. Biol.* 25:169–190.

Tindale, N. B. 1965. Some population changes among the Kaiadilt people of Bentinck Island, Queensland. *Rec. South Aust. Mus.* (Adelaide) 14:297–336.

Tindale, N. B. 1974. *Aboriginal tribes of Australia.* Berkeley: University of California Press.

Vorkapich, M. 1976. Population density and Great Basin ecology. In *The perception of evolution.* Ed. L. Mai. *Anthropology, UCLA,* 71: unpublished.

Williams, B. J. 1969. *Contributions to anthropology: Band societies.* Bulletin 228. Ottawa: National Museums of Canada.

Chapter 7

The Variation and Adaptation of Social Groups of Chimpanzees and Black and White Colobus Monkeys

Akira Suzuki

Shifting to studies of nonhuman primates, Dr. Suzuki invites our attention once again to the generality of basic mechanisms of adaptation and simultaneously indicates the greater simplicity of the nonhuman primate model relating social organization to ecology. There will be certain obvious parallels with the previous chapter on Australian aborigines, and the relative simplicity of the nonhuman primates does not preclude a considerable degree of complexity as compared to simplistic models of social organization.

In considering black and white colobus monkeys and chimpanzees living in the same general habitat, Suzuki shows that different social organizations reflect different basic adaptations to an environment. On the other hand, his descriptions of chimpanzees living under a variety of habitat conditions reveals the flexibility of the chimpanzee social organization which permits the same basic social organization to function adequately in diverse habitats with minimal adjustments, which are built into the system itself.

Even the colobus groups demonstrate more variability than might have been expected based on the relative uniformity of their habitat. Although the one male unit may have been typical, some troops contained as many as three adult and subadult males. The chimpanzee social organization demonstrated an even greater flexibility of group composition and some variants were more common in the forest whereas others predominated in the savanna woodland. More importantly, the chimpanzees living in the savanna woodland showed seasonal changes in social structure apparently related to the changing pattern of resources and habitat utilization in the wet and dry seasons. The chimpanzee variability is maintained in a more generalized basic social organization which is apparently a tolerable social organization in multiple habitats. Comparison with the generalized adaptations of the Aborigines discussed in the previous chapter is invited.

Introduction

In primate species there are few works discussing such problems as social variability in a species and the influence of environments on social organization. Working on the assumption that each species should have a species specific society which is genetically determined, many researchers have generalized from a few observations of individual groups to the species as a whole, rather than studying variations in social organization.

However, it is very important for the study of social evolution in primates to clarify the relationship between inter- and intraspecific social variability and environmental variables. With this in mind, I have carried out ecological and sociological studies on wild-living chimpanzees and black and white colobus monkeys since 1964 [chimpanzees: in western Tanzania, 1964–1965, 1970 (Suzuki, 1966, 1969, 1971a); in the Budongo forest, Uganda, 1967–1968, 1970–1971, 1972–1973 (Suzuki, 1971b); colobus monkeys: in the Budongo forest, 1967–1968, 1970–1971, 1972–1973].

This chapter summarizes the results of these studies from the viewpoints of variability and adaptation of primate society. The social compositions and organizations of these species show variations which can be related to changes in the environment and population structure, and may be important in discussions of the evolution and adaptation of society in primate species.

Habitat Descriptions

Budongo Forest, Uganda

The vegetation of the Budongo forest is described as *Cynometra-Celtis* medium-altitude semideciduous forest, which is situated outside tropical rain forest (Langdale-Brown, Osmaston, and Wilson, 1964). Eggeling (1947) divided the vegetation of the Budongo Forest into five types: (1) colonizing woodland forest, (2) colonizing *Maesopsis* forest, (3) mixed forest, (4) *Cynometra* forest and (5) swamp forest. Figure 1 gives a vegetation map of the study area. In the parts of the mixed forest areas, *Chrysophyllum* dominated mixed forest and is shown separately. This particular forest type develops in high humidity areas in valleys and is rich in terms of food production for chimpanzees. Dry *Combretum* savanna and grassland are found on the outskirts of the forest, which was formed following human cultivation.

Western Tanzania

The other study area is in the Kigoma and Mpanda districts of Tanzania. The area is bounded on the west by the eastern shores of Lake Tanganyika, on the north by the Malagarasi River, on the south by the Lugufu River, and on the east by the Ugala Swamp River, a tributary of the Malgarasi River.

The flora of West Tanzania is described as open woodland with *Isoberlinia, Julbernardia,* and *Brachystegia* vegetation. The dominant species are Caesalpininaceae trees, called "Miombo"; consequently this type of vegetation is often called "Miombo woodland." In contrast to the flora in Mimosaceae dry savanna, that of West Tanzania can be called a Caesalpiniaceae savanna type. I have previously referred to this type of vegetation simply as "savanna woodland" (Suzuki, 1969).

The area is situated on the eastern side of the Western Rift Valley. There are many streams on the wall of the Rift Valley, and semideciduous or evergreen riverine forests develop along each stream.

I divided the savanna woodland vegetation in the study area into three major types: (1) grassland and savanna, (2) open forest or woodland, and (3) riverine forest and thicket. These were subdivided into a total of 20 vegetation types. The three major vegetation divisions accounted for 30%, 60%, and 10% of the total area, respectively.

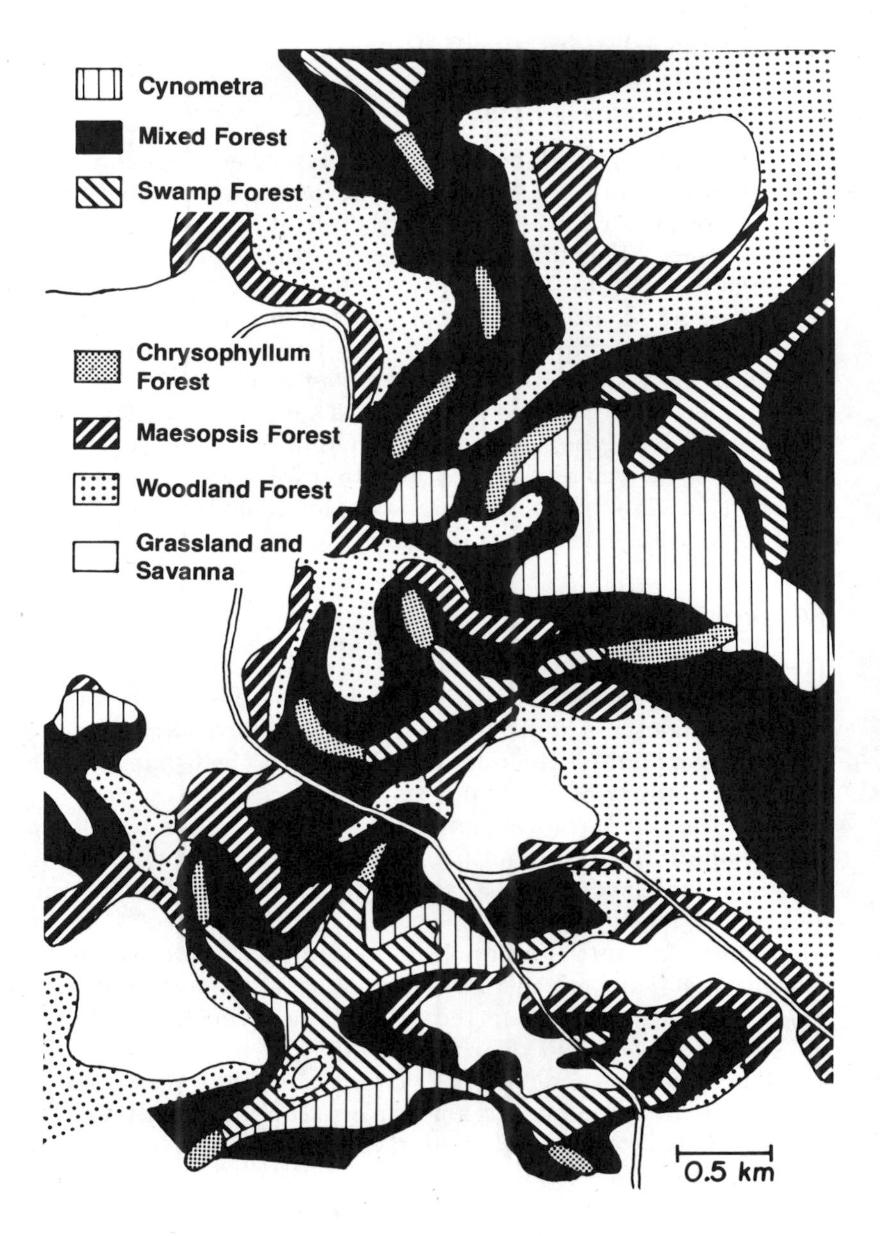

Figure 1. Vegetation map of the Budongo forest.

Species and Population Densities in the Two Areas

Six species of primates other than prosimians are found in the Budongo forest. They are chimpanzees (*Pan troglodytes*), black and white colobus monkeys (*Colobus guereza*), redtail monkeys (*Cercopithecus ascanius*), blue monkeys (*C. mitis*), vervet monkeys (*C. aethiops*), and baboons (*Papio anubis*). The home ranges of the species overlap widely. In the forest area, three species of forest-living monkeys (black and white colobus monkeys, blue monkeys, and redtail monkeys), keep in close relationship and are often associated together in mixed-species parties wherever their home ranges overlap. Baboons and vervets have the core areas of their home ranges outside of the forest and have larger home ranges than the three species of forest-living monkeys. Chimpanzees have very large home ranges which cover several home ranges of all other monkey species.

Table 1 shows the number of individuals in a group or social unit of each species, the area of the home range of a group or social unit, and population density is highest in the blue monkey (86.9 individuals per km^2), intermediate in the black and white colobus monkey (48.6 ind/km^2), and lowest in the baboon (11.1 ind/km^2), the vervet monkey (10.9 ind/km^2), and the chimpanzee (4.1 ind/km^2). Only one group of vervet monkeys was found in the study area. I suspect that the population density of the vervet monkeys here does not necessarily reflect normal levels, since they have suffered from hunting by local farmers and may have been forced to move into the study area from the southern part of

Table 1. Group Sizes, Home Ranges, and Population Densities of Primates in the Budongo Forest, Uganda.

Primate	*Number of Individuals per Group* ($\bar{X}$)	*Range (Individuals)*	*Home Range (km^2)*	*Population Density* $\frac{Individuals}{km^2}$
Black and white Colobus	7	(2–13)	0.14	48.6
Blue monkey	20	(16–23)	0.23	86.9
Redtail monkey	20	(14–27)	0.28	71.4
Vervet monkey	12		1.10	10.9
Baboon	35	(30–40)	3.15	11.1
Chimpanzee	80		19.31	4.1

the forest. The observed population density of chimpanzees in this study is similar to that obtained by Reynolds and Reynolds (1965) (3–4 individuals per km^2).

In the savanna woodland of the western part of Tanzania, the area of home range of the chimpanzee social unit was estimated to be about 200 km^2. The density of animals within the home range is 0.2–0.25 ind/km^2, but the estimate of the population density of chimpanzees in the study area as a whole is closer to 0.5 ind/km^2 since their home ranges overlap widely (Suzuki, 1969). The typical size of the chimpanzee home ranges exhibits great variability. In the Budongo forest it is 20 km^2, whereas in the Kasakati basin of Tanzania (a savanna woodland) it is approximately 200 km^2, and in the driest area in the western part of Tanzania it has been estimated at 470–506 km^2 (Kano, 1972).

Population Structure of Black and White Colobus Monkeys in the Budongo Forest

Black and white colobus monkeys are generally known to live in one-male groups and to be strongly territorial (Marler, 1969). The composition of the 25 groups studied in the Budongo forest is given in Table 2, and their home ranges are shown in Figure 2. The average home range size is 0.144 km^2, which is very similar to the 0.137 km^2 estimated by Marler (1969) in the same study area some years earlier. Thus, during this time home-range size seems to have remained stable in the Budongo forest. Each colobus group has one or more special resting or safety trees into which the animals can run if they meet danger. These trees are marked in Figure 2.

Most of the groups of this species consist of one male and several females, but 5 of the 25 groups are multi-male groups (*see* Table 2). Intergroup relationships have been observed to take a variety of forms, including not only antagonistic interactions, but also friendly interactions. The smallest group (No. 27, Table 2) consisted of just two individuals, an adult male and adult female.

New groups seem to be formed generally by the fission of larger (13–15 individuals) groups. Two cases of group fission were observed during 1967–1973. In 1967, a group of 14–15 individuals split to form two groups (C14 and C15). When these two groups were observed in 1973 they were composed of nine and six individuals, respectively. In 1967, another fission was

Table 2. Compositions of *Colobus guereza* Groups in the Budongo Forest, Uganda.

Group Number	Total	Adult Males	Subadult Males	Adult Females with Young	Other Adult Females	Young Adult Females	White Infant	0–1 Years	2–3 Years	4–5 Years
1	7	1	——	2	1	——	1	——	1	1
2	7	1	——	2	1	1	——	——	2	——
3	8	1	——	2	1	1	——	1	2	——
4	7	1	——	2	1	1	1	1	——	——
5	8	1	——	2	——	1	——	1	2	1
6	4	1	——	1	1	——	——	1	——	——
7	9	1	1	1	2	2	——	1	1	——
8	4	1	——	1	——	——	——	1	1	——
9	5	1	——	1	1	1	——	——	1	——
10	5	1	——	1	1	1	——	——	1	——
11	4	1	——	——	2	1	——	——	——	——
12	8	1	——	1	2	3	——	1	——	——
13	10	2	——	2	2	1	1	——	1	——
14	9	1	——	1	4	1	——	——	——	1
15	6	1	——	2	——	——	——	2	——	——
16	*a*									
17	*a*									
18	8	1	——	1	3	1	——	2	——	——
19	4	1		1	——	——	1	——	——	——
20	6	1	——	2	——	1	——	1	1	——
21	13	2	1	4	1	——	1	3	1	——
22	10	2	1	2	2	——	1	——	1	1
23	8	1	——	1	3	——	——	——	1	2
24	*a*									
25	*a*									
26	6	1	——	1	2	1	——	1	——	——
27	2	1	——	——	1	——	——	——	——	——
28	9	1	1	2	1	1	——	1	1	1
29	6	1	——	1	1	1	——	1	——	——
Total (25 groups)	173	28	4	36	33	21	6	17	19	9
Average/Group	6.92	1.12	0.16	1.44	1.32	0.84	0.24	0.68	0.76	0.36

[a]No census data available.

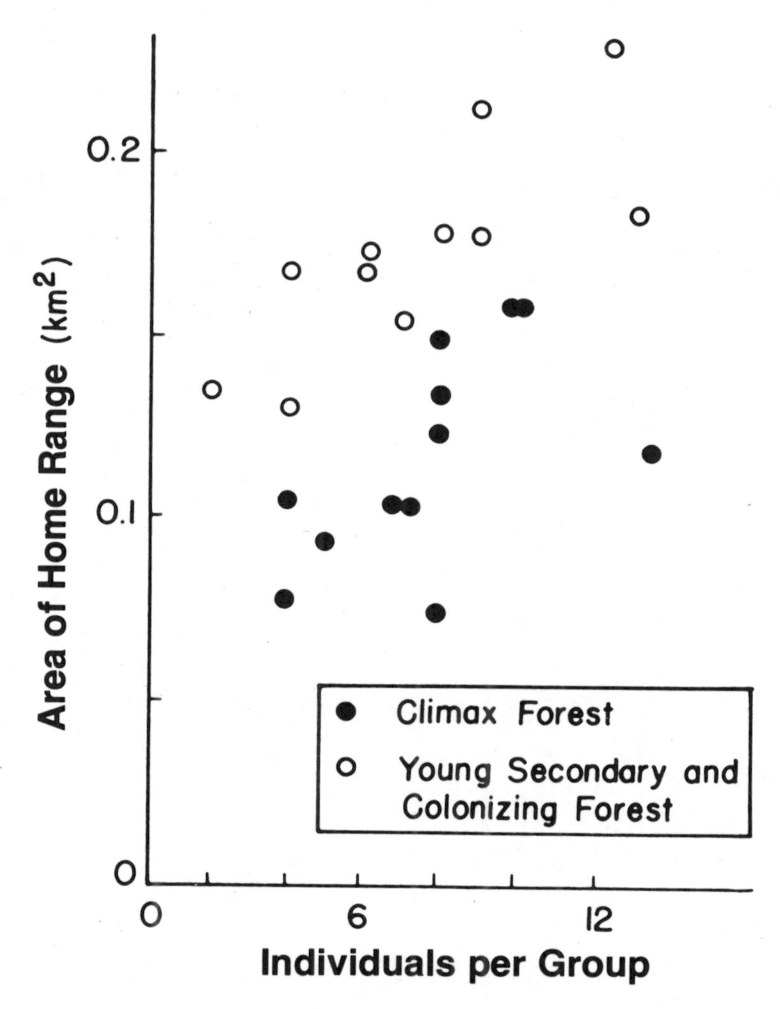

Figure 2. Home ranges of black and white colobus monkeys in the Budongo Forest.

observed when the C11 and C13 groups were formed from one large group. The C14 and C15 groups were observed many times in close proximity when feeding and resting. This very friendly relationship between the two groups is thought to be related to the fact that they were originally a single group.

Although their origins are not known, groups C21 and C22 were also observed to have a friendly relationship, the members of both groups mixing together socially with great ease. On 12

September 1972, five members of group C21 (an adult male, two females, and two juveniles) transferred temporarily to group C22. The home ranges of the two groups overlapped in their main feeding areas.

Flexibility in the social structure (*See* Rowell, this volume, Chapter 1) can be seen in the formation of two groups from three existing groups. Two aggregations of animals (10 and 13 individuals, respectively) were formed from existing groups C12, C18, and C20. This fusion may have been temporary as a total of four groups (C12, C18, C20, and C1) were staying in a small area on the forest edge after a rainstorm.

It is worth noticing that groups which showed friendly relationships were all living in higher density areas with more overlapping home ranges. The groups occupied forest margin areas where young stands of colonizing *Maesopsis* and *Celtis* existed. It is suggested that such an intergroup relationship has developed because groups on the forest margin cannot move to other areas despite high population densities; instead, they establish larger home ranges that overlap extensively with those of neighboring groups. Figure 3 shows home-range areas, plotted against group size for different forest types.

It can be seen that the area per individual is greater in woodland and *Maesopsis* forest than in other forest types. I divide the groups into two types, Type A (see Table 3), those occupying mainly climax forest (*Cynometra*, mixed, and swamp forests); and Type B (see Table 3), those occupying young secondary and colonizing forest (*Maesopsis* and woodland forests) (see Table 3 and Figure 3). The mean population density for each type shows a clear difference of 61.97 ind/km^2 in the former and 38.42 ind/km^2 in the latter. The areas of home ranges seem to depend primarily on the amount of food available in each type of forest. *Cynometra*, mixed, and swamp forests do not have many deciduous trees and constantly produce food, mainly leaves, throughout the year; while *Maesopsis* and woodland forests are deciduous and shed their leaves during the dry season. At that time the colobus obtain part of their diet from the seeds of *Albizia* and other trees or in the thicket zone of the forest edges.

It is significant that differences in social behavior are found between groups in the climax forests (*Cynometra*, mixed, and swamp forests) and those in young colonizing forests (*Maesopsis* and woodland forests); the latter groups having more friendly relationships with neighboring groups.

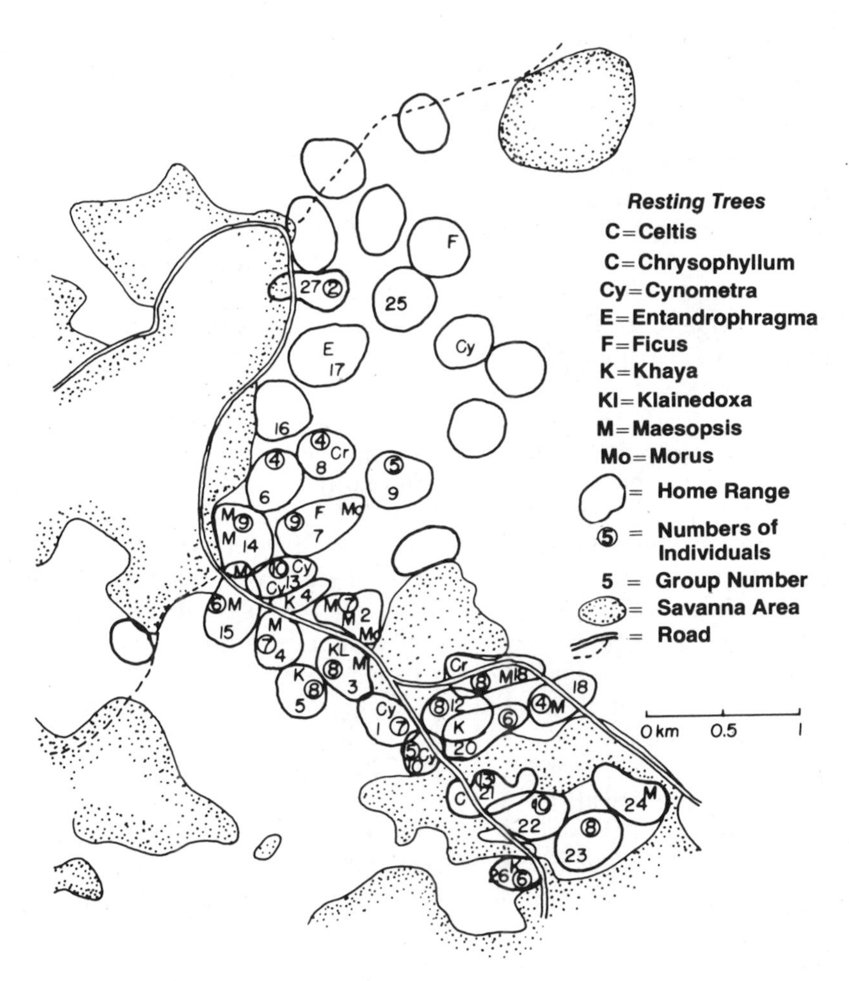

Figure 3. The correlation of home ranges, group sizes, and vegetation types of 21 groups of *Colobus guereza*.

Grouping Patterns of Chimpanzees

From observations of 43 individual processions at Filabanga in the western part of Tanzania in 1965 and some other observed cases, Itani and Suzuki (1967) presented a hypothesis that chimpanzees have large-sized groups or social units which consist of 30 to 80 individuals including several adult males. They also pointed out that parties of chimpanzees are formed and dissolved freely within the large-sized group, and that such

Table 3. Home Ranges and Population Density of Colobus guereza **in the Budongo Forest.**

Group Number	Number of Individuals per Group	Home Range (km^2)	Population Density $\frac{Individuals}{km^2}$	Forest Type[a] A			Forest Type[a] B		Forest Type Classification
				M	CY	S	MA	W	
1	7	0.103	67.96	X	X				A
2	7	0.155	45.16	X					B
3	8	0.123	65.04	X					A
4	7	0.103	67.96	X					A
5	8	0.073	109.59	X	X		X		A
6	4	0.168	23.81					X	B
7	9	0.179	50.28	X				X	A
8	4	0.105	38.10	X					A
9	5	——	——						
10	5	0.093	53.76	X	X				A
11	4	0.078	51.28	X					A
12	8	0.150	53.33	X		X			A
13	10	0.158	63.29	X					A
14	9	0.213	45.25	X			X	X	B
15	6	0.170	35.29					X	B
16	?	0.125	——						
17	?	0.125	——						
18	8	0.178	44.94	X			X	X	B
19	4	0.130	30.77				X	X	B
20	6	0.173	34.68	X		X		X	B
21	13	0.183	71.04	X			X		B
22	10	0.158	63.29	X		X			A
23	8	0.134	59.70	X		X			A
24	?	0.158	——						
25	?	0.213	——						
26	6	?	——						
27	2	0.135	14.81	X			X	X	B
28	9	?	——						
29	6	?	——						

[a]A = Climax forest
M = Mixed forest
CY = *Cynometra* forest
S = Swamp forest

B = Young secondary and colonizing forest
MA = *Maesopsis* forest
W = Woodland forest

changeability in grouping is the most characteristic feature of chimpanzee society. Following this, further studies of chimpanzees in several areas have recognized the existence of large groups of chimpanzees as social units (Nishida, 1968; Teleki, 1975; Sugiyama, 1968).

The social unit of chimpanzees, observed in the Budongo forest and named the Picnic Site group (P Group), consisted of about 80 individuals. Table 4 shows the membership of this social unit in 1967–1968. Individual chimpanzees, however, are generally observed in nomadic groups of less than ten individuals in the forest area.

A big nomadic group which consists of males and several females, juveniles, and infants moves throughout the home range and forms the nucleus of the social unit. Table 5 shows some examples of these large nomadic groups observed in 1967–1968. In this kind of nomadic group, 1–20 adult males generally make up the nucleus.

From 1 August to 31 October 1967, a total of 87 nomadic groups (2 or more individuals) were observed in the Budongo forest, Uganda. Sixty of these 87 (69.0% were homogeneous groups (all male groups, all adult groups, or all mother groups) and numbered 11 individuals or less. Only 16 of the groups (26.7%) observed of less than 11 individuals were of a heterogeneous or mixed composition, while all 11 groups of 11 or more individuals were of a mixed composition. During the observation period 21 lone individuals were observed (19 males and 2 unidentified).

The data are further analyzed in Table 6 to show the distribution of members according to age-sex class among different types of groups. Forty percent of mothers with 0–1-year-old infants were in mother groups which were generally scattered around the feeding trees. When the infants were 2-3 years-old, the mothers spent more time in mixed groups. After

Table 4. Members of a Social Unit of Chimpanzees, the Picnic Site Group, in the Budongo Forest during 1967–1968.

Males		*Females*	
Age Class	*Number of Individuals*	*Age Class*	*Number of Individuals*
Adult	18	Adult with young	17
Subadult	6	Adult without young and subadult	10
Juvenile	4		
3–4 years	3	3–4 years	1
2–3 years	3	2–3 years	2
1–2 years	0	1–2 years	3
0–1 years	2	0–1 years	3
Temporal immigrant males	10	Sex unknown, 3–4 years	1
		Sex unknown, 2–3 years	0
		Sex unknown, 1–2 years	3
		Sex unknown, 0–1 years	4

Table 5. Examples of Compositions in Large Nomadic Groups of Chimpanzees in the Budongo Forest.

	End of Celtis Food Season 1967						*Changing Time from Maesopsis to* Pseudospondias *Season* 1968		
Primate Classification	*1 Nov.*	*2 Nov.*	*6 Nov.*	*8 Nov.*	*9 Nov.*	*20 Nov.*	*28 May*	*29 May*	*30 May*
Adult and young males	16		21	20	20	15	19	22	23
Adult and young females	1		2	3	3	2	8	19	14
Mothers with infants	——		——	(3)[a]	(2)[a]	(1)[a]	(4)[a]	(13)[a]	(9)[a]
Oestrous females	(1)[a]		(2)[a]	——	(1)[a]	——	(5)[a]	(6)[a]	(5)[a]
Juveniles	1		4	5	1	3	5	15	7
Infants	——		——	3	2	1	4	13	9
Total (Individuals)	18	42[b]	27	31	26	21	36	69	53

[a]Numbers in parentheses indicate the number of females in each situation.
[b]Age/sex determinations not possible because of observation conditions.

Table 6. Types of Nomadic Groups in which Individuals Were Observed in Budongo Forest, 1 Aug.–31 Oct., 1967 (Number of individuals observed).

Classification	*Mixed Group*	*All Males*	*All Adults*	*All Mothers*	*All Juveniles*	*Lone Individual*	*Total*
Males	130 (41.1%)	115 (36.4%)	50 (15.8%)	2 (0.6%)	——	19 (6.8%)	316
Young adult males	12 (62.2%)	4 (15.4%)	——	8 (30.8%)	1 (3.8%)	1 (3.8%)	26
Mothers	28 (57.1%)	——	——	17 (37.8%)	——	——	45
Oestrus females	16 (46.2%)	9 (32.1%)	——	3 (10.7%)	——	——	28
4–5-year-old juveniles	5 (23.8%)	6 (28.6%)	——	10 (47.6%)	——	——	21
2–3-year-old infants	24 (64.9%)	—— (2.7%)	1 (2.7%)	11 (29.7%)	1 (2.7%)	——	37
1-year-old infants	14 (58.3%)	——	——	10 (41.7%)	——	——	24
Total observed individuals	229	125	60	61	2	20	497
Total groups	27	37	9	13	2	20	108
Range of number of individuals in each group type	3–31	2–9	2–11	2–11	0–1	0–1	

the age of four or five years, juvenile males become independent from their mothers and were sometimes observed in all-male groups. The data for adult males show that 41.4% (130) were in mixed groups, 36.4% (115) in male groups, 15.8% (50) in adult groups, and 6.0% (19) cases were lone individuals. Half of the nomadic groups including females with infants were observed in mixed groups. Oestrous females were usually in adult groups or mixed groups.

The variability in group composition is further highlighted by observations of the groups of chimpanzees which visited in the fig tree, B1 (in Table 7), from mid-September to early October, 1972. At first, a large nomadic group visited the tree before the fruit ripened. After a second visit by a big nomadic group as the first fruits ripened on 18 September, several mothers and their infants came to feed on the fruit and stayed near the tree. Only a few males joined this feeding group. In the forest, mothers and infants generally stayed near the feeding trees, while the males or adults of the nomadic group moved around looking for new feeding trees. Mothers and infants

Table 7. Months when Fruits Were Ripe on Each Food Tree in the Budongo Forest.

Classification of Tree		*1967*				*1968*					*1970*		*1971*		*1972*		*1973*					
		5	*7*	*9*	*11*	*1*	*3*	*5*	*7*	*9*	*10*	*12*	*2*	*2*	*9*	*11*	*1*	*3*	*5*	*7*	*9*	*11*
Trecuria africana	1		+++						++											+	+	
	2			+																		
	3			+																		
Ficus mucuso	A1										+					+						
	2										+					+						
	3				++						+					+++						++
	4															+						
	5				+																	
	6																					
	7																					
	8																+					+
	9																					
	10																					
	11				+++											+						+
	12				+	++										++						+
	13				++																	
	14										+											
	15										+											
	16										++											
	17					+																
Ficus sp.	B1															+++						
	2										++											
	3				++																	
Ficus sp.	C1																					
	2					+	+			+												
	3																					
Ficus sp.	D1										+++					++						
	2														++							
	3														+							
	4											+						+				++
	5						++															
	6																					
	7			++						++												
Ficus sp.	E1																					
	2																					
Ficus sp.	F1											+										

learned of the new feeding trees through the loud calls of the adult males and other adults in the nomadic groups.

Observations of chimpanzees living in the savanna woodland area of the western part of Tanzania in 1964–1965 indicate a clear contrast to the preceding situation described for the Budongo. The compositions of nomadic groups in the savanna

Table 8. Summary of Observations of Nomadic Groups of Chimpanzees in Different Habitats in East Africa.

Type of Group	Budongo Forest		Gombe	Mahali Mts.	Kasakati
	Suzuki (This article)	Reynolds and Reynolds (1965)	Goodall (1965)	Nishida (1968)	Suzuki (1969)
Mixed	25.0%	37.2%	30.0%	50.4%	77.9%
Adults	8.3%	30.2%	18.0%	17.5%	4.8%
Mothers	12.0%	16.8%	24.0%	7.2%	2.8%
Males	34.3%	15.8%	10.0%	8.0%	1.0%
Lone males	20.4%	0	18.0%	9.5%	12.5%
Lone females	0	0	0	7.4%	1.0%
Number observed	108	215	350	538	104

woodlands indicate that 78% of the observations were of mixed groups. Table 8 shows the distribution of the nomadic groups in different habitats. In the savanna woodland area, the grouping of chimpanzees was more compact and concentrated, while in forest areas the aggregations tended to be dispersed and not so compact.

In the savanna areas, the nomadic ranges of each group have not been completely determined, but I have estimated them to be about 200 km^2 or more (Suzuki, 1969; Kano, 1972). The home ranges in the savanna area are thus more than ten times those found in the forest area. Although they contain riverine forests that are ecologically similar to the semideciduous forest of the Budongo forest, riverine forest accounts for only 10% of the home range of the chimpanzees in the savanna area. If food production in the riverine forest of the savanna area is the same per unit as that of comparable forest in the Budongo, it will not be enough to support all individuals that are found living in the savanna area.

In fact, the chimpanzees are forced to rely on the open savanna areas in order to obtain sufficient quantities of food. Suzuki (1969) summarizes the food sources of the chimpanzees in the savanna. These data indicate marked seasonal changes in diet composition.

A most important point is that in the dry season, when food supplies are low in the forest area, the savanna-living chimpanzees move out in large numbers onto the open land and use the rich food sources of the savanna woodland (Suzuki, 1969). Characteristic food sources in the savanna woodland during the dry season include hard seed of Caesalpinaceae and Papilionaceae, the fleshy half-dried fruits of *Vitex*, *Uapaca*, *Parinari*, and *Garcinia* and the fruits of *Strychnoa* and the seeds of *Brachiaria* grains. It may also be noted that during the dry season in the savanna area chimpanzees found a lot of dried fruits such as *Canthium*, *Tarenna*, and *Uragoga* in the thickets. Thus during the five months of the dry season (May to October), the main food resources for the chimpanzee are in open forest or woodland.

These facts show that the chimpanzees can only live in this area by depending on open areas in the dry season to obtain foods which are not available in sufficient quantities in the riverine forest. It also means that the chimpanzees must move over the savanna woodland, which is an extremely open habitat in the dry season, and quite different environment from the riverine forest.

The number and types of foods taken by chimpanzees in each food season are fewer in the forest habitats than in the savanna woodland. In the forest habitat chimpanzees restrict themselves to a few types of abundant foods for long periods (a month or more). The number of different food types is usually less than 10 per month in the Budongo forest, while in the savanna area it is 10–20 types. These findings are consistent with the results of examination of the droppings of the chimpanzees: 7–8 kinds of food are usually found in the savanna area, whereas 3–5 kinds are more typical in the Budongo forest. Thus, the chimpanzees in the savanna habitat eat a wider variety of food species than do those in the forest.

Figure 4 shows the distribution of the favorite food species of chimpanzees, in the home range of the P group in the Budongo forest. Table 8 shows when each of the *Ficus* trees was in fruit, and it can be seen that each type of tree does not necessarily bear fruit every season. Consequently, chimpanzees are unable to obtain *Ficus* fruits in all seasons, and the quantity available varies across the habitat, even though they occupy a large home range. Again, each of the species is distributed only in parts of the chimpanzee home range, making it necessary for the chimpanzees to maintain large home ranges.

The differences in the nomadic life-styles of the chimpanzees in the two habitats appear to give rise to differences in the grouping patterns and dispersion of individuals. It would seem, for example, that the mixed type of aggregation is adaptive to a life of moving long distances along the riverine forests. In this type of more open habitat, the chimpanzees need the security of the group in case of danger, as well as the presence of males to

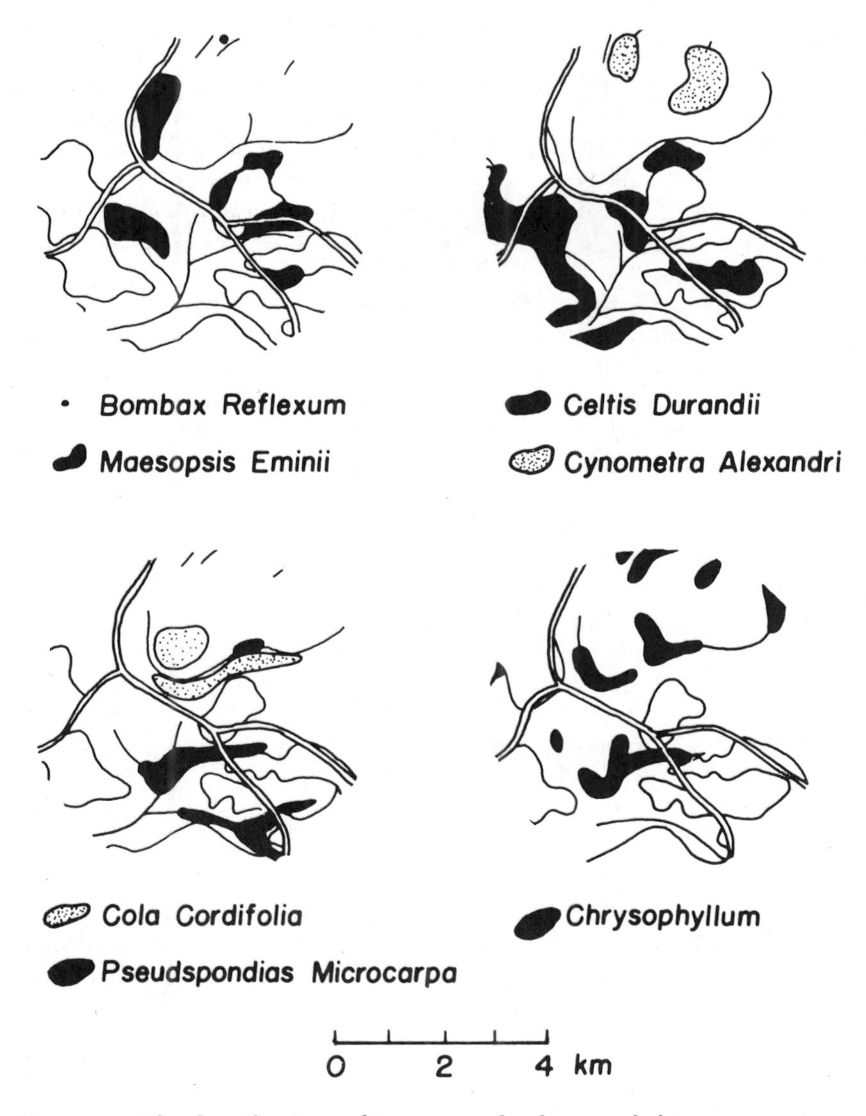

Figure 4. The distributions of important food trees of chimpanzees in the Budongo forest.

find the best course of travel and the food resources.

In the forest habitats, however, they do not need to travel so widely. Consequently, mothers and infants are generally able to stay in small areas, while the males move over a wider area. The chimpanzees' flexible social system can thus be seen to allow them to efficiently exploit ecologically different habitats.

Discussion

Life-Style Form and Primate Social Organization

Most living primates can generally be classified as either leaf-eating or fruit-eating with the exception of some prosimians and New and Old World monkeys that are insectivorous or omnivorous. Generally primates prefer fruits; even the leaf-eating monkeys often consume large quantities of fruits. However, meat-eating and hunting behavior have been observed in both chimpanzees and baboons, although both of them are primarily vegetarian (Washburn and DeVore, 1961; Goodall, 1965; Teleki, 1973, 1975; Suzuki, 1971a, b, 1975). Even black and white colobus monkeys, which are generally thought to be leaf-eaters, have been observed feeding on termites on the ground. In the Budongo forest, for example, I observed many groups of colobus monkeys coming down to the ground in order to feed on termites in the morning. These observations have elsewhere led me to argue that all primates, even leaf-eaters, are basically omnivorous (Suzuki, 1976).

Among the prosimians, for example, only the herbivorous species live in social groups. Generally speaking, the more insectivorous species live a nocturnal, solitary life with a very loose social organization, while herbivorous species are diurnal and group living, with few exceptions. These contrasts suggest that patterns of social life are fundamentally related to the species' mode of obtaining food. Thus, the group-living characteristic of most of the anthropoid primates correlates with the change to a predominantly vegetarian life-style.

It can be argued that a vegetarian life correlates with a larger physical size, a larger group size, and a larger ranging area. Consequently, in order to understand such features as group size and home-range size, it is necessary to examine not only the distribution of the food species in the animals' home range throughout all seasons, but also the animals' habits and the methods employed in obtaining their food. Of course, a genetically determined species-specific social ability is fundamentally important to the species' particular form of social organization. Nonetheless, as Clutton-Brock's (1974) study of red colobus (*Colobus badius*) and black and white colobus (*Colobus guereza*) suggests, clearly an ecological life-style can affect social organization in important ways.

Ecological Influences on Social Organization

We have seen above how the quality of the home range affects social life: Figure 3 showed that there was a correlation between vegetation type and the area of the home ranges of the black and white colobus monkeys. The necessity of having large home ranges means that they tend to overlap with each other, and this in turn tends to influence and affect intergroup relationships.

Comparable differences were found for chimpanzees in different habitats. In the Budongo forest, for example, the relationship between *P* and *K* groups was usually friendly. At Kasoge in Tanzania, on the other hand, Nishida (1968) reported that *K* and *M* groups were invariably antagonistic with each other. It is also interesting that group organization and intergroup relationships of mountain gorillas which Schaller (1963) observed are different from those observed by Fossey (personal communication by Itani). The characteristic relationship between groups may be a basic species-specific character, but it is clearly the case that it is influenced by such social and ecological factors as the age of the group, the process of group formation and fission, and the extent to which individuals transfer between groups, as well as the ecological reasons mentioned above. Thus, to generalize from a few, separate cases of observation to specific species characters is to oversimplify a complex situation.

I have similarly argued that among the chimpanzees the mixed type nomadic group containing several males is especially advantageous for long-range wandering in open savanna environments during the dry season. In the forest, however, dispersal of the individuals may be necessary for a frugivore so that animals can obtain enough fruit.

The differences in the nomadic pattern and grouping of chimipanzees between the forest and savanna habitats are good examples of how social organization is affected by the environment and the species' way of life. These points are very important for understanding the origins of hominid society, since they show that we must first attempt to understand both the environments in which the protohominids lived, as well as their life-styles.

References

Clutton-Brock, T. H. 1974. Primate social organization and ecology. *Nature* 250:539–542.

Eggeling, W. J. 1947. Observations on the ecology of the Budongo rainforest, Uganda. *J. Ecol.* 34:28–87.

Goodall, J. 1965. Chimpanzees of the Gombe Stream Reserve. Pages 425–473 in *Primate behavior*. Ed. I. DeVore. New York: Holt, Rinehart and Winston.

Itani, J., and Suzuki, A. 1967. The social unit of chimpanzees. *Primates* 8:355–381.

Kano, T. 1972. Distribution and adaptation of chimpanzee in the open country on the Eastern Shore of Lake Tanganyika. *Kyoto Univ. Afr. Stud.* 7:37–129.

Langdale-Brown, I., Osmaston, H. A., and Wilson, J. G. 1964. *The vegetation of Uganda and its bearing on land-use.* Kampala: The Government of Uganda.

Marler, P. 1969. Colobus guereza: Territoriality and group composition. *Science* 163:93–95.

Nishida, T. 1968. The social group of wild chimpanzees in the Mahali Mountains. *Primates* 9:167–224.

Reynolds, V., and Reynolds, F. 1965. Chimpanzees of the Budongo forest. Pages 368–424 in *Primate behavior.* Ed. I. DeVore. New York: Holt, Rinehart and Winston.

Schaller, G. B. 1963. *The mountain gorilla—ecology and behavior.* Chicago: University of Chicago Press.

Sugiyama, Y. 1968. Social organization of chimpanzees in the Budongo Forest Uganda. *Primates* 9:225–258.

Suzuki, A. 1966. On the insect-eating habits among wild chimpanzees living in the savanna woodland of Western Tanzania. *Primates* 7:481–487.

Suzuki, A. 1969. An ecological study of chimpanzees in a savanna woodland. *Primates* 10:103–148.

Suzuki, A. 1971a. Carnivority and cannibalism among chimpanzees. *Journal of the Anthropological Society of Nippon* 79:30-48.

Suzuki, A. 1971b. On the problems of conservation of the chimpanzees in East Africa and the preservation of their environment. *Primates* 12:415–418.

Suzuki, A. 1975. The origin of hominid hunting: A primatological perspective. Pages 259–278 in *Socioecology and psychology of primates.* Ed. R. H. Tuttle. The Hague and Paris: Mouton.

Suzuki, A. 1976. Cannibalism among chimpanzees. *Science* 6(8). Pages 18-29 (in Japanese).

Teleki, G. 1973. *The predator behavior of wild chimpanzees.* Lewisburg, Pa.: Bucknell University Press.

Teleki, G. 1975. Primate subsistence patterns: Collector-predators and gatherer-hunters. *J. of Hum. Evol.* 4:125–184.

Washburn, S. L., and DeVore, I. 1961. Social behavior of baboons and early man. Pages 91-105 in *Social life of early man.* Ed. S. L. Washburn. New York: Viking Fund Publications in Anthropology, No. 31, Wenner-Gren Foundation.

Chapter 8

Activity Patterns in Howler and Spider Monkeys: An Application of Socio-Bioenergetic Methods

Anthony M. Coelho, Jr.
Claud A. Bramblett
Larry B. Quick

In this chapter, Coelho, Bramblett, and Quick further emphasize the notion of evolution of the tolerable. At Tikal, both spider monkeys and howler monkeys are found in overlapping home ranges. Spider monkeys organize themselves into large heterosexual groups, which fragment and reunite quite frequently. Howlers, on the other hand, typically organize themselves into one-male groups. In this particular habitat, then, both howlers and spiders organize themselves in very different fashions, but both seem to have found adaptations which are suitable to this ecological situation.

Normally one might expect that sympatric primates with different social organizations were also making differential use of the habitat. If food were the limiting resource and both used the same food supplies, one might expect competition to result in the exclusion of one species as the other reached the carrying capacity of the habitat. Of course, the losing competitor might be sustained through continual immigration but some character displacement would nonetheless be expected.

At Tikal, however, Coelho, Bramblett, and Quick found that the howlers and spider monkeys were both self-sustaining populations exploiting virtually the identical food supply. Differences in group organization were not related to differential utilization of the food supply but, nonetheless, there seemed little critical competition for food based on an analysis of the quality and quantity of food resources available at Tikal. They concluded that with ramon fruit as the principal food substance in the diet, there was a more than ample food supply available to support the entire primate population. Food was therefore not the limiting factor in the habitat, but the precise population limiting factors could not be identified. Although the impact of human activities may make Tikal an extraordinary locale from an ecological perspective, it must nevertheless be taken as an example of the relative independence which may exist between social organization and the specific parameters of any immediate given habitat. Both species have social organizations which permit adequate adjustment to the existing conditions regardless of the apparent diversity of the two social systems.

The amounts and types of activity an animal engages in have important biological and ecological consequences. Performance of activity requires expenditure of energy, and expenditure of energy requires energy intake in the form of food calories if homeostasis is to be maintained. The relationship between food calorie requirements of a species and the ability of a habitat to provide those requirements is of considerable interest to individuals studying primate evolution, biology, and ecology. However, the amount and distribution of energy expenditures and requirements among members of free ranging primate populations at present or in the past is largely unknown. One factor contributing to this situation is the reluctance of researchers to acquire, process, and make available data on the duration of time spent by primates in the performance of various behaviors, as well as the distribution of these behaviors through time in terms of diurnal and annual cycles. Data on duration of activity in conjunction with body weights of animals are necessary in order to provide estimates of caloric requirements and expenditures.

The calorie is a biologically and ecologically meaningful unit for assessing the relationship of an organism to its habitat. The calorie provides a way of measuring the impact of an animal's behavior on the habitat as well as a way of assessing the extent to which a habitat is limiting or providing opportunities and potentials for an animal's behavior. Consequently, studies of primate socio-bioenergetics, or energy budgets, provide a means of (1) modeling and studying the dynamics of species and their habitats, (2) modeling, comparing, and quantifying variations in behaviors between and among species as well as, (3) assessing the significance of these behaviors from different perspectives.

Socio-bioenergetics is a term employed by Coelho (1974) to denote the study of energy expenditure as it is related to an animal's membership and interaction within a social unit. The term is employed to differentiate the study of energy budgets at the level of the individual organism and population from the study of the biochemical aspects of energy transformation at the cellular level, which Lehninger (1965) defines as bioenergetics.

Socio-bioenergetics data are herein presented on howler and spider monkeys living in Tikal, Guatemala, to provide a description of their diurnal energy budgets and a comparison of activity between these sympatric species. Their activity is presented in terms of (1) calorie expenditure associated with performance of three activities, and (2) distribution of these activities through time. In addition, a detailed description of the methods, assumptions, and equations used in estimating energy expenditure in nonhuman primates is presented. These data are used to test the hypothesis that there are differences in the diurnal patterns of energy expenditure between sympatric species of *Alouatta villosa pigra* and *Ateles geoffroyi* living at Tikal, Guatemala.

Methods and Materials

Observation Techniques

Data were collected in a 5 km^2 study site, located in the center of the 567 km^2 Tikal National Park by a research team of 7 persons during the 66-day period beginning June 1, 1973. This particular study site was chosen for a number of important reasons:

1. The Tikal National Park is a government protected reserve and a famous Maya archaeological site, which has been studied since 1853 by European archaeologists and intensively by American archaeologists since the early 1900s. Consequently, the topography of the site and locations of archaeological ruins are extremely well documented and detailed on available maps. In addition, intensive year-round research in the area has been carried out since 1956

with the inauguration of the Tikal excavation project. As a result of the many excellent maps available from the University Museum of the University of Pennsylvania and through the use of obvious structures as landmarks, it was possible to trace the movement of the primate population with great accuracy and without spending many months in the field mapping specific geographic markers.

2. The flora of the area have been studied extensively by several botanists, who, in addition to collecting and classifying botanical samples in the area for more than 40 years, have major portions of their samples at the herbarium at The University of Texas at Austin. Thus, identification of plants of the area was greatly simplified.
3. On the basis of an early reconnaissance of the site by Dr. Clarence Ray Carpenter and several other individuals (including one of the authors, C.A.B.), it was ascertained that the primates of Tikal were accessible and could be habituated very easily to human observation within a very short time period.
4. The site has been inhabited for at least 2,000 years by humans. We were able to call upon current inhabitants to act as local informants to assist, on the basis of their experience, in choice and initial locations of sites for maximum probability of encountering howler and spider monkeys, and to provide us with historical documentation of the area.

Of 2,318 hours spent in the forest, 1,147 were in contact with monkeys (49.5% contact time). Three independent teams of observers alternated among separate howler (*Alouatta villosa pigra*) and spider (*Ateles geoffroyi*) monkey groups to obtain data used in this analysis. Multiple teams enabled field hours to be rotated and gave relief to team members for meals, fatigue, and so on, while maintaining continuity of data acquisition. Each observation team was composed of at least two persons: an observer who watched subject animals and dictated information to a second person who then recorded information on a standardized data sheet (Coelho *et al.*, 1976; Bramblett, 1976). This data sheet, with a modified diary format, allows organization of data necessary to answer the question, "Who is doing what to whom and for how long?" Time was logged at the beginning of an observation day or session in 24-hour clock notation, with subsequent times recorded from a 100-second-face stopwatch.

Duration of an activity was later calculated from the start of a behavior to its completion. The minimal unit of time for any act was arbitrarily set at 5 seconds. Behavioral categories were formulated during periods of observer training and site familiarization. Codes and abbreviations were assigned to designate each age/sex category and act (see Coelho *et al.*, 1976).

The comments portion of the sheet is used to make diary entries that describe noteworthy events in anecdotal form, similar to data used by other field researchers. These comments and anecdotes supplement and greatly expand the observation records, but do not enter into the bioenergetic computations. Comments also include the plant species used, the direction and distance of travel, landmarks, tree number, weather, or any other appropriate supplementary information.

Each research team maintained a daily diary, which was a summary of the day's activities and observations, including where and when the team were not in contact with monkeys.

In accordance with our objective of obtaining a cross-sectional survey of the actual activity and energy expenditure pattern of individuals, focal-animal technique of observation was not used. Individuals were identified by age, sex, and body-size category and, except for howlers, no individual was knowingly observed for more than one day. The small population size of howlers living in the area resulted in repeated samplings of the same individuals. Moreover, unlike the spiders, it was possible to identify individual howlers and follow them for up to a week at a time. Longitudinal observation of howlers was terminated initially by the howlers being able to lose observers, but after learning the direction and nuances of Tikal's arboreal pathways as a result of experience and tape markers, it became increasingly easier to remain with them. Within one month we had a large sample of howler data and we began to concentrate on sampling and acquiring data on spiders; however, both species were sampled throughout the entire summer. In the sampling procedure we attempted to obtain relatively equal amounts of data on each age-sex category and at all hours of daylight. Consequently, all data were recorded and tabulated by age-sex category rather than by individual. We felt that this type of data was necessary to cope with the problem of animals remaining out of sight for periods of time as they traveled through the dense, arboreal foliage. (It is difficult to be sure if the animal exiting from foliage on the north side of a tree is the same animal who entered the foliage only a few seconds earlier on the south side of the tree.) Without the aid of tags or collars, or some other form of unambiguous identification, we felt that sex and body-size categories could be used for data recording and tabulating and would be most appropriate to meet our objectives of sampling a large number of individuals.

Several observation techniques were possible because of the large number of observers. Shadowing of progressions of howler and spider monkey groups from approximately 0530 to 1730 hours comprised a major portion of the data (14,063 observations). Multiple observer teams were able to observe different groups of howler and spider monkeys in widespread areas of the habitat, that is, simultaneous observations. During the 13-hour census-gathering periods, all researchers formed picket lines that encircled important feeding areas and congregation sites. Removal of trees to make roadways in the park provided a break in the arboreal paths that cross from one part of the central park area to the other. Picket lines were also conducted that placed observers at the positions of canopy contact above the roads to observe movement of animals from one side of the central park area to the other. Group composition, direction of movement, and time of each progression by an individual or group of monkeys were recorded as the group passed over an observer's assigned segment of the picket line. Intensive surveillance of a popular congregation area (Palace Reservoir) by a single observer on the Maler Palace was carried out over the entire 2½-month study period. The monkeys were observed from above canopy level with a minimum of disturbance. Large temple structures in the park were also utilized to observe group movements and to locate vocalizations.

Each observer team carried rolls of brightly colored plastic surveyor's tape and marked trees utilized by the monkeys with strips of this tape. Date, time, observer team, species, and group under observation were recorded on the tape and provided an accurate and unambiguous record of multiple use of particular trees and areas. This system proved useful, not only for us but also for subsequent researchers (Schlichte, personal communication) who discovered out tape records a year later as they observed some of the same animals using some of the same areas.

Large-scale archaeological maps were used for orientation and location. Arboreal pathways, feeding trees, and other important locations in the habitat were surveyed by pacing with a compass from one charted landmark to another and then were recorded on the site maps.

Food samples dropped by animals were identified by vernacular names and then converted to appropriate scientific nomenclature (*see* Coelho *et al.*, 1976, 1977b). Nutritional composition values for each food item were obtained and presented by Coelho *et al.* (1977b).

Extensive habitat descriptions and field methods employed by our research team are presented by Coelho *et al.* (1976, 1977a) and Bramblett (1976).

Energy expenditure rates for each animal are based on body weight (Table 1), amount of time observed performing different acts, and the cost per unit of time to perform each act (Coelho, 1974; Coelho *et al.*, 1976). The formulae presented by Kleiber (1961), Tucker (1970), and Coelho (1974) are used to calculate energy expenditure. These formulae are based on the correlation ($r = +0.98$) between metabolic body size ($kg^{0.75}$) and metabolic rate (Kleiber, 1961). Studies by Benedict (1938), Brody (1945), and Kleiber (1961) indicate that the fuel to energy conversion process is similar in all mammals. Empirically derived regression equations provide constants used in estimating caloric expenditure associated with basal metabolic rate as well as general activity expenditure. Methods of energy estimation, their underlying assumptions, and application to mammalian ecology are reviewed by Moen (1973).

During the course of field research in Tikal, howler and spider monkeys were routinely observed during the 12-hour period between 0530 and 1730. On the basis of durational data gathered in this time period, an animal's total energy expenditure for a 24-hour period is calculated by the equation:

$$C = \frac{B}{2} + \sum_{i=1}^{n} A_i \qquad \text{(Eq. 1)}$$

Table 1. Body Weights and Individual Categories for Each Species.[a]

	Howler (*kg*)	*Spider* (*kg*)
Juvenile	3.40	3.41
Infant	1.09	1.14
Adult male	10.91	8.64
Adult female	9.09	8.56

[a]Weights derived from visual estimates and values of actual weighings by Jorg-Schlicthe (personal communication).

Where C = kcal/animal/24 hrs

B = kcal/animal/24 hours of basal metabolic rate (basal metabolic rate was assumed for 12 hours of nonobservation between 1730 and 0530)

A_i = kcal/animal/12 hours observed activity (between 0530 and 1730; thus $n = 12$). Each hour's value is the sum of expenditure associated with performance of three behavior categories in order to obtain a value for each A_i; however, several computational steps precede the execution of this equation and are presented in this text.

Behaviors previously used and presented by Coelho *et al.* (1976) were grouped into three major activity categories: (1) rest, (2) move, and (3) feed. The category *rest* includes all forms of inactivity observed. *Feed* includes all forms of observed drinking and feeding. The category *move* includes all forms of motion other than feeding. The cost of resting is estimated to be .06986 kcal/kg$^{0.75}$/min. Cost of feeding is estimated to be .118103 kcal/kg$^{0.75}$/min and is higher than values previously presented by Coelho *et al.* (1976) because this activity category, as used in this paper, now encompasses suspensory feeding as well as ingestion and chewing of food items. The calorie cost of this activity is based on a mean of the estimated calorie cost of each individual type of feeding behavior reported in Coelho *et al.* (1976). Cost of movement is equated to locomotion and is estimated by Tucker's (1970) equation according to procedures which are presented in the text.

Unlike laboratory conditions, where primates are visible almost 100% of the time, one problem of field study is that it is not always possible to account for every second of an animal's time during an observation period. This problem arises as a result of occasional periods of poor visibility due to dense foliage, lack of light, or an animal going out of sight of the observer. An observer may be in contact or close proximity with an animal for an entire hour but may, for example, not be able to account for and record more than 45 minutes of an animal's activity. Consequently, it is necessary to adjust the recorded values by proportionally scaling the observed duration so that each hour of observation is represented by an hour of animal activity. Techniques used to prepare data used in this paper are the following.

STEP 1 The behavioral data are grouped by species, body weight category, act, and time of day that behaviors were performed.

The 12-hour observation period between 0530 and 1730 is divided into 12 equal units and labeled consecutively from 0600 to 1700, corresponding to the hours of observation. All values are tabulated using SPSS (Nie *et al.*, 1975) computer programs. Mean values for each act and age/sex category are calculated and shown in Table 2.

Table 2. Mean Duration (min/hr) of Observed Juvenile Spider Activity.[a]

Time of Day	*Rest*	*Move*	*Feed*	*Sum*	*Time Out of Sight*
0600	8.75	33.88	2.38	45.01	14.99
——	——	——	——	——	——
——	——	——	——	——	——
1700	——	——	——	——	——

[a]Artificial data.

STEP 2 Each durational value is adjusted by proportional scaling so that each hour of observation was represented by 60 minutes of animal activity, thereby compensating for amount of time out of sight, as the following example of resting illustrates:

$$T_{r_6} = \frac{P_{r_6}}{\sum_{i=1}^{n} S_{i_6}} \times 60 \qquad \text{(Eq. 2)}$$

where T = duration of an act (r = rest) observed during a time period (for example, 0600) adjusted for visibility

P = actual amount of time animal is observed performing a specific activity (such as rest = r) at a given time of day (6 = 6 o'clock period)

S_i = Sum of time spent performing activities (i = rest, move, feed, thus n = 3) during 0600 period by a particular animal (values from Table 1).

Thus

$$\frac{8.75}{45.01} \times 60 = .1944 \times 60 = 11.66 \text{ min resting}$$

STEP 3 Each adjusted value is again tabulated and tabled as before.

STEP 4 The resulting adjusted T values are entered into a series of equations in order to estimate each animal's caloric expenditure per unit of observation time.

$$A_i = D_i \times W^{0.75} \times T_i \qquad \text{(Eq. 3)}$$

where A_i = calories expended by an animal of W body weight performing act D_i for T_i duration
W = body weight of an actor in kg
D_i = caloric expenditure rate (kcal/kg$^{0.75}$/min) for a particular act (*see* text and Coelho *et al.*, 1976)
T_i = amount of time spent performing the act (D_i) adjusted for visibility.

In the case of our average juvenile spider, the following value results from calculating the cost of resting observed in the 0600 hour period:

$$A_{r_6} = D_r \times W^{0.75} \times T_{r_6}$$

$= .069086 \times (3.41)^{0.75} \times 11.66$
$= .069086 \times 2.51 \times 11.66$
$= 2.021$ kcal expended in resting during the 0600 time period.

STEP 5 The preceding procedure is repeated for each act, actor, and time period except locomotion (or in the case of this paper all activities categorized as movement). Caloric expenditure of the activity move or locomotion is calculated by Tucker's (1970) formula:

$$\text{kcal/kg/km} = .100\ (10)^E \qquad \text{(Eq. 4)}$$

where $E = 1.67\ W^{-0.126}$
W = body weight (kg) of an actor.

Thus, for our average spider juvenile the following values result for the activity category "move" on the basis of observations made in the 0600 time period.

$E = 1.67\ (3.41)^{-0.126}$
$E = 1.67\ (.85679)$
$E = 1.4308$
kcal/kg/km $= .100\ (10)^{1.4308}$

$= .100\ (26.9665)$
$= 2.6965.$

STEP 6 Caloric expenditure per period of observation is computed as follows:

$$A_m = L \times W \times V \times T_m \qquad \text{(Eq.5)}$$

where A_m = caloric expenditure resulting from moving during an observation period
L = rate of expenditure (kcal/kg/km) from Tucker (1970)
W = body weight (kg)
V = speed of locomotion (km/hr)
T_m = duration of time spent moving at V during the period of observation.

The value resulting from $(V \times T_m)$ is obviously an estimate of distance traveled since speed (V) varies. Distance may be substituted for $(V \times T_m)$ if it is known. For the purposes of this computation, spider monkeys are assumed to travel at 5 km/hr and howler monkeys at 4 km/hr (these values are higher than those previously used by Coelho, 1976, 1977a, b). The velocity rates assumed for each species are based on recorded timings of these species traveling over an estimated distance. The values chosen are deliberately high so as to provide an overestimate of caloric cost.

The following values result for the activity category "move" on the basis of observations made on our average juvenile spider monkey in the 0600 time period.

$$A_{m_6} = L \times W \times V \times T_{m_6}$$
$$= 2.6965 \times 3.41 \times (5/60) \times 45.17$$
$$= 9.1951 \times .0833 \times 45.17$$
$$= 34.5980 \text{ kcal.}$$

STEP 7 Total caloric expenditure for a time period is calculated by summing caloric expenditure associated with each act performed in that time period; for example, the average juvenile's expenditure during the 0600 time period is equal to:

$$A_6 = \sum_{i=1}^{n} A_{i_6} \qquad \text{(Eq. 6)}$$
$$= A_{r_6} + A_{m_6} + A_{f_6}$$
$$= 2.02 + 34.60 + .94$$
$$= 37.56 \text{ kcal.}$$

The above procedures are repeated for each time frame and each animal.

STEP 8 Tikal's howler and spider monkey population was not routinely observed between 1730 and 0530; consequently, caloric expenditure during this time period was assumed to be equal to basal metabolic rate. Kleiber's (1961) equation is used to estimate basal rate:

$$B = 70\ W^{0.75} \qquad \text{(Eq. 7)}$$

where B = basal metabolic rate/24 hours
W = body weight (kg).

Thus, for our juvenile spider the following values result.

$$B = 70\ (3.41)^{0.75}$$
$$= 70\ (2.51)$$
$$= 175.7 \text{ kcal/24 hours.}$$

STEP 9 The expenditure value for the 12-hour night period is ½ the B value or 87.85 kcal/12 hours.

STEP 10 Each animal's total caloric expenditure (C) for a 24-hour period is calculated by entering values obtained from equations 2 through 6 for each time period; and the value from equation 7 into equation 1.

$$C = \frac{B}{2} + \sum_{i=1}^{n} A_i$$

$$= \frac{175.7}{2} + \sum_{i=1}^{12} A_i$$

$$= 87.85 + (37.56 + A_7 + A_8 \ldots)$$
$$= 87.85 + 360.36$$
$$= 448.21 \text{ kcal/24 hours.}$$

Implementation of the previously described procedure results in estimates of caloric expenditure for each age/sex category, act, and time period. Using these procedures and computer programs (Nie *et al.*, 1975) approximately 14,000 individually timed and recorded observations were reduced to 36 mean activity duration values (3 activity category values/hr x

12 hr) per age/sex/species category (4 age/sex/body weight categories x 2 species). These grouped values are subsequently used in conjunction with appropriate statistical formulae (Nie *et al.*, 1975) to test the proposed hypotheses.

Results

Energy Requirements and Habitat Productivity

Daily caloric expenditure estimates for each animal and species category are presented in Table 3. The *C* values (total caloric expenditure/24 hr) in Table 3 are considerably higher than previously reported by Coelho *et al.* (1976, 1977a, b). The increases are due largely to grouping of many behaviors into only three activity categories. Procedures used in this paper differ from procedures previously utilized by Coelho *et al.* (1976, 1977a, b) which use 11 distinct body-weight categories/species and 48 behavior categories. Grouping of body weights and behaviors results in a simplification of methods and procedures but considerable inflation (and somewhat less accuracy) in estimates of energy expenditure. In particular, grouping all nonresting and nonfeeding behaviors into the category move results in considerable inflation of *C* values. The category move, which is herein equated with locomotion, is one of the most expensive activities (kcal/min) performed by an animal because it assumes transport of the entire body mass over distance. Moreover, body weights (Table 1) used in the adult age/sex categories are larger than those previously reported by Coelho *et al.* (1976) and represent estimates of large specimens of each species. The *C* value of an 8.64 kg male spider monkey (Table 3) is approximately 1/3 the value (2,800 kcal/24 hr) recommended by the Food and Nutrition Board, National Academy of Sciences–National Research Council (1968) for a 70 kg human. Despite these very generous estimates of *C* values, it is clear that the size, growth potential, and activity of Tikal's howler and spider monkey populations are not being limited to their present level by problems of caloric insufficiencies. Coelho *et al.* (1976, 1977a, b) report that the 5 km^2 study site is capable of producing 50,137,500 kcal/24 hr of ramon fruit pulp and seed. Ramon is a major food source for both species, it is available on a year-round basis, and represents only one of more than 100 food items available in the habitat to the howler and spider monkeys. Thus, if Tikal's monkey population had access to only ramon fruit, a

Table 3. Daily Energy Expenditures of Tikal's Cebidae Population.

	Basal Metabolic Rate (kcal/12 hr)	*Activity Expenditure (kcal/12 hr)*	*Total Individual (kcal/24 hr)*	*Number in Tikal's Population*	*Total Daily Expenditure (kcal/24 hr)*
Howler					
ADM	210	493	703	6	4,221
ADF	183	453	636	7	4,455
JUV	87	247	335	8	2,680
INF	37	124	161	4	647
Spider					
ADM	176	669	845	33	27,899
ADF	175	649	824	75	61,874
JUV	87	360	448	67	30,030
INF	38	162	201	50	10,079
TOTAL				250	141,889

population more than 100 times larger could be supported by this single favored food item. Coelho *et al.* (1976, 1977a, b) indicate that the diet available to Tikal's monkeys adequately meets their nutrient requirement of fats, proteins, carbohydrates, vitamins, and amino acids. Moreover, it is unlikely that the diet is toxic since these food items are used by local natives and the Institute for Nutrition for Central America and Panama to provide balanced diets for humans (*see* Coelho *et al.*, 1976, 1977a, b).

Comparative Activity Patterns

Values presented in Figure 1 illustrate a clear lack of association between howler and spider monkeys in terms of diurnal pattern of energy expenditure. Tests (r_s) between these species at the level of individual categories also fail to demonstrate any significant associations (Table 4). Differences in caloric expenditure between these species are particularly evident in all age/sex categories except infants for resting and moving estimates (Tables 5, 6, and 7). Lack of differences between howler and spider feeding activity when compared on the basis of time reported feeding is also described by Richard (1970). Lack of species difference in the activity is interesting since it might be expected that the smaller-bodied spider monkeys would spend less time and energy feeding than large-bodied howler monkeys.

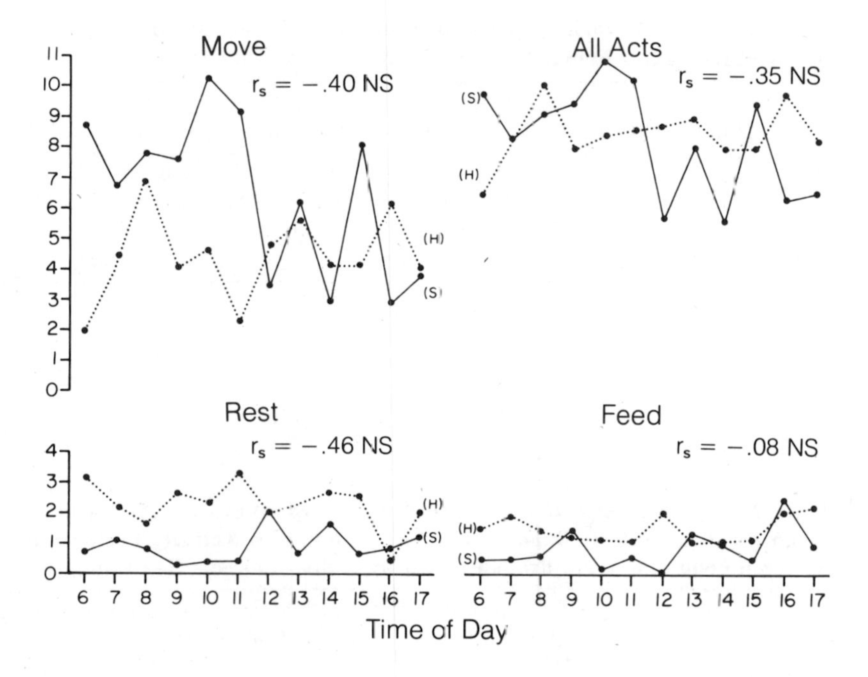

Figure 1. Percent of each species total activity expenditure (kcal/12 hr) broken down by time of day and act. Category "All Acts" represents percent of total energy expended in each hour (6 to 17). Time period 6 represents energy expended in $Move_6 + Rest_6 + Feed_6$ (for example, howlers expend approximately 6.5% of their energy at 6:00 (All Acts) 3% $Rest_6$ + 2% $Move_6$ + 1.5% $Feed_6$).

Table 4. Correlation (r_S) between Howler and Spider Monkeys on the Basis of Percentage of Calories that Each Hour (0600 to 1700) Represents of Total Daily Energy Expenditure.

	Juveniles	*Infants*	*Adult Males*	*Adult Females*
Rest	-.4266	.4460	.0769	-.5315
Move	-.2867	-.1331	.2517	-.5455
Feed	.1678	.1261	.2238	-.4755
Hourly total	-.4448	-.0420	.1648	-.3643

Note: No r_S significant at .05 two-tailed.

Table 5. Mean Caloric Expenditure Based on Totals of an Adult Male, Adult Female, Juvenile, and Infant Monkey.

	Rest		*Move*		*Feed*		*Total*	
	Howler	*Spider*	*Howler*	*Spider*	*Howler*	*Spider*	*Howler*	*Spider*
$\overline{X}$	31.88	17.49	58.05	119.01	19.93	16.98	109.87	153.49
sd	7.37	10.03	19.64	48.24	5.69	12.78	14.11	33.89
t value	3.83		-3.88		.69		3.94	
Prob.	.001		.001		ns		.001	

Table 6. Comparison[a] of Howler and Spider Monkeys on the Basis of Mean (kcal/hr) Caloric Expenditure Associated with Activity Performed Between 0600 and 1700, Broken Down by Individual Age/Sex Category.

	Juveniles		*Infants*		*Adult Males*		*Adult Females*	
	Howler	*Spider*	*Howler*	*Spider*	*Howler*	*Spider*	*Howler*	*Spider*
Rest								
$\overline{X}$	4.59	3.12	1.83	1.99	14.18	6.35	11.28	6.04
sd	1.55	1.41	0.65	1.29	3.29	5.45	2.88	3.99
t value	2.45		-.40		4.27		3.68	
Prob.	.023		.695		.001		.001	
Move								
$\overline{X}$	11.88	23.58	6.72	10.44	19.96	43.94	19.51	41.08
sd	4.47	9.46	1.98	7.25	7.42	22.85	7.79	20.45
t value	-3.88		-1.72		-3.46		-3.42	
Prob.	.001		.110		.004		.004	
Feed								
$\overline{X}$	4.15	3.33	1.83	1.14	6.98	5.47	6.98	7.02
sd	1.68	2.54	.71	1.13	3.12	6.68	2.32	5.34
t value	0.93		1.79		.71		.03	
Prob.	.366		.090		.489		.979	
Total Activity								
$\overline{X}$	20.62	30.03	10.38	13.58	41.12	55.75	37.77	54.15
sd	3.23	6.56	1.59	5.80	5.33	15.92	5.36	13.82
t value	-4.46		-1.84		-3.02		-3.83	
Prob.	.001		.089		.01		.002	

[a]paired *t*-test (n = 12)

Table 7. Comparison[a] of Howler and Spider Monkeys on the Basis of Mean Percentage Each Act Represents of Each Hour (0600 to 1700) of Energy Expenditure.

	Juveniles		*Infants*		*Adult Males*		*Adult Females*	
	Howler	*Spider*	*Howler*	*Spider*	*Howler*	*Spider*	*Howler*	*Spider*
Percent rest								
$\overline{X}$	23.94	11.63	18.96	24.16	35.95	15.50	31.45	13.65
sd	12.20	7.00	10.00	28.6	12.50	18.70	12.70	11.80
t value	3.04		-.59		3.15		3.56	
Prob.	.007		.561		.005		.002	
Percent move								
$\overline{X}$	55.64	75.48	63.35	65.04	47.16	73.23	49.91	71.15
sd	14.50	16.00	10.50	33.00	12.60	22.30	13.90	20.80
t value	-3.18		-.17		-3.53		-2.94	
Prob.	.004		.869		.003		.008	
Percent feed								
$\overline{X}$	20.42	12.89	17.69	10.80	16.89	11.27	18.64	15.20
sd	8.40	11.50	6.30	13.70	6.50	15.30	6.20	13.60
t value	1.83		1.59		1.17		.80	
Prob.	.08		.133		.259		.437	

[a]paired *t*-test ($n = 12$)

However, data obtained in this study (Tables 5, 6, and 7) clearly indicate the spider monkeys expend greater amounts of energy in moving activity and overall activity than howlers. The result is that spiders require more food calories per day in order to meet their present activity requirements. Within-species comparisons indicate statistically significant (p=1.05) associations in pattern of diurnal energy expenditure (Tables 8 and 9) and hourly (Tables 10 and 11) patterns of energy expenditure.

In all instances there are a greater number of significant associations among howler monkeys than among spider monkeys. In particular, diurnal activity patterns of adult male and female howlers were significantly associated while those of spider monkey adults were not. This may possibly reflect differences in social organization and group compositions between howler and spider monkeys. Howlers at Tikal travel and live in small, tightly organized, bisexual social units; whereas spiders (Table 12) are characterized by fragmenting social units having a variety of different compositions which may change through the day (Coelho *et al.*, 1977a, b). The greatest number of associations between spider individuals is found in the relationships of juveniles, infants, and adult females. These findings

Table 8. Correlation (r_S) among Howler Monkeys on the Basis of Mean Percentage of Calories That Each Hour (0600 to 1700) Represents of Total Energy Expenditure Resulting from Activity.

	Infants	*Adult Males*	*Adult Females*
Juvenile			
Rest	.5874[a]	.6434[a]	.8042[a]
Move	.9091[a]	.5524	.7902
Feed	.5664	.4825	.2797
Hourly Total	.9422[a]	.5184	.8123[a]
Infant			
Rest		.2028	.5105
Move		.3287	.7347[a]
Feed		.1399	.4615
Hourly Total		.3776	.6725[a]
Adult Male			
Rest			.7063[a]
Move			.6364[a]
Feed			-.5171
Hourly Total			.6445[a]

[a]Significant at .05 two-tailed.

Table 9. Correlation (r_S) among Howler Monkeys on the Basis of Percentage That Each Activity Represents of Each Hour of Energy Expenditure.

	Infants	*Adult Males*	*Adult Females*
Juvenile			
Rest	.5804[a]	.6154[a]	.8252[a]
Move	.9021[a]	.5874[a]	.8392[a]
Feed	.5804[a]	.5315	.3923[a]
Infant			
Rest		.2378	.5874[a]
Move		.4196	.6993[a]
Feed		.2448	.4974
Adult Male			
Rest			.5664
Move			.6573[a]
Feed			.2837

[a]Significant at .05 two-tailed.

Table 10. Association (r_S) among Spider Monkeys on the Basis of Percentage of Calories That Each Hour (0600 to 1700) Represents Represents of Total Energy Expenditure Resulting from Activity.

	Infants	*Adult Males*	*Adult Females*
Juvenile			
Rest	.3566	.7622[a]	.7482[a]
Move	.6340	.3636	.7972[a]
Feed	.3827	.4685	.8182[a]
Hourly Total	.6853[a]	.5378	.8462[a]
Infant			
Rest		.1469	.5245
Move		.1611	.7426[a]
Feed		.1711	.4694
Hourly Total		.2347	.7203[a]
Adult Male			
Rest			.2657
Move			-.0280
Feed			.2657
Hourly Total			.1156

[a]Significant at .05 two-tailed.

Table 11. Association (r_S) among Spider Monkeys on the Basis of Percentage Each Activity Represents of Each Hour of Energy Expenditure.

	Infants	*Adult Males*	*Adult Females*
Juvenile			
Rest	.4063	.7133[a]	.8056[a]
Move	.6982[a]	.3853	.8616[a]
Feed	.3404	.5009	.8511
Infant			
Rest		.1226	.5719[a]
Move		.1751	.8161[a]
Feed		.0070	.3678
Adult Male			
Rest			.3117
Move			.1119
Feed			.1748

[a]Significant at .05 two-tailed.

Table 12. Size of Spider Social Units.[a]

Number of Individuals in Social Unit[b]	*Frequency of Occurrence (Percent)*
1	32.2
2	27.5
3	14.2
4	9.8
5	4.8
6	4.8
7	1.8
8	1.2
9	0.7
10–20	1.8
21–30	0.2
31+	1.0

[a]Values based on counts of individuals occuring while observers were following animals.

[b]The term social unit is used because spider monkey fission and fusion behavior made it difficult for even seven observers to ascertain what constituted a group. Coelho *et al.*, (1977b) discuss the problem of spider monkey foraging parties and grouping behavior as it is related to distribution of animals in the canopy.

support our subjective impressions and quantitative data on group composition which suggest that a majority of spider monkey social units were composed of adult females, juveniles, and older infants; whereas adult males seemed to travel independently between female foraging parties (Coelho *et al.*, 1976, 1977a, b).

Values illustrated in Figure 1 suggest that there are species differences in diurnal peaks of energy expenditure resulting from feeding activity. These values raise the question as to whether or not these different diurnal patterns are ecologically significant. It can be hypothesized that differential distribution of feeding periods is an indication that these sympatric species are partitioning the habitat on the basis of time of day the site is used. If this is so, then it raises the question of why are they partitioning the habitat? Competition for resources, which are assumed to be limited, is one possible explanation. However, Coelho *et al.* (1976, 1977a, b) present data indicating that these populations are living well below the carrying capacity of the habitat.

Coelho *et al.* (1977b) document extensive overlap in area utilized by these sympatric species. They indicate that these species share many identical sleeping and feeding sites. However, Coelho *et al.* (1977b) present data indicating that the density of Tikal's monkey population is very low, possibly as a result of successive disease epidemics. Therefore, it seems unlikely that these species need to compete for space in the habitat.

Tikal's howler and spider monkeys are principally frugiverous and rely on and seem to prefer many of the same food items, especially ramon (Coelho *et al.*, 1976, 1977a, b). Both of these species were observed feeding in many of the same tree locations, both independently in time and simultaneously. However, Coelho *et al.* (1976, 1977a, b) indicate that the habitat at Tikal is capable of supporting a population many times larger on the basis of the productivity of the ramon fruit, the preferred food source of both species. Moreover, ramon is only one of more than 100 food species which are available to Tikal's monkeys. Consequently, it seems unlikely that these sympatric species are competing for food energy [or the other nutritional elements presented by Coelho *et al.* (1977b)].

If these species do not need to compete for food calories or space, then why are they exploiting the same habitat differentially in terms of time, but not food or space, and why do they differ in their forms of social organization and structure? The problem is especially enigmatic for hypotheses and explanations of the relationship of ecology and social organization which suggest that social structure and organization are determined by ecological influences. In some instances proposed hypotheses and explanations have confused correlation or association with causality. For example, it is assumed that certain forms of social structures (for instance, small social groups containing only one adult male) represent adaptations to scarce and widely distributed food resources. Although this may explain the existence of one-male groups in some habitats (Hall, 1965), it does little to explain their existence in other habitats (Aldrich-Blake, 1970).

Ecological theories, of both the group selectionists and the sociobiologists, emphasize the importance of environmental stressors and limitations to the point of almost completely overlooking environmental potentials. Ecological theories in primatology present natural selection as the principal evolutionary force responsible for existing social organizations, and in terms of the way it was understood in Darwinian days; namely, as the "struggle for existence" and "the survival of the fittest." Thus it is tacitly assumed that a form of social structure or social organization exists in a particular habitat because it is or was the "best" solution to some environmental limitation or problem. There are few if any data available to suppport these assertions.

Natural selection is primarily a process of differential reproduction and not primarily a process of elimination. Natural selection does not favor the fittest individuals in terms of most dominant, largest, fastest, or any other aspect of the fang and claw interpretations of struggle, or competition. Natural selection does not favor the fittest unless the fittest are defined as those that have more offspring. Simpson (1964:221) argues that this usually means individuals best adapted to the conditions in which they find themselves, or those individuals who will be best able to adapt to other conditions existing in the future, or those best able to meet opportunity. "Moreover the correlation between those having more offspring, and therefore really favored by natural selection, and those best adapted or best adapting to change is neither perfect nor invariable; it is only approximate and usual" (Simpson, 1964:221).

It is possible that the observed existence of small stable social units of howlers (6 or 7 animals, including more than 1 adult male and adult female) and the simultaneous occurrence of spider social units (which fuse into large sleeping and foraging parties of more than 70 animals on some occasions or fission into small social units containing 1 to 4 animals) is not the direct result of adaptation or selection for these forms of behavior. The behavior, social organization, and social structure of Tikal's howler and spider monkeys may represent an example of opportunism. Simpson (1964:160) explains opportunism as a convenient label for tendencies in evolution or "that what can happen usually does happen; changes occur as they may and not as would be hypothetically best; and the course of evolution follows opportunity rather than plan."

If ecology influences social organization and structure, then it may be as a result of the opportunities in the environment and the ability of organisms to seize them. In the case of Tikal's howlers and spiders, the available food calories, nutrients, sleeping sites, and habitat are far in excess of what these animals require, and do not seem to be especially patchy in distribution (Coelho *et al.*, 1977b). It can be suggested that, given these extensive opportunities, a number of different ways of life are possible. Howler and spider monkeys seem to have the behavioral plasticity necessary to adjust to living in small or large social units and populations.

Perhaps variations in activity, social organization, social structure, and behavior are not the result of selection acting to produce particular forms which are adaptive to particular environmental conditions. Instead, it can be hypothesized that these variations in lifeways are artifacts or noise resulting from the process of selection acting on other biological systems such as differential disease susceptibility. It can also be hypothesized that variations exist because there were a number of lifeways which were not maladaptive to conditions which prevailed at some time in the past. Kummer describes this as evolution for the tolerable:

> Discussions of adaptiveness sometimes leave us with the impression that every trait observed in a species must by definition be ideally adaptive, whereas all we can say with certainty is that it must be tolerable since it did not lead to extinction.... Ecological conditions can put a premium on a certain type of society, but it cannot tell the species how to create such a society (Kummer 1971:90).

As long as none of them become maladaptive or confer special advantage to the organisms, they may all survive.

> Evolution works on the materials at hand; ... organisms as they exist at any given time and the mutations that happen to arise in them. The materials are the results of earlier adaptations plus random additions and the orienting factors in change is adaptation to new opportunities. If this view of evolution is correct, then we must expect to find similar opportunities exploited in different ways. The problems involved in performing certain functions should have multiple solutions. (Simpson, 1964:164–165)

Variations in activity, social structure, organization, and behavior are significant not because they may or may not represent the result of evolutionary forces acting in the past but because they represent what can be selected on in the future. The adaptiveness of any form (gene, trait, species, social structure, and so on) is ultimately measured in terms of its ability to continue through time and not by how and why it came to exist in the first place. Adaptation is not just the result of evolution, it is the future of evolution.

Acknowledgments

We acknowledge the assistance of Dr. and Mrs. Clarence Ray Carpenter, Dr. Charles Darby, and Dr. Charles Douglas, who along with Dr. and Mrs. Claud A. Bramblett conducted an initial reconnaissance of the Tikal National Park; Dr. Luis Lujan, director de Instituto de Anthropologia y historia; Mr. Amilcar Guzman, administrator of the Tikal National Park; the park staff and residents of the Tikal area; Mrs. Sharon S. Bramblett, Mrs. Lynne Quick, Ms. Caroline Johnson, and Ms. Kathrine Dolan for assistance in gathering data; Drs. Cyrus L. Lundell and Marshall Johnston for their assistance in clarifying the taxonomic nomenclature of the Tikal botany and making available samples in the Lundell herbarium; Drs. Dennis Puleston and R. E. W. Adams for sharing information based on their many years of observations during their archaeological study of Tikal; Ms. Linda S. Coelho for assistance in data preparation, processing, and for illustrating the botany; Drs. Robert M. Malina, S. Chad Oliver, Phil Grant, and Claud Desjardins for their comments and suggestions.

This study was funded by a grant from the University of Texas to cover the cost of gasoline and vehicle maintenance expenses and grant NSF-USDPGU-1598, which provided funds for work-study students to keypunch the data and fieldnotes. In addition, funds for computer time were provided by the departments of anthropology at the University of Texas at Austin and Texas Tech University and funds for the writeup, analysis, and illustration were provided by Southwest Foundation for Research and Education, General Research Grant 5-507-RR05519-13.

References

Aldrich-Blake, F. P. G. 1970. Problems of social structure in forest monkeys. Pages 79-102 in *Social behaviour in birds and mammals.* Ed. J. H. Crook. London: Academic Press.

Benedict, F. G. 1938. *Vital energetics: A study of comparative basal metabolism.* Washington, D.C.: Carnegie Institute, Publ. 503:1–215.

Bramblett, C. A. 1976. *Patterns of primate behavior.* Palo Alto, Ca.: Mayfield.

Brody, S. 1945. *Bioenergetics and growth.* New York: Reinhold.

Coelho, A. M., Jr. 1974. Socio-bioenergetics and sexual dimorphism in primates. *Primates* 15:263–269.

Coelho, A. M., Jr., Bramblett, C. A., Quick, L. B., and Bramblett, S. S. 1976. Resource availability and population density in primates: A socio-bioenergetic analysis of the energy budgets of Guatemalan howler and spider monkeys. *Primates* 17:63–80.

Coelho, A. M., Jr., Bramblett, C. A., and Quick, L. B. 1977a. Social organization and food resource availability in primates: A socio-bioenergetic analysis of diet and disease hypothesis. *Amer. J. Phys. Anthrop.* 46:253–264.

Coelho, A. M., Jr., Coelho, L. S., Bramblett, C. A., and Quick, L. B. 1977b. Ecology, population characteristics and sympatric association in primates: A socio-bioenergetic analysis of howler and spider monkeys in Tikal, Guatemala. *Yrbk. Phys. Anthropol.* 20:96–135.

Food and Nutrition Board–National Academy of Sciences. 1968. Recommended Dietary Allowances Publication 1694 National Academy of Sciences, Washington, D.C.

Hall, K. R. L. 1965. Behavior and ecology of the wild patas monkeys, *Erythrocebus patas* in Uganda. Pages 32–119 in *The baboon in medical research.* Ed. H. Vagtborg. Austin: University of Texas Press.

Kleiber, M. 1961. *The fire of life: An introduction to animal energetics.* New York: Wiley.

Kummer, H. 1971. *Primate societies: Group techniques of ecological adaptation.* Chicago: Aldine-Atherton.

Lehninger, A. L. 1965. *Bioenergetics: The molecular basis of biological energy transformations.* New York: Benjamin.

Moen, A. N. 1973. *Wildlife ecology: An analytical approach.* San Francisco: Freeman.

Nie, H. H., Bent, D. H., and Hull, C. H. 1975. *SPSS: Statistical package for social sciences.* New York: McGraw-Hill.

Richard, A. 1970. A comparative study of the activity patterns and behavior of *Alouatta villosa* and *Ateles geoffroyi. Folia Primat.* 12: 241–263.

Simpson, G. G. 1964. *The meaning of evolution.* New Haven: Yale University Press.

Tucker, V. A. 1970. Energetic cost of locomotion in animals. *Comp. Biochem. Physiol.* 34:841–846.

Chapter 9

Home Range Size, Population Density and Phylogeny in Primates

T. H. Clutton-Brock
Paul H. Harvey

Although we have emphasized the multiple theoretical sources of variability in primate social organization and described some of the variability seen in the same habitats and constancies seen across diverse habitats, none of us believes that there is absolutely no relationship between ecology and social organization. Despite the multiple sources of variability, certain aspects of ecology must surely influence some aspects of behavior in imposing severe limits on the range of acceptable expressions. One may search for such relationships by running simple correlations among all the available measures of behavior and all the available measures of ecology. Such an approach is doomed almost from the start by our poor definitions of both ecology and behavior and, perhaps more importantly, because most of our available data consist of what must be considered single samples for most species or even genera. In a study of social organization, each field study of a troop, no matter what the duration, must be regarded as an N of one. The single measure then is compared with other single measures and comparisons are made without any knowledge of within sample variance. This limitation, along with the many other sources of variance already discussed, then holds little promise for shotgun correlational approaches.

An alternative is to deduce theoretical relationships which should be more tightly constrained based on the logical assumptions made about selected variables. Clutton-Brock and Harvey have used this approach and hypothesized relationships among measures of home range, day range, population density, and biomass based on a specific logical framework.

In general, they concluded that heavier groups have larger home ranges and longer day ranges than lighter groups, and folivorous species have shorter day ranges than omnivorous species. In all taxa studied, they found population density and biomass to be positively correlated. We may summarize their findings by saying that the larger the primate, the lower its population density, and the greater its ranging distances.

The apes pose an interesting problem in these data, for they show lower density and biomass than many other primates and have larger home ranges. This would seem to indicate that the apes are adapted to feeding on clumped, unpredictable and widely dispersed food supplies, while more folivorous species would be expected to feed on more predictable, more evenly dispersed food supplies and, consequently, have shorter day ranges. Of course, there are always problems with correlational models, but this one does allow for the generation of testable hypotheses about the relationship of ecology to certain aspects of physique and behavior.

Introduction

Species differences in ranging behavior and population density among primates are related to differences in the density and distribution of their food supplies (Crook and Gartlan, 1966; Jolly, 1972; Eisenberg, Muckenhirn and Rudran, 1972). Recent studies have demonstrated that species living in heavier groups (that is, either groups of large size or composed of individuals of greater weight) tend to have larger home ranges than species living in lighter groups, and species of large individual body size show lower population densities than those of smaller size (Clutton-Brock and Harvey, 1977a). In addition, species feeding to a greater extent on foliage have smaller home ranges and higher population densities than those feeding on fruit, flowers, or animal matter, probably because their food supplies are more abundant (Clutton-Brock and Harvey, 1977a, b).

However, all analyses of interspecific variation in ranging behavior have combined species belonging to different phylogenetic groups. Consequently, they have obscured the possibility that relationships exist as a result of differences between major taxonomic groups rather than within them. For example, a relationship between home-range size and body weight in an imaginary order could be claimed if four families varied in body weight and home-range size but no relationship occurred within each family (Figure 1a); if body size and home-range size were related within families as well as between them (Figure 1b); or if home-range size were related to body weight and no differences existed between families (Figure 1c). Such differences are important because they are likely to influence both our explanation of the relationship between ranging behavior and ecological variables, and our statistical treatment of the data.

In this analysis, we investigate whether differences in home range and population density within four taxonomic groups of primates (Prosimii, Ceboidea, Cercopithecoidea and Hominoidea) are related to group weight and diet type in the same way as across the order as a whole, and whether relationships differ between taxonomic groups.

Methods

The data set used in this sample is substantially the same as that used in previous analyses (Clutton-Brock and Harvey, 1977a, b) (*see* Table 1). References are not reproduced here since they are extensive, but they will be published in full elsewhere (Clutton-Brock, Harvey and Rudder, in preparation). Weights were calculated from data for wild-caught animals wherever possible. Population group size is the number of individuals sharing a common home range, defined as the total area used by a consistent social group. Estimates of proportion of foliage eaten were based on several different methods of measurement (including time spent feeding, visual measurement of intake, and analysis of stomach contents—*see* Clutton-Brock and Harvey, 1977a) and a considerable amount of error was probably introduced for this reason. As in previous analyses, where several reliable studies of the same species were available, we calculated means for all available estimates. Several anomalies should be mentioned: for the Lorisidae, where females occupy separate home ranges from males, we have taken average *female* home-range size as our measure wherever this was available on the

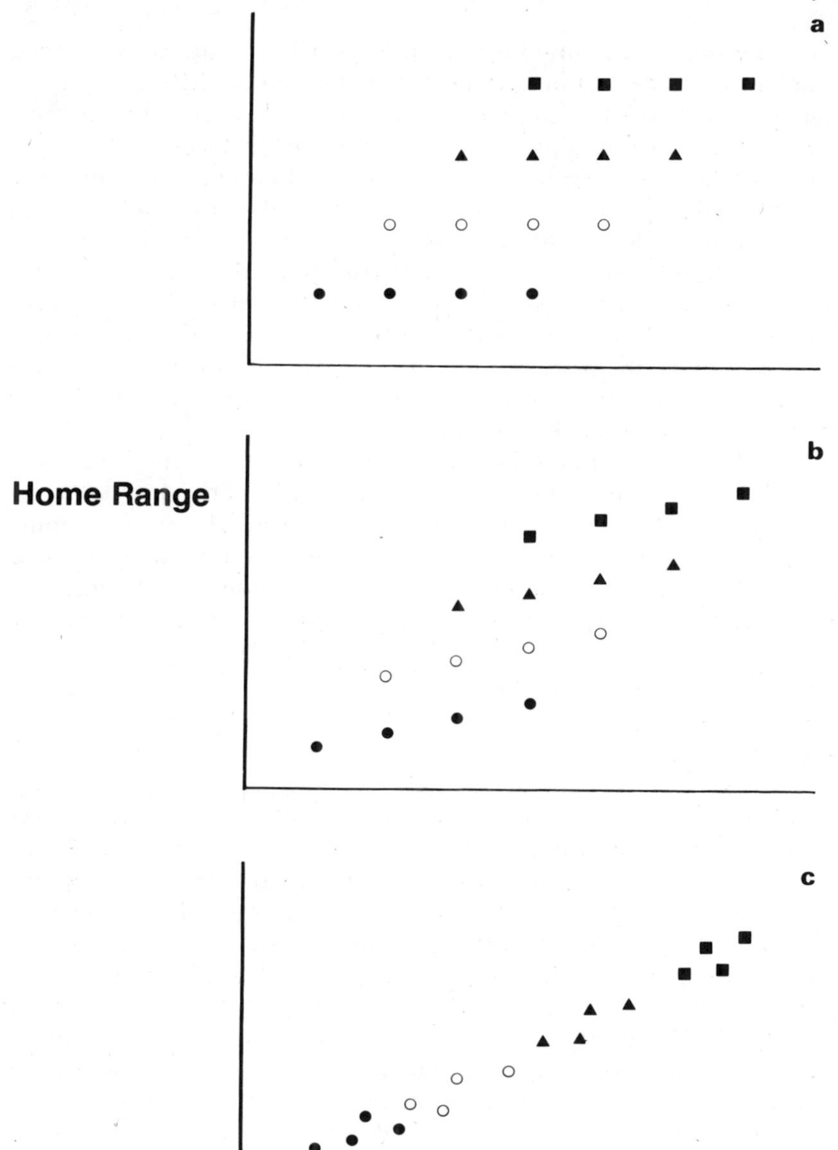

Figure 1. Hypothetical relationships between body size and home-range size in an imaginary order with four families and four species in each family.

Table 1. Mean Values of Body Weight, Group Size, Percent of Foliage in the Diet, Population Density, and Home-range Size for 47 Primate Species Used in This Analysis.

	Adult Male Weight (kg)	Adult Female Weight (kg)	Population Group Size[a]	Percent of Foliage in the Diet[a]	Population Density (animals/km²)	Home-range Size (km²)
Prosimii						
Lemur catta	2.9	2.5	18	34	233	0.07
Lemur fulvus	2.1	2.1	9.5	71	1125	0.01
Lemur mongoz	1.8	1.8	2.6	1.5	350	0.01
Lepilemur mustelinus	0.61	0.64	1	51	270	0.002
Indri indri	12.5	12.5	3	57	12.5	0.23
Propithecus verreauxi	3.7	3.5	6.5	40	103	0.08
Loris tardigradus	0.3	0.3	1	0	100	0.01
Perodicticus potto	1.2	1.1	1	0	8	0.1
Galago alleni	0.26	0.26	——	0	15	——
Galago elegantulus	0.3	0.3	——	0	15	——
Galago demidovii	0.06	0.06	1	0	50	0.01
Ceboidea						
Callicebus moloch	1.1	1.5	3	26	24	0.04
Callicebus torquatus	1.1	1.1	3	4	13	0.20
Alouatta villosa	7.4	5.7	17	49	70	0.38
Cebus capucinus	3.8	2.7	10	15	17	0.86
Ateles belzebuth	6.2	5.8	——	7	13.5	——
Ateles geoffroyi	6.2	5.8	12	19	——	0.60
Cercopithecoidea						
Macaca sinica	5.7	3.6	——	0	100	——
Macaca fascicularis	5.9	4.1	23	16	50	0.32
Macaca mulatta	9.0	6.5	21.5	19	44	2.3
Cercocebus albigena	9.0	6.4	15	5	27	4.1
Cercocebus galeritus	10.2	5.5	26	13	100	0.55
Papio anubis	21.0	12.0	34	17	4	24.3
Papio hamadryas	21.5	9.4	——	7	1.8	——
Theropithecus gelada	20.5	13.6	——	45	46	——
Cercopithecus aethiops	5.0	3.75	15	12	112	0.37
Cercopithecus cephus	4.1	2.9	10	8	25	0.35
Cercopithecus mitis	7.4	5.3	25	21	42	0.73
Cercopithecus pogonias	4.5	3.0	15	2	23	0.78
Cercopithecus neglectus	7.0	4.0	5	9	34	0.15
Cercopithecus nictitans	6.6	4.2	20	28	30	0.67
Cercopithecus ascanius	4.2	2.9	35	16	140	0.28
Miopithecus talapoin	1.4	1.1	70	2	40	1.2
Erythrocebus patas	10.0	5.6	20	2	0.3	52.0
Presbytis melalophos	6.7	6.6	9.3	37	74	0.21
Presbytis entellus	18.4	11.4	19	54	57	3.4
Presbytis senex	8.5	7.8	8	60	154	0.12
Presbytis johnii	14.8	12.0	13	71	90	0.23
Presbytis obscura	7.6	6.6	10	48	31	0.29
Colobus satanas	12.0	9.5	12.5	37	30	0.60
Colobus guereza	11.8	9.25	11	82	104	0.15
Colobus badius	10.5	5.8	50	76	298	0.75
Hominoidea						
Hylobates lar	5.7	5.3	4	34	4.7	0.54
Symphalangus syndactylus	10.8	10.6	3.8	45	4.2	0.23
Pongo pygmaeus	69.0	37.0	2	22	2	5.0
Pan troglodytes	41.6	31.1	2	28	2	7.0
Gorilla gorilla	160.0	93.0	10	86	1.8	6.2

[a]See text.

grounds that female ranging is likely to be related to the distribution of food while male ranging may be related to the distribution of females (*see* Clutton-Brock and Harvey, 1976, 1978). For both the chimpanzee and the orangutan we have taken the size of individual female home ranges since recent studies suggest that females occupy independent home ranges.

Analysis of relationships within taxonomic groups employed multiple regression analysis (Snedecor, 1956). Standardized partial regression coefficients were calculated to compare the relative importance of each of the two independent variables as predicators of the dependent variable. Analyses of covariance on the multiple regressions were used to compare regression lines between taxonomic groups (Snedecor, 1956). As estimates of comparative home-range size and population density in particular species, we used deviations from the regression line of the relevant taxonomic group.

The same reservations concerning broad cross-species comparisons mentioned in our previous studies (Clutton-Brock and Harvey, 1977a, b) apply here. Methodological differences between studies or differences in study duration may introduce considerable error. Particular studies may have been carried out in aberrant study areas or groups and may produce unrealistic results. It is often difficult to construct realistic species averages in the many cases where intraspecific variation is marked (*see* Clutton-Brock, 1977). Finally, we did not examine the possibility that nonlinear relationships exist between variables. Though these problems weaken cross-species comparisons, they do not invalidate them. Although considerable error may be present, there is usually no reason to suppose that this might be responsible for creating a relationship where none existed. However, the heterogeneous nature of the data allows us to place no reliance either on negative results or on comparisons between individual species.

Results

Home-Range Size

Within all four taxonomic groups home-range size was positively related to group weight and negatively to the proportion of foliage eaten (Table 2) though only in the Cercopithecoidea (the group for which most data were available) were relationships

Table 2. Standardized Partial Regression Coefficients for Home Range Regressed on Percent of Foliage in the Diet (b_1') and Group Weight (b_2'). For Each Comparison Significant Tests (t_1 and t_2) Are Also Shown.

Phylogenetic Group	*Sample Size*	b_1'	b_2'	*Elevation*	t_1	t_2
Prosimii	9	-.638	1.025	-3.242	1.579	2.539†
Ceboidea	5	-.762	1.217	-1.355	3.866*	6.175†
Cercopithecoidea	22	-.630	0.719	-5.198	4.844**	5.527**
Hominoidea	5	-.956	1.353	0.114	2.418	3.423*

*$p < 0.1$
†$p < 0.05$
**$p < 0.001$

significant. Comparisons of standardized regression coefficients (b_1, b_2) suggested that the positive relationship with group weight was stronger than the negative one with the proportion of foliage eaten.

No significant heterogeneity in the slopes of the multiple regressions was evident between groups ($F_{6.29} = 1.306$, not significant), though differences in elevation were highly significant $F_{3,35} = 5.301$, $p < 0.001$). Relative to diet and group weight, the Prosimii and the Cercopithecoidea have smaller home ranges than the Ceboidea and Hominoidea (see Table 2).

Population Density

Results for population density paralleled those for home-range size. In all groups, population density was positively related to the proportion of foliage eaten and in all cases except one (the Ceboidea) it was negatively related to individual weight (see Table 3). Relationships between population density and diet were only significant in the Ceboidea and Cercopithecoidea, while those between density and body weight were only significant in the Cercopithecoidea and Hominoidea. Except in the Hominoidea, density was more closely related to differences in diet than to those in body weight.

As in the case of home-range size, the slopes of relationships did not differ between taxonomic groups ($F_{6,34} = 1.034$, not significant) while differences in elevation were highly significant ($F_{3,40} = 5.460$, $p < 0.001$). Relative to their diet and group weight, the Ceboidea and Hominoidea show lower population densities than the Prosimii and Cercopithecoidea.

Table 3. Standardized Partial Regression Coefficients for Population Density Regressed on Percent of Foliage in the Diet (b_1') and Individual Body Weight (b_2'). For Each Comparison Significance Tests (t_1 and t_2) Are Also Shown.

Phylogenetic Group	Sample Size	b_1'	b_2'	Elevation	t_1	t_2
Prosimii	11	0.824	-0.431	3.024	2.030*	1.062
Ceboidea	5	0.932	0.087	1.710	4.667†	0.436
Cercopithecoidea	25	0.797	-0.590	4.296	4.704**	3.477**
Hominoidea	5	0.271	-1.063	1.831	2.165	8.488†

*$p < 0.1$
†$p < 0.05$
**$p < 0.001$

Discussion

The analysis provides evidence that home-range size and population density are related in similar ways within primate families to group (or individual) weight and diet as they are across the order as a whole: home-range size is positively related to group weight and negatively to the proportion of foliage eaten, while population density is positively related to the proportion of foliage eaten and negatively to individual weight. Relationships are only consistently significant in the Cercopithecoidea, but this might be expected since it is the only group for which more than ten data points are available. Home-range size tends to be more closely related to group weight than diet while for population density the situation is reversed, with the proportion of tonnage in the diet a better predictor than individual weight.

The study confirms the common impression that the densities of apes are relatively low compared to those of monkeys. Part of this difference may be the product of increased human interference: it is evident that both orangs and gorillas have suffered extensive persecution throughout much of their range for many years (MacKinnon, 1974; Fossey, 1974; Goodall, 1977). And part, too, may be the product of the extension of the chimpanzee and gorilla into habitats where the biomass of available food supplies is relatively low compared to deciduous/evergreen or evergeen forest (see Izawa, 1970; Fossey, 1974; Fossey and Harcourt, 1977). However, there are several indications that, when both these effects have been taken into account, population densities (and biomasses) of apes may be lower than expected. The biomass of gibbon and siamang populations in comparatively undisturbed areas (Chivers, 1973, 1974) is low compared to that of simian species of similar body size. Even in relatively undisturbed African rain forests, the biomass of chimpanzees is low compared to the biomasses shown by many of the sympatric monkeys (for example, *Cercopithecus* spp.; Struhsaker, 1975, 1978).

One possible explanation is that the apes are primarily adapted to exploiting widely dispersed food which is intermittently available (MacKinnon, 1974; Struhsaker, 1975) and that they depend extensively on ripe (versus unripe) fruit (Wrangham, in preparation). Differences in relative home-range size may also be related to their body size. Although the functional significance of anatomical differences between monkeys and apes has been extensively discussed, perhaps the most obvious morphological difference between the two groups is usually ignored. Apes are larger than monkeys. The most likely advantages of large size in primates are:

1. Reduced food requirements relative to body weight (Kleiber, 1961), permitting larger species with a given digestive rate to feed on nutritionally poorer foods than smaller species (Janis, 1976)
2. Reduced costs of locomotion per unit weight (Schmidt-Nielsen, 1972a, b), at least for terrestrial species, and the ability to travel further per unit of time
3. Increased potential for defense against predators.

If body size is an adaptation to the species' feeding niche, we might expect folivores and species needing to travel considerable distances when food is short to show increased body size. This appears to be the case. In primates, the body size is significantly related to the proportion of foliage eaten (b_1) and to home-range size (b_2) ($b_1 = 0.255$, $t = 5.235$, $p < 0.001$; $b_2 = 0.447$, $t = 9.977$, $p < 0.001$; $n = 42$). The argument can, however, be reversed. Body size may determine both diet (Hladik, 1975) and home-range size (since it contributes a large proportion of the variance in group weight). Since we should expect the products of evolution to be complexes of functionally interrelated traits, the tautology should not embarrass us though it is important to distinguish between the two arguments: (1) Differences in body size evolved for some reason unrelated to diet. Large species were subsequently enabled to feed extensively on poor quality foods and were forced to adopt large home ranges. (2) Body size, diet, and ranging behavior evolved as a complex of functionally interrelated traits. Although among contemporary primate populations body size constrains diet and ranging behavior (since it is ontogenetically more stable), permanent shifts in food quality, density, or distribution will be likely to have repercussions on body size. To us, the second argument seems more parsimonious.

Returning to the apes, we have suggested that their tendency to show large home ranges and low population densities occurs because they rely on clumped, unpredictable, and widely distributed food supplies and may be associated with their dependence on ripe fruit (Struhsaker, 1975; Wrangham, in preparation). This accords with their pattern of social organization: both the orang and the chimpanzee have flexible grouping patterns, individuals ranging widely and aggregating at food sources (Wrangham, 1975, 1977; MacKinnon, 1974; Rodman, 1973). It is even possible that the relatively large brain size and advanced logical capabilities characteristic of apes (Jolly, 1972) may also represent adaptations to exploitation of a complex food supply.

The large body size typical of pongids could have preadapted the ancestors of the gorilla for radiation into a terrestrial, folivorous niche where reduced feeding interference permits the development of large feeding groups: it is possible that the gorilla may have had ecological analogues both in Madagascar (*Megaladapis*) and in southeast Asia (*Gigantopithecus*). The Hylobatidae represent an apparent exception since their home-range sizes and body weights are considerably smaller than those of the great apes. One possible explanation is that the African apes require large ranges in order to maintain access to enough different tree species to provide them with ripe fruit in all seasons of the year (*see* Clutton-Brock, 1975). In southeast Asia, a primate which depends on ripe fruit may be able to obtain adequate food supplies from a smaller area because of the high tree species diversity typical of this region (Richards, 1966; Chivers, 1974), and gibbons may have evolved to fill this niche while the orangutan became adapted to feeding on more dispersed and unpredictable fruit sources.

The relevant data to test the hypothesis that apes are better adapted to feeding on clumped and unpredictable food supplies than monkeys is missing. If the hypothesis is correct, we would predict that apes should show increased spatial and temporal variance in home range utilization and that pongid populations should extend into habitats too variable for monkeys feeding on similar diets. We hope that future studies may fill this gap.

Differences in relative home-range size and population density between New and Old World monkeys probably occur because no terrestrial open-country species occur in the New World. Differences in species diversity (*see* Richards, 1966; MacArthur, 1972) may also be involved: not only is high primate species diversity likely to produce low densities (and large home ranges) within particular species, but high-tree species diversity may be associated with a low density of utilizable food sources. This accords with the tendency for primate species in Madagascar (which shows relatively low species diversity compared to mainland Africa) to exist at relatively high densities and, with the exception of *Indri*, to have relatively small home ranges.

The presence of pronounced differences between taxonomic groups in home-range size and population density relative to diet and group (or individual) weight raises important questions about similar differences in other biological variables. Phylogenetic differences in the sizes of particular organs, including the brain (Jerison, 1973), and in the duration of particular growth processes (Rudder, personal communication) are common and are frequently explained as examples of phylogenetic inertia (*see* Wilson, 1975)—the effects of common genetic constraints inherited by all members of a taxonomic group. This seems a less likely explanation of phylogenetic differences in variables so evidently malleable as home-range size and population density, which are probably explicable in terms of differences in habitat and food distribution, and, by inference, suggests that phylogenetic differences in growth processes may also be adaptive.

Summary

First, previous analysis has shown that species differences in mean home-range size are positively related to group weight and negatively to the proportion of foliage in the diet, while differences in population density are negatively related to individual weight and positively to the proportion of foliage eaten.

Within four phylogenetic groups (Prosimii, Ceboidea, Cercopithecoidea, and Hominoidea) the available data suggest that home-range size and population density are similarly related to diet and group (or individual) weight as across the order as a whole. However, marked differences in relative home-range size and population density exist between taxonomic groups: the Prosimii and Cercopithecoidea show relatively small ranges and high densities compared to the Ceboidea and Hominoidea.

Finally, differences between Hominoidea and Cercopithecoidea may reflect a tendency for apes to depend on widely dispersed and erratic food supplies or to feed extensively on ripe (versus unripe) fruit. Differences between Ceboidea and Cercopithecoidea may arise both because no ceboid species have invaded terrestrial niches, and because of high diversity in both primate and tree species in the neotropics.

Acknowledgments

We should like to thank Ben Rudder for allowing us to use data on body weights extracted by him from the literature and all those who have allowed us to use their unpublished data. We are particularly grateful for help, advice, support or criticism from Professor J. Maynard Smith F.R.S., Dr. B. C. Bertram, S. D. Albon, Dr. R. W. Wrangham, Dr. I. S. Bernstein, Dr. E. O. Smith and the other participants of the present symposium.

References

Chivers, D. J. 1973. Introduction to the socio-ecology of Malayan forest primates. Pages 101–146 in *Comparative ecology and behaviour of primates.* Eds. R. P. Michael and J. H. Crook. London: Academic Press.

Chivers, D. J . 1974. The siamang in Malaya: A field study of a primate in a tropical rain forest. *Contrib. Primat.* 4:1–355.

Clutton-Brock, T. H . 1975. Ranging behaviour of red colobus (*Colobus badius tephrosceles*) in the Gombe National Park. *Anim. Behav.* 23:706–722.

Clutton-Brock, T. H. 1977. Some aspects of intraspecific variation in feeding and ranging behaviour in primates. Pages 539–556 in *Primate ecology: Studies of feeding and ranging behaviour in lemurs, monkeys and apes.* London: Academic Press.

Clutton-Brock, T. H., and Harvey, P. H. 1976. Evolutionary rules and primate societies. Pages 195–237 in *Growing points in ethology.* Eds. P. P. G. Bateson and R. A. Hinde. Cambridge: University Press.

Clutton-Brock, T. H., and Harvey, P. H. 1977a. Species differences in feeding and ranging behaviour in primates. Pages 557–584 in *Primate ecology: Studies of feeding and ranging behaviour in lemurs, monkeys and apes.* Ed. T. H. Clutton-Brock. London: Academic Press.

Clutton-Brock, T. H., and Harvey, P. H. 1977b. Primate ecology and social organisation. *J. Zool.* 183:1–39.

Clutton-Brock, T. H., and Harvey, P. H. 1978. Mammals, resources and reproductive strategies. *Nature* 273:191–195.

Clutton-Brock, T. H ., Harvey, P. H., and Rudder, B. In preparation. Adaptive aspects of life history variables in primates.

Crook, J. H ., and Gartlan, J . S. 1966. Evolution of primate societies. *Nature* 210:1200–1203.

Eisenberg, J. F., Muckenhirn, W. A., and Rudran, R. 1972. The relation between ecology and social structure in primates. *Science* 176: 863–874.

Fossey, D. 1974. Observations on the home range of one group of mountain gorillas (*Gorilla gorilla berengei*). *Anim. Behav.* 22:568–581.

Fossey, D., and Harcourt, A. H. 1977. Feeding ecology of free-ranging mountain gorilla (*Gorilla gorilla berengei*). Pages 415–448 in *Primate ecology: Studies of feeding and ranging behaviour in lemurs, monkeys and apes.* Ed. T. H. Clutton-Brock. London: Academic Press.

Goodall, A. 1977. Feeding and ranging behaviour of a mountain gorilla group (*Gorilla gorilla berengei*) in the Tshibinda-Kahnzi region (Zaire). Pages 450–479 in *Primate ecology: Studies of feeding and ranging behaviour in lemurs, monkeys and apes.* Ed. T. H. Clutton-Brock. London: Academic Press.

Hladik, M. 1975. Ecology, diet and social patterning in old and new world primates. Pages 3–35 in *Socioecology and psychology of primates.* Ed. R. H. Tuttle. The Hague: Mouton.

Izawa, K. 1970. Unit groups of chimpanzees and their nomadism in the savanna woodland. *Primates* 11:1–46.

Janis, C. 1976. The evolutionary strategy of the equidae and the origins of rumen and caecal digestion. *Evolution* 30:757–774.

Jerison, H. J . 1973. *Evolution of the brain and intelligence.* New York: Academic Press.

Jolly, A. 1972. *The evolution of primate behavior.* New York: Macmillan.

Kleiber, M. 1961. *The fire of life.* New York: Wiley.

MacArthur, R. H. 1972. *Geographical ecology.* New York: Harper & Row.

MacKinnon, J. 1974. The behaviour and ecology of wild orang-utans (*Pongo pygmaeus*). *Anim. Behav.* 22:3–74.

Richards, P. W. 1966. *The tropical rain forest.* Cambridge: University Press.

Rodman, P. S. 1973. Population composition and adaptive organisation among orang-utans of the Kutai Nature Reserve, East Kalimanatan. Pages 383–413 in *Primate ecology: Studies of feeding and ranging behaviour in lemurs, monkeys and apes.* Ed. T. H. Clutton-Brock. London: Academic Press.

Schmidt-Nielsen, K. 1972a. Locomotion: Energy costs of swimming, flying and running. *Science* 177:222–228.

Schmidt-Nielsen, K. 1972b. *How animals work.* Cambridge: University Press.

Snedecor, G. W. 1956. *Statistical methods.* Iowa City: Iowa University Press.

Struhsaker, T. T. 1975. *Behavior and ecology of red colobus monkeys.* Chicago: University of Chicago Press.

Struhsaker, T. T. 1978. Food habits of five monkey species in the Kibale Forest, Uganda. Pages 225–248 in *Recent advances in primatology,* vol. 1. Eds. D. J . Chivers and J. Herbert. London: Academic Press.

Wrangham, R. W. 1975. Behavioural ecology of chimpanzees in Gombe National Park, Tanzania. Ph.D. thesis, University of Cambridge.

Wrangham, R. W. 1977. Feeding behaviour of chimpanzees in Gombe National Park, Tanzania. Pages 503–538 in *Primate ecology: Studies of feeding and ranging behaviour in lemurs, monkeys and apes.* Ed. T. H. Clutton-Brock. London: Academic Press.

Wrangham, R. W. In preparation. The social behaviour of the great apes.

Wilson, E. O. 1975. *Sociobiology: The new synthesis.* Cambridge, Mass.: Belknap Press.

Chapter 10

Habitat, Economy, and Society: Some Correlations and Hypotheses for the Neotropical Primates

John F. Eisenberg

In the previous chapter, Clutton-Brock and Harvey limited themselves to certain specified hypotheses which they then examined across the full range of available primate data. In this chapter Eisenberg takes a different sort of global view and focuses on the ecological conditions influencing neotropical primates. He, too, notes that smaller species have lower biomass, that home range increases with group weight, and that folivores have higher biomass than omnivores, as well as smaller day ranges. Considerable variability is also noted and Eisenberg examines the sources of such variability by including considerations of species diversity, geographic history, and phylogenetic history.

*Considering the geographic distribution of New World primates in general, and the areas of sympatry and allopatry, Eisenberg suggests that some genera (*Cebus *and* Alouatta*) might be regarded as pioneer species with generalized adaptations to a diversity of habitats. Other genera may be regarded as more specialized with social organizations more finely attuned to specific ecological situations.*

The New World primates are also regarded as having a phylogenetically ancient contributor to social organization in obligate monogamy. He theorizes that this results as the percentage of the weight of offspring increases with respect to adult weight, and where the next pregnancy follows soon after delivery. These life history parameters thus result in a particular obligate social organization which has limited flexibility in responding to ecological pressures unrelated to those which selected for small adult body size, large offspring size (the weight of either single or multiple offspring combined), and rapid reproductive rates.

In reviewing the data on primate distribution in the neotropics, Eisenberg notes that whereas demographic data is often difficult to obtain there is good evidence for systematic variation in the diversity of species present in different areas. An examination of the impoverished areas reveals that these are the black water river areas and this suggests a generally impoverished environment may exist in such river basins based on geology and soil compositions. Eisenberg notes that in areas where primates maintain high biomass, there is an increase in species diversity. The corollary thus might be that those primates that do live in the black water river areas will show lower biomass in that area than in other areas, despite the lack of additional potentially competitive primate species. The distribution and habitat use of primates in the neotropics may also be regarded as a mechanism to reduce competition, but in the more impoverished environments, habitat specialization may not be feasible.

In examining the various proposed correlations between ecology and social organization, Eisenberg notes that generalizations at the generic level ignore the considerable diversity which occurs among members of the same genus. Moreover, he indicates that there is far more variability from troop to troop than most of us would have guessed.

Dr. Eisenberg's chapter thus touches on a number of the repetitive themes uncovered in our discussions during the conference and emphasizes some of the neglected variables which must be considered in relating the broad subject of ecology to social organization. Intraspecific variability, intergeneric influences, and the whole phylogenetic and geologic history of the area must be considered. The time scale under consideration is certainly beyond any brief field study and the diversity of knowledge required goes beyond most single investigators. The complexity of the task may seem overwhelming, and few of us may have Dr. Eisenberg's ability to assume such a broad global approach to the topic, or his knowledge of so many mammalian forms for comparison. If correlations between cer-

tain factors in ecology and social organization can be demonstrated for the poorly known New World primates, then surely, with literally dozens of field workers concentrating on a few Old World primates living in a diversity of habitats, we should be able to hope to understand how ecology limits and influences the social behavior of at least a few of the better known species. Broader, all-encompassing theories may have to wait for a future time when we will have sufficient detailed information on a sufficient sample for representative primate species. We may have to build slowly, taxonomic level at a time, geographic region by geographic region, but we may yet hope to achieve the goals set by the first theoreticians working in this area.

Introduction

Investigations of neotropical primate behavior and ecology have lagged behind comparable studies in the paleotropics. It would appear that the balance is finally being redressed somewhat, given the recent efforts of Glander (1975), Baldwin and Baldwin (1972), Neville (1972), Dawson (1977), Izawa (1976), and Coelho (1974). For some years the neotropical primates of Panama on Barro Colorado Island have been under study by the staff and visiting researchers of the Smithsonian. Following the pioneer studies on *Ateles* and *Alouatta* by Carpenter (1934, 1935), ethological and ecological studies, often combining both a field and captive colony approach, were completed for the following Panamanian species: *Alouatta palliata*, Altmann (1959), Chivers (1969), Smith (1977), Milton (1978); *Ateles geoffroyi*, Eisenberg and Kuehn (1966), Dare (1974); *Ateles fusciceps*, Eisenberg (1976); *Cebus capucinus*, Oppenheimer (1968); *Saimiri sciureus oerstedi*, Baldwin and Baldwin (1972); *Aotus trivirgatus*, Moynihan (1964); *Saguinus oedipus geoffroyi*, Muckenhirn (1967), Dawson (1977), Moynihan (1970). The ecological studies by the Hladiks were pursued on Barro Colorado Island and generated a number of interesting hypotheses concerning niche separation (Hladik and Hladik, 1969). Recently Eisenberg, R. Rudran, and J. Robinson have attempted to develop new study areas for *Ateles*, *Alouatta* and *Cebus* in Venezuela (see Figure 1).

Under the auspices of the Pan American Health Organization, the ILAR Primate Committee of the National Academy of

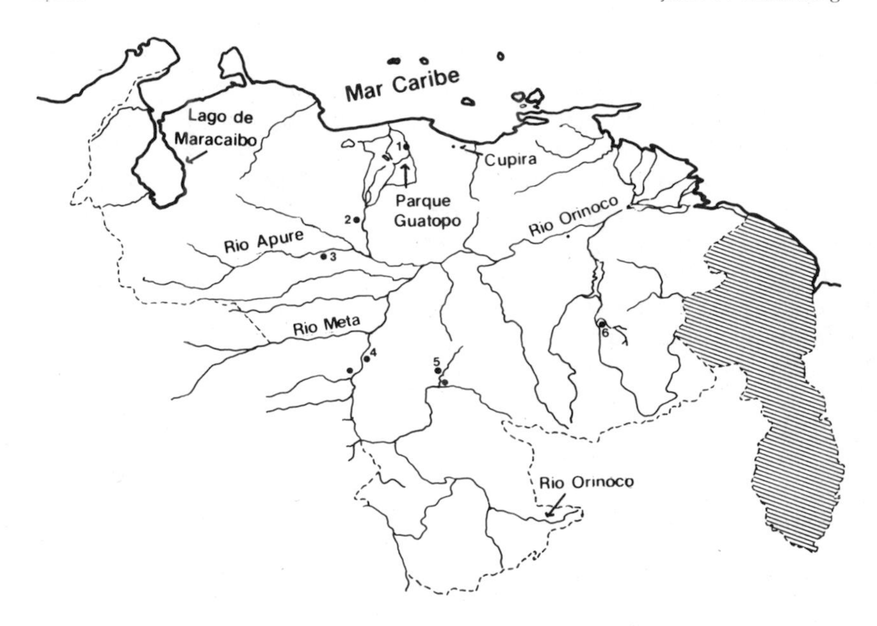

Figure 1. Survey sites and long-term study areas (1–3) in Venezuela. The crosshatched area is that part of Guyana west of the Essequibo River. The three intensive study areas in Venezuela are: 1. Guatopo National Park; 2. Fundo Pecuario Masaguaral; 3. Hato El Frio. Additional survey areas include: 4. Puerto Ayacucho; 5. San Juan de Manapiare; and 6. Canaima.

Sciences sponsored a series of primate censusing operations in Colombia, Peru, Bolivia, and Guyana. Basic data concerning habitat destruction, relative abundance, and distribution of species were summarized in the reports by Thorington and Heltne (1976), Neville (1975), Struhsaker *et al.* (1975), Heltne *et al.* (1976), Muckenhirn *et al.* (1976), and Bernstein *et al.* (1976). Some of the results of these censusing efforts have been utilized in my preliminary calculations concerning relative biomass and density contributions by a given species (*see* Appendix of this chapter). Hernandez-Camacho and Cooper (1976) have summarized primate distributions in Colombia, and Handley (1976) has published the distributional data for the Smithsonian-Venezuela Mammal Project. The data from Handley, together with the censusing results from Muckenhirn *et al.*, and the Hernandez-Camacho/Cooper distribution maps, have been used by me in determining some ecological factors influencing sympatry in northern South America.

A Consideration of Some Problems in Data Collection and Evaluation

Density estimates and all derivatives such as biomass conversions ultimately are only as strong as the reliability of the census methods. In comparing data from various authors collected under varying conditions with differing standards of reliability, one must be somewhat selective. The very best density estimates are based on long-term residence in the study area, and a careful enumeration of troops, their composition, and their home ranges. The result of such labors will be a density estimate based on known troops for a defined "base area." The base area, or area of reference is somewhat arbitrary: (1) it may be a grove of trees comprising ten hectares in a cleared field; (2) it could be a larger forested area grading to an entirely different vegetation form, the latter being unoccupied by the species under study; (3) finally, the study area could be a mosaic of different physiognomic classes of vegetation which are used at different levels by the various primates ranging within the area. When studying forest primates in a new area it may not be at all obvious to the observer when one is actually moving from one habitat subtype to another.

Censusing primates by means of a walking transect generally involves two important decisions: first, one must choose the length of the transect for each sample which will realistically give a consistent count; second, one must define the detection width of the transect in terms of the vegetation structure and the species one is attempting to census. Green (1978) critiques the latter problems, while Heltne and Thorington (1976) examine the whole problem of forest primate censusing in a broad context. I do not wish to review these problems again but instead will offer the results of efforts by myself and colleagues in Venezuela and Panama (*see* Appendix). In the Appendix, I present two types of density estimation: an estimate by mapping and repeated or long-term study, and an estimate derived from transect walking. Since populations wax and wane through the years, I have included successive annual estimates where possible. Not all of the columns are filled, since the methods of individual workers over a span of 17 years cannot be made uniform in the here and now. It is encouraging to note however, that where we have a match between a transect and an actual count within a 20-month interval, the actual crude density and the transect density estimates positively covary. This correlation also tells us that when walking a transect *Alouatta seniculus*, in open habitats, may be slightly overestimated; that *Alouatta*, *Ateles*, and *Saguinus* in heavily forested habitats are usually underestimated; and that *Cebus* is usually somewhat overestimated. The overestimates for *Cebus* during transect walking have been commented on previously by Eisenberg and Thorington (1973). The positive correlation between transect density estimates and actual crude density estimates lends credence to the validity of utilizing transect data for relative biomass estimates when sympatric species for a given censusing locus are compared. Such use of transect data expressed as numbers per plot of a defined area can actually represent an index of abundance and be converted to a density estimate (Caughley, 1977).

This brings us then to the different forms of recording density. Table 1 adopted from intermediate to long-term studies of primate populations includes three classes of density:

1. *K* density is that value derived from the study of one or more troops for a sufficient time to determine the home range in terms of some area measurement. (A volumetric measure would be better but is as yet impractical.) This *K* density probably is near the carrying capacity for the species in question within the particular environment where the measurements were carried out. Variations in *K* density within one species from one habitat to the next probably reflect real variations in carrying capacity for the particular habitat.
2. Ecological density results from a study where the number of troops or individuals is expressed in terms of the area which may be termed suitable habitat.
3. Crude density refers to counts expressed in terms of the total base area of the survey whether or not the base area contains suitable habitat for the species in question. Crude density and ecological density become almost synonymous if the habitat is so uniform as to constitute ideal habitat for the species in question. This situation is seldom realized in practice. Thus, crude density estimates tend to be lower than ecological density and the latter lower than *K* density. All three expressions of density are useful but must be compared separately.

Table 2 transforms the data from Table 1 into three forms of biomass estimates. An inspection of the columns will indicate that the biomass levels for a given species will vary for any category of density estimation. Such variations within species may be the result of external perturbations such as hunting (Izawa, 1976) or forest clearing (Bernstein *et al.*, 1976). Clearing of forests adjacent to a study plot can even result in temporary elevations in density and biomass (*see* Baldwin and Baldwin, 1976a). If we leave aside external disturbances, then variations in density and biomass within a given class of data for a species probably reflect differences in the carrying capacity of the habitat. Variations in carrying capacity can result from regional variations in plant productivity, presence or absence of competitors, or relative toxicity of plant foods. These three influences on carrying capacity will be examined in the next section.

When one compares home range and biomass values among the primate species in Table 2 several general trends are evident. If one considers absolute size, the smaller species contribute a lower biomass than do the larger species. If one considers the average biomass of the group (group weight) the home-range size increases as the group weight increases. Partial folivores such as *Alouatta* attain higher biomasses than do frugivores, and both partial folivores and sap-feeding genera such as *Callithrix* and *Cebuella* utilize smaller home ranges per unit group weight than do specialists for feeding on fruits and insects (*Saguinus*, *Saimiri*, and *Cebus*). These general conclusions are in agreement with the trends noted by Clutton-Brock and Harvey (1977) when considering mainly prosimians, pongids, and cercopithecines. The New World primates appear to reflect a set of feeding adaptations which parallel the Old World species. The available data are still quite sparse but the initial analyses conform to trends which might be anticipated based on a knowledge of relative size and trophic position. The smaller genera such as *Saguinus*, *Callithrix*, and *Cebuella* may reach high *K* densities, but crude density estimates are very low. This could result from censusing errors, but it may well reflect that when one considers a relatively large area as the base area for the ecological study then the habitat is a genuine mosaic from the standpoint of the smaller species. It is then a question of relative resource patch size and the relative distance between patches which determines the carrying capacity for small species.

Table 1. Some Primate Densities Resulting from Intensive Studies and

Species	Locality	Home Range (ha)	Troop Size
Saguinus oedipus geoffroyi	Panama	26	6.93
S. o. oedipus	Colombia	7.8–10	3–13
S. leucopus	Colombia	——	4.8
Cebuella pygmaea	Peru	0.3	4–5
Callithrix jacchus	Brazil	0.6–1.3	5–13
Callicebus moloch	Colombia	0.4	3.1
C. moloch	Colombia	3.29–4.18	3.9
C. torquatus	Peru	20	3–4
Saimiri sciureus	Colombia	——	42–43
S. sciureus	Colombia	——	25–35
S. sciureus oerstedi	Panama	17.5–40	23–27
Pithecia pithecia	Surinam	4–10	3.5
P. monachus	Colombia	>50	3.0
Cebus albifrons	Colombia	~300	15–20
Cebus apella	Colombia	——	.6–12
C. apella	Colombia	——	18–20
C. capucinus	Panama	90	~15
C. capucinus	Costa Rica	50	15–20
C. capucinus	Panama	32–40	27–30
Ateles geoffroyi	Guatemala	——	——
Ateles geoffroyi	Panama	110–115	12–14
A. geoffroyi	Costa Rica	——	~20
A. belzebuth	Colombia	260–390	17–22
A. paniscus	Peru	330	20
Lagothrix lagotricha	Colombia	>500	42–43
Lagothrix lagotricha	Colombia	——	25–70
Lagothrix lagotricha	Colombia	——	>12
Alouatta palliata	Panama, BCI '33	44–76	4-35
A. palliata	Panama, BCI '59	——	3–45
A. palliata	Panama, BCI '67	7.9–11.6	11–18
A. p. villosa	Costa Rica	——	3–24
A. p. villosa	Panama	3.2–6.9	7–28
A. p. villosa	Costa Rica	9.9	13
A. p. pigra	Guatemala	125	3–6
A. seniculus	Colombia	——	3–6
A. seniculus	Venezuela	3.2	4–15
A. seniculus	Guatopo I	——	6.3
A. seniculus	Masaguaral-W	3.5	8.5
A. seniculus	Masaguaral-E	——	——

Actual Troop Counts.

K Density (No./ha)	*Ecological Density (No./km²)*	*Crude Density (No./km²)*	*Authority*
——	23	——	Dawson, 1977
0.30–1.8	——	——	Neyman, 1977
——	——	15	Green, 1978
5.6	——	——	Castro & Soini, 1977
9	——	——	Stevenson, 1978
4.3	——	——	Mason, 1966, 1971
0.8	——	——	Robinson, 1977
0.2	15	——	Kinzey, *et al,* 1977
——	——	4.2–5	Izawa, 1976
——	19–31	——	Klein & Klein, 1975
1.3	——	——	Baldwin & Baldwin, 1976a
0.4	——	0.8–7	Buchanan, in press
0.1	——	3	Izawa, 1976
——	——	5	Izawa, 1976
——	——	8	Klein & Klein, 1975
——	——	——	Izawa, 1976
0.16	——	12	Oppenheimer, 1968
——	——	6	Freese, 1976
1.3	70–90	——	Baldwin & Baldwin, 1976a
——	45	——	Coelho, 1974
0.10	——	——	Dare, 1974
——	——	6–9	Freese, 1976
——	12–15	——	Klein & Klein, 1976
——	——	——	Durham, 1971
——	——	4.3	Izawa, 1976
——	——	~16	Izawa, 1976
——	——	9–14	Bernstein, *et al.*, 1976
~0.5	——	32	Carpenter, 1965
——	——	52	Carpenter, 1965
0.6–0.8	60	——	Chivers, 1969
——	——	18–25	Freese, 1976
>4	——	——	Baldwin & Baldwin, 1976a
1.3	——	——	Glander, 1975
0.04 ?	5	——	Coelho, 1974; Schlichte, 1978
——	——	12–29	Klein & Klein, 1975
1.6	86	——	Neville, 1976
——	——	16–20	Eisenberg, unpubl.
——	(87)	41	Eisenberg & Kleiman, unpubl.
——	48	<20	Eisenberg & Green, unpubl.

Table 2. Biomass, Home Range, and Diet for Some Neotropical Primates.

Species	Unit Weight (gm)	Group Weight (kg)	K Biomass (kg/km²)	Ecological Biomass (kg/km²)	Crude Biomass (kg/km²)	Trophic[a] Category
Saguinus o. geoffroyi	600	4.158	——	13.8	——	I;F
S. o. oedipus	600	4.800	63	——	——	I;F
S. leucopus	600	2.88	——	——	9.0	I;F
Cebuella pymaea	150	0.68	84	——	——	I;F;G
Callithrix jacchus	450	4.05	252	——	——	I;F;G
Callicebus moloch	900	2.79	387	——	——	I;F
C. moloch	900	3.51	72	——	——	I;F
C. torquatus	1200	4.20	24	18	——	I;F
Saimiri sciureus	800	34.0	——	——	3.68	I;F
S. sciureus	800	24.0	——	20	——	I;F
S. s. oerstedi	800	20.0	104	——	——	I;F
Pithecia pithecia	1100	3.85	44	——	4.29	I;F
P. monachus	1100	3.30	11	——	3.30	I;F
Cebus albifrons	2600	45.5	——	——	13.0	I;F
C. apella	2600	23.4	——	——	20.8	I;F
C. apella	2600	49.4	——	——	——	I;F
C. capucinus	2600	39.0	41.6	——	31.2	I;F
C. capucinus	2600	45.5	——	——	15.6	I;F
C. capucinus	2600	74.1	338	208	——	I;F
Ateles geoffroyi	5000	——	——	225	——	F;L
A. geoffroyi	5000	65.0	50	——	——	F;L
A. geoffroyi	5000	100	——	——	37.5	F;L
A. belzebuth	5000	97.5	——	67.5	——	F;L
A. paniscus	5000	100	——	——	——	F;L
Lagothrix lagotricha	5200	221	——	——	22.36	F;L
L. lagotricha	5200	247	——	——	83.2	F;L
L. lagotricha	5200	62.4	——	——	59.8	F;L
Alouatta pulliata	5500	107.3	275	——	176	F;LL
A. palliata	5500	132.0	——	——	286	F;LL
A. palliata	5500	82.5	385	330	——	F;LL
A. p. villosa	5500	74.3	——	——	118.3	F;LL
A. p. villosa	5500	96.3	2200	——	——	F;LL
A. p. pigra	5500	24.8	22	27.5	——	F;L
A. p. villosa	5500	71.5	715	——	——	F;LL
A. seniculus	5500	24.8	——	——	112.8	F;LL
A. seniculus	5500	46.7	880	473	——	F;LL
A. seniculus	5500	34.7	——	——	99.0	F;LL
A. seniculus	5500	46.8	——	479	225	F;LL
A. seniculus	5500	——	——	264	110	F;LL

[a] I = Insects, F = Fruit, L = Some Leaves, LL = >35 Percent Leaves in Diet, G = Gums or Sap.

Productivity and Diversity

The ceboid primates include two families: the Callitrichidae and the Cebidae. Sixteen genera and 61 species are currently recognized, with 5 genera and 32 species assigned to the Callitrichidae. With such a rich assemblage of forms one could profitably ask questions concerning the degrees of sympatry, niche breadth, and polyspecific associations.

If we examine the number of sympatric species for a given locality, we find it ranges from 2–11.[1] At the extreme continental ranges of neotropical primate distribution, generally only one species is present, *Alouatta*. Currently *Alouatta* reaches approximately 22°N latitude and 28° 30′S latitude and has the broadest range of any neotropical genus. The genus *Ateles* may have extended further north than *Alouatta* in the state of Tamaulipas to 23°N (Alvarez, 1963), but *Ateles* does not show an extended southern range (Krieg, 1948). *Cebus* also shows an extended range as a genus, but ranks second to *Alouatta* in breadth of distribution (see Figure 2).

Plant productivity, whether it be leaves, fruits, blossoms, or woody structures, is a function of rainfall and temperature, provided soil fertility and porosity are relatively constant (Walter, 1973). In order to test the hypothesis that plant productivity influences primate diversity or the number of sympatric species in a given area, three analyses were performed for lowland areas lying between 14°N and 15°S latitude. Primate diversity was examined as a function of: (1) the total duration of drought in an annual cycle,[2] (2) the total annual precipitation in millimeters, (3) the difference in mean monthly maximum and minimum temperatures. The data clearly show that drought duration is an excellent predictor of primate diversity (*see* Figures 3 and 4). The shorter the time of annual drought, the greater the probability that primate diversity will also be high. Precipitation and temperature show less perfect correlations, but as a rule of thumb less annual precipitation means a lower primate diversity, and a greater difference between the mean monthly high and low temperature also reflects a lower primate diversity.[3] There are some localities that do not conform to the preceding trends and these invariably are either at the northern and southern extremes of the primate range or are localities which show a reduced diversity for zoogeographical reasons (*see* the next section).

Given then that primate diversity appears to parallel environmental indicators of plant productivity, I predicted that as the primate diversity increased the density of total primate numbers for an area should increase. An analysis of the data indicates that the correlation is valid except for those areas showing faunal impoverishment for zoogeographic reasons (*see* Figure 5). A more refined correlation can be derived however, if we use numerical

Figure 2. The limits of current primate distributions in Central and South America. Numbers 1–6 indicate study sites by the author from 1960 to the present. = *Ateles;* = *Alouatta;* = *Cebus capucinus;* = *Cebus apella;* = *Saimiri sciureus;* = *Aotus trivirgatus.* The dotted lines demarcate the Tropic of Cancer and the Tropic of Capricorn. Records from Handley (1966), Krieg (1948), and Pope (1966).

density and biomass figures for areas of long-term study. When these calculations have been performed, it is clear that a high diversity of primates (4–5 species) generally means a crude biomass of primates exceeding 200 kg/m^2. High biomass values can be achieved with low diversity in some localities where faunal impoverishment is demonstrable for zoogeographic reasons (*see* Table 3).

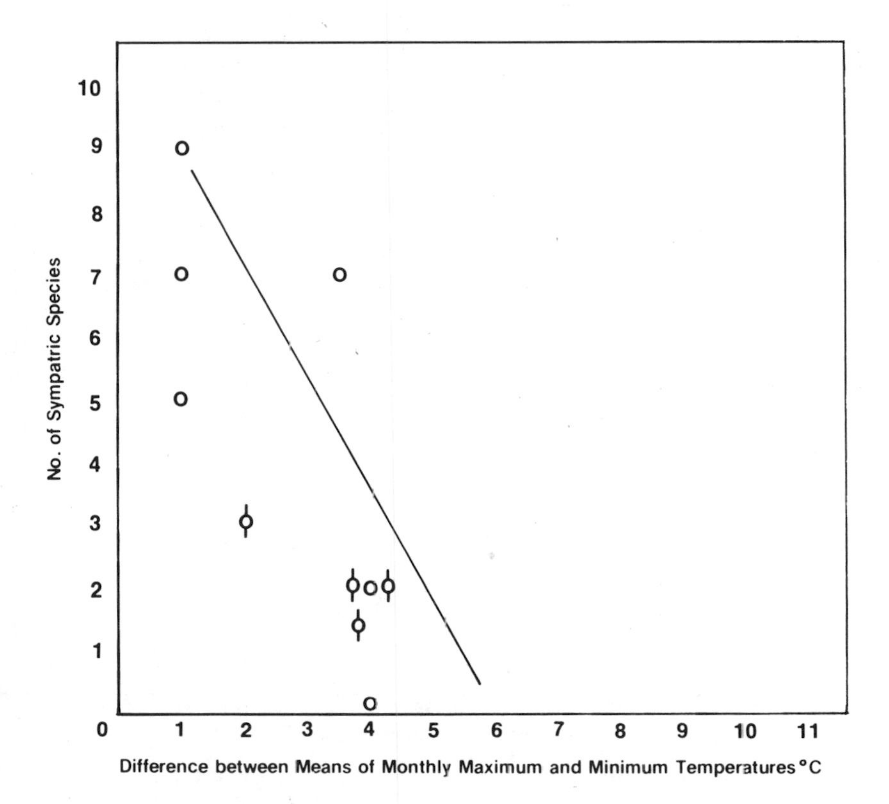

Figure 3. Number of sympatric primate species as a function of the difference between the mean monthly maximum and mean monthly minimum temperature. All stations between 14°N and 15°S latitude below 1,000 m. Temperature records from World Weather Records, U.S. (1951–1960) Department of Commerce. ϕ = Venezuelan and Colombian Llanos.

One final analysis of diversity was performed. Utilizing the data from the U.S. National Museum and Muckenhirn *et al.* (1976), it was possible to determine primate diversity indices for some 26 locations in Venezuela and Guyana. Looking at each species in turn, the total number of instances of sympatry was divided by the number of stations at which the species was recorded. This so-called overlap index will be high where a species was noted at few stations but always associated with a large number of sympatric species. Conversely, when the index

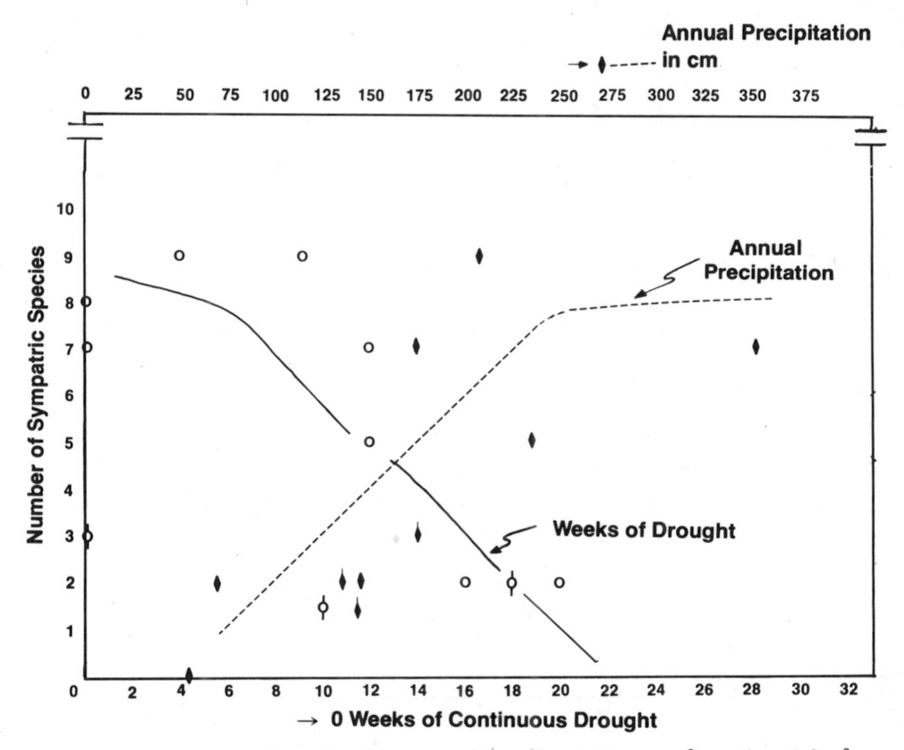

Figure 4. Double plot of primate species diversity as a function (a) of the average duration of annual drought in weeks (o); (b) of the mean annual precipitation in centimeters (⧫). Prolonged annual drought periods exceeding 2 to 2½ months usually indicate a tropical deciduous or semideciduous forest. All stations between 14°N and 15°S latitude at an elevation less than 1,000 m. Weather records as noted in Figure 3. Drought as defined by Walter (1973). (ϕ) = Areas which are marginal habitats.

value is low the species is broadly distributed but many times occurs alone or with few sympatric forms (*see* Figure 6). *Cebus* and *Alouatta* have the lowest indices and may be thought of as pioneer species over a range of habitats. *Ateles* may show an artificially low index in many areas for several reasons. First, it would appear that *Ateles* is sensitive to deforestation, and second, this species is preferentially hunted in many localities and thus shows a reduced density. The broad habitat tolerance by *Alouatta* and *Cebus* is born out in zoogeographical terms when the total range for these genera is considered (*see* Figure 2).

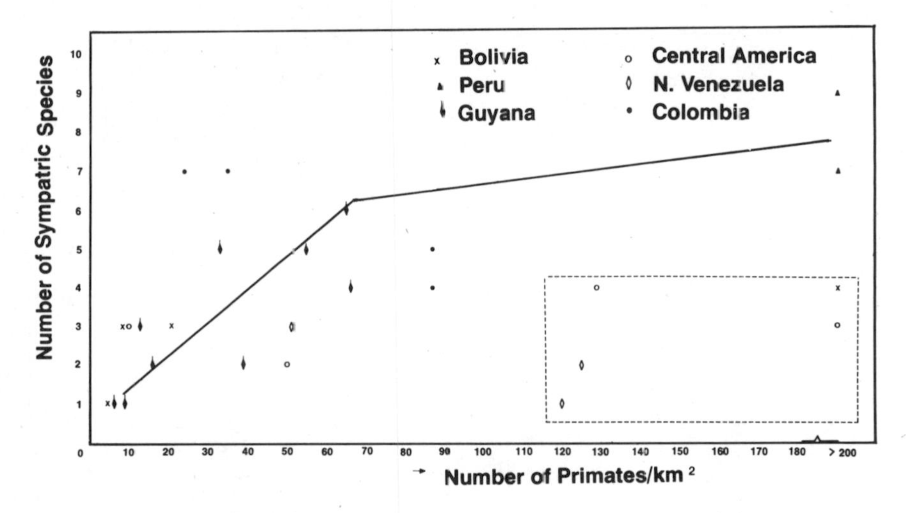

Figure 5. Number of sympatric primate species per station plotted as a function of absolute density of primates per km². The points enclosed by dotted lines are areas of faunal impoverishment resulting from zoogeographical factors (*see* Appendix Tables I and II).

Table 3. Diversity, Density, and Crude Biomass for Some Primate Communities.

Locality	*X̄ Number/km²*	*kg/km²*	*Number of Sympatric Species*
1. Areas near the limits of primate ranges			
Costa Rica	35	171	3
Guatemala	50	280	2
2. Areas showing depauperate faunas			
Northern Venezuela			
(a) Guatopo	45.1	181	3
(b) Llanos (W)	42	231	1
(c) Llanos (E)	45	175	2
3. Areas relatively undisturbed			
Colombia			
La Macarena	61	226	4
Cerro Bran	87	304	5
Panama			
B.C.I.	129	418	5

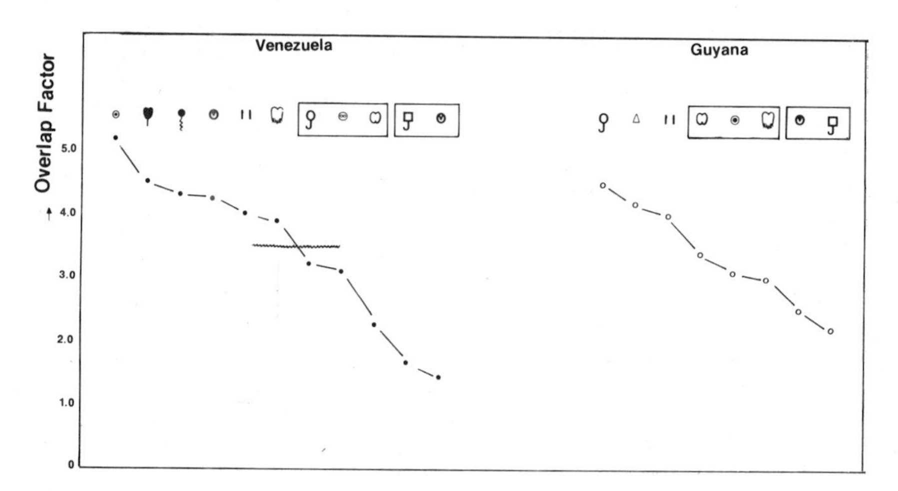

Figure 6. Tendency of primate species to be found with other sympatric species. A high overlap index indicates a tendency to be associated with many species at every locus of occurrence. A low value indicates many instances of either occurrence alone or with one or two sympatric species. Note that *Cebus* and *Alouatta* are consistent low scoring species in both Venezuela and Guyana. Symbols as in Figure 7. Data in part from Handley (1976) and Muckenhirn *et al.* (1976).

Some Zoogeographical Barriers to Diversity

As one proceeds up the isthmus of Panama, primate diversity drops sharply from six species in western Panama to three sympatric species in western Costa Rica. This loss results from a failure of *Saguinus, Aotus,* and *Saimiri* to extend very far beyond the Panama/Costa Rica boundary area. This does not appear to be the result of drastic climatic or vegetational changes. *Cebus* drops out in eastern Honduras and only *Ateles* and *Alouatta* carry on north to the state of Veracruz, Mexico. The failure of *Saguinus, Aotus,* and *Saimiri* to extend to Honduras may be a result of faunal imbalances deriving from the relatively recent formation of the isthmian land bridge. To the south in Argentina *Cebus, Aotus,* and *Alouatta* reach to approximately 27°S latitude on the Rio Paraña.

The Orinoco Basin of Venezuela was partially under water in the Pliocene and lay hemmed in between the eastern cordillera of the Andes and the Guiana Highlands to the south. Today northern Venezuela still shows signs of recent faunal colonization. Although the mammal fauna of the Maracaibo Basin resembles that of Panama and northern Colombia, to the east of the Andes the fauna is impoverished over much of the lower Orinoco system—only *Cebus* and Alouatta occur in frequent sympatry. *Ateles* is sporadically distributed in the montane forests of northern Venezuela and its "colonization" efforts seem to have been somewhat retarded.

The Guiana Highland remnant south of the Orinoco presents some genuine puzzles. The major rivers which drain the areas (the Caroni and the Caura) are black water rivers. Extensive black water rivers in the tropics are known from South America, West Africa, southeast Asia, and Borneo. By far the largest system occurs in South America and drains the Guiana Highlands in southern Venezuela and Guyana. The major south drainage flows into the Rio Negro emptying into the Amazon. From the Guiana Highlands the north and western drainage includes the Ventuari, Choroni, and Caura which enter the Orinoco. The Essequibo leaves the Guiana Highlands and terminates in the Caribbean.

Rivers in northern South America are broadly divided into three types: black water, white water, and clear or "crystal." As Sioli (1975) points out, the color of the river system indirectly indicates the predominant rock and soil substrates which it drains *upstream*. Black water rivers include white water and crystal tributaries, but the predominant input comes from black water streams which drain Archean rocks with a high silica content. Sioli (1975) has compared the water in these three categories of rivers and demonstrated that black water has a low pH and a low suspended nutrient content. Fittkau *et al.* (1975) point out that primary and secondary productivity is vastly reduced in black water streams when compared to white water. Janzen (1974) further points out that the vegetation over vast areas drained by black water streams tend to reflect an overall adaptation by the plants to reduced nutrients. As a consequence the leaves and fruits are often impregnated with alkaloids which have been developed by the plants as a protective device in response to the selective pressures of phytophagous vertebrates and invertebrates. Upstream black water systems generally show reduced densities of fish faunas and all associated predators of fish. The vegetation associated with white sand soils and sandstone in turn support a reduced insect fauna and all attendant insect predators.

Of course the further downstream one penetrates in a black water system, the less likely one is to be continuously surrounded by white sands, low nutrient soils, and specialized vegetation. In reality the habitat drained by black water streams is a mosaic of soil and vegetation forms. However, the upstream condition is often dominated by nutrient poor soils and a distinctive vegetation complex.

Since the sources of the major black water rivers lie in the Guiana Highlands, and further since these highland are believed to have served as Pleistocene forest refugia (Simpson, 1971), it should come as no surprise that some vertebrates as well as invertebrates may have come to adapt to such forest types. McKey (1978) argues that the black colobus population (*Colobus satanus*) which he studied in Cameron shows specific adaptations for a black water forest climax. Kinzey *et al.* (1977), offers evidence that *Callicebus torquotus* is adapted for forests on nutrient poor soils whereas *C. moloch* cannot invade this habitat when the two species are in macrosympatry.

Given then that the forests on nutrient poor soils in the Guiana Highlands and geologically related areas are a mosaic, and further given that species may have evolved specific adaptations to such forests, it would appear that the carrying capacity for many vertebrates is reduced. Even though hardy colonizers such as *Cebus* and *Alouatta* are to be found in selected valleys within the Guiana shield area, the primate diversity appears to be low. Indeed, *Pithecia pithecia* seems to be the only persistent invader of the highlands together with the two aforementioned species (*see* Figure 7).

In southern Venezuela the situation is not at all simple. In the Amazon lowlands of the Territerio Federal Amazonas at least nine primate species may again be found in macrosympatry. Some of the rivers taking origin from the highest parts of the southern escarpment of the Guiana Highlands, which includes carboniferous deposits, are not black water but crystal or white such as Mavaca. This latter river was an extremely rich collecting site for primates and other mammals in the late sixties (Handley, 1976).

To summarize then, the dominant vegetational complex over the southeastern portion of Venezuela is distinctive and termed "Guianian." The nature of the soil suggests low productivity and may imply reduced faunal diversity. As such this zoogeographical area might be a potent barrier to dispersal although one species, *Pithecia pithecia,* appears to have colonized from the east. *Cebus* and Alouatta, our durable pioneers, are often present in sympatry with *Pithecia.* This vast area is a habitat mosaic with pockets of savanna and white water drainage systems, but

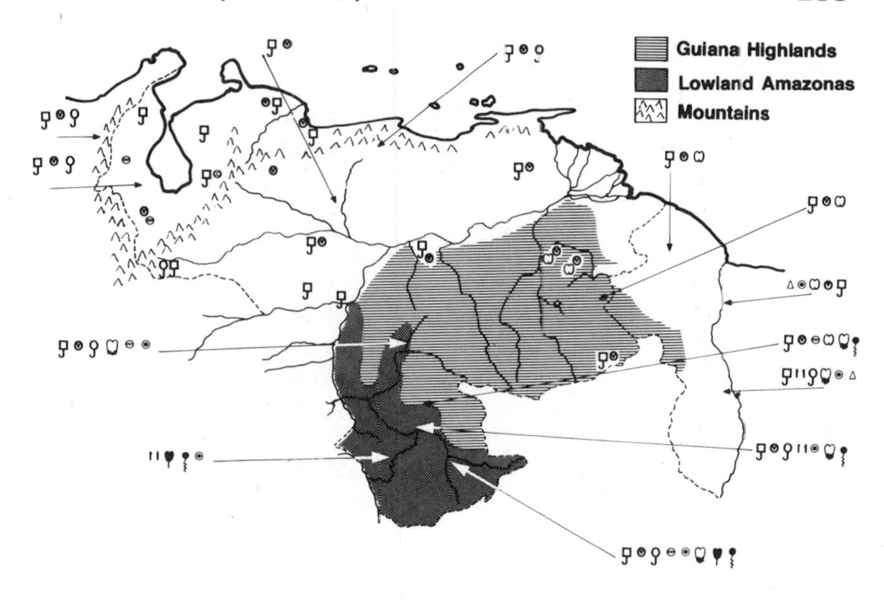

Figure 7. Sympatric primate distributions for 26 stations in Venezuela and Guyana. = *Alouatta seniculus;* = *Cebus albifrons* or *C. nigrivittatus;* = *Ateles;* = *Cebus apella;* = *Aotus trivirgatus;* = *Saimiri scuireus;* = *Callicebus;* = *Saguinus;* = *Chiropotes sanatas;* = *Pithecia pithecia;* = *Cacajao melanocephalus.* (Symbols adapted from Muckenhirn *et al.*, 1976. Data from Handley, 1976; Muckenhirn *et al.*, 1976; and Hernandez-Comacho and Cooper, 1976.)

its influence as a general barrier to colonization from Guyana should not be underestimated (see Figure 7).

Feeding Habits and Biomass

When the census data for neotropical primates are converted to biomass values (see Tables 2 and 4), it is possible to graph the range of values in terms of the rank order of the species' average size (see Figure 8). An analysis demonstrates that *Alouatta* averages the highest biomass but shows great variations. In general the largest primates show the highest biomass levels but *Saimiri* is a partial exception (see Appendix Table II). *Saimiri* is often censused in riverine gallery forest where it occurs in sympatry with one or two other species. When density calculations are made for *Saimiri* they are usually made with reference to this particular edaphic form of vegetation. Thus the density

Table 4. Actual Biomass (kg/km²) and Percentage of Total Biomass Contributions for Primates from Eight Localities. [a]

Locality	Alouatta	Ateles	Cebus sp.	Cebus apella	Saimiri	Aotus	Saguinus	*Number of Sympatric Species*	*Authority*
Guatemala	27.5 (11%)	252 (89%)	——	——	——	——	——	2	Coehlo, 1974
Costa Rica	118.5 (69%)	37.5 (21%)	15.6 (10%)	——	——	——	——	3	Freese, 1976
Panama	363 (86%)	5.0 (<1%)	45 (10%)	——	——	<4 (1%)	<3 (<1%)	5	Eisenberg & Thorington, 1973
Venezuela									
(a) Llanos (W)	231 (100%)	——	——	——	——	——	——	1	Eisenberg, Rudran & Green, unpubl.
(b) Llanos (E)	110 (63%)	——	65 (37%)	——	——	——	——	2	Eisenberg, Rudran & Green, unpubl.
(c) Guatopo I	88 (49%)	22.5 (13%)	70.2 (39%)	——	——	——	——	3	Eisenberg, Rudran & Green, unpubl.
Colombia									
Cerro Bran	121 (36%)	165 (49%)	39 (12%)	——	——	? (<1%)	9 (3%)	5	Green, 1978
La Macarena	112.8 (44%)	67.5 (29%)	——	20.8 (8%)	20 (8%)	——	——	4	Klein & Klein, 1976

[a] See Tables 1 and 2.

234

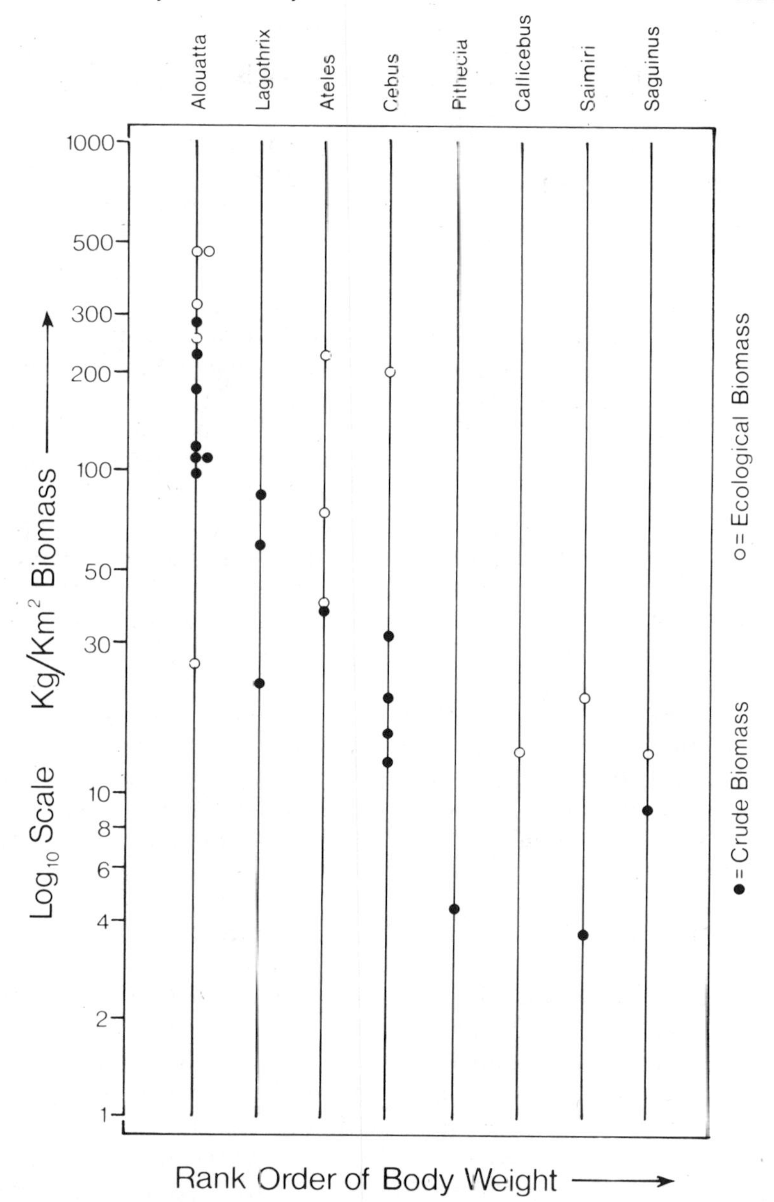

Figure 8. Biomass of primate species arranged in descending order (left to right) of mean adult body weight. Each point refers to a biomass value for a given geographic locality (*see* Appendix I). *Saimiri* may reach high biomass levels in gallery forests where it shares the habitat with one or two sympatric species. *Alouatta* has a great range of biomass concentrations but may reach the highest values for selected localities. (Derived from Table 2).

estimates inadvertently are representative of a correction for optimum habitat type. This form of density estimate is usually not performed for the other species showing a range of habitat tolerance. On the average *Alouatta* contributes the maximum percentage to a given local primate biomass (for example, 51%) but there is a great deal of variation. Causes for this variation will be examined in the next section.

The highest biomass levels are achieved by those species of mammals which utilize a significant proportion of herbaceous vegetation in their diet (Eisenberg and Thorington, 1973). Since the degree of herbivority which a mammal can achieve is in part a function of absolute size (Eisenberg, 1978), it is not surprising that *Alouatta, Lagothrix,* and *Brachyteles* are not only the largest primates but also the most folivorous with *Alouatta* typically utilizing 40–60% young leaves in its diet.

From the preceding considerations alone one would expect *Alouatta* to be a dominant contributor to the total primate biomass unless its numbers had been vastly reduced through epidemics; it is especially vulnerable to yellow fever (Galindo and Srihongse, 1967; Trapido and Galindo, 1956). Nevertheless, *Alouatta* is dependent on ripe fruits for a significant portion of its diet (Smith, 1977; Milton, 1978). An analysis of existing data suggests that the density of *Alouatta* may decrease in some areas where the number of frugivorous, sympatric primates is high. Thus, *Alouatta* may definitely be influenced by frugivore competitors under some habitat conditions (*see* Appendix Table II).

The Diversity of Niches and Competition

In their Panama studies the Hladiks clearly demonstrated that on Barro Colorado Island the five sympatric primate species maintained a respectable niche separation. *Aotus trivirgatus* as the only nocturnal primate is removed from direct competition with all other primate species in the Neotropics. *Saguinus oedipus* with its small size and insectivorous habits, plus its ability to use early second growth, is removed from direct competition with the larger genera. *Cebus* is also a genus which can use second growth and, together with its foraging for invertebrates, may compete with *Saguinus* under some circumstances. *Ateles, Alouatta,* and *Cebus* do broadly overlap with respect to some of the fruits which they feed on in common.

Some current primate distributions suggest active, competitive exclusion as a result of an initial geographical separation (*see* Moynihan, 1976). Although the tiny *Cebuella* occurs in sympatry with the larger species of *Saguinus*, it would appear that the genus *Saguinus* effectively occupies the "marmoset" niche north of the Amazon, while *Callithrix* occupies the niche to the south. Although it is true that *Callithrix* and *Saguinus* exist in macrosympatry in some areas of the northwestern Amazon basin, the whole problem deserves special scrutiny.

In a similar fashion the genus *Pithecia* seems to be abundant wherever the genera *Cacajao* and *Chiropotes* are absent or rare, respectively. These observations hold in southern Colombia when *P. monachus* does not overlap the ranges of *Cacajao melanocephalus* (Hernandez-Camacho and Cooper, 1976). The observations are the same for *Pithecia pithecia* and *C. melanocephalus* in southern Venezuela (Handley, 1976). It should be noted, however, that *Cacajao melanocephalus* and *Chiropotes satanus* exist in microsympatry over a broad geographical range.

The genus *Saimiri* appears to be a specialist for riverine habitats in Guyana (Muckenhirn *et al.*, 1976). This microhabitat preference, together with its small size, may reduce its direct competition with species of the genus *Cebus* even when foraging together. It would appear that the presence of *Saimiri* reduces the carrying capacity for the sympatric *Saguinus midas* (Muckenhirn *et al.*1976; Appendix Table II).

The influence of microhabitat preferences and competition are perhaps best understood for the dyad *Cebus*[4] and *Alouatta*. Broadly speaking, over a range of habitats as the density of *Alouatta* increases so does the density of *Cebus* (*see* Figure 9). This probably reflects the relative plant productivity for a given region. Yet there are areas where *Cebus* biomass exceeds that of *Alouatta*. An analysis of data from our recent studies shows that early second growth favors *Cebus* in Venezuela and furthermore that where *Cebus* is absent, *Alouatta* may reach very high biomass concentrations (*see* Figure 10).

The relative decline in *Cebus* numbers on Barro Colorado Island over a period of some 12 years and the concomitant increase of *Alouatta* substantiates the notion that *Cebus* exploits second growth quite well. Knight (1975) has analyzed the forest types on B.C.I. and presented convincing evidence that several areas on the island have recovered from extensive cutting over the last 50 years. This observation is consistent with those of

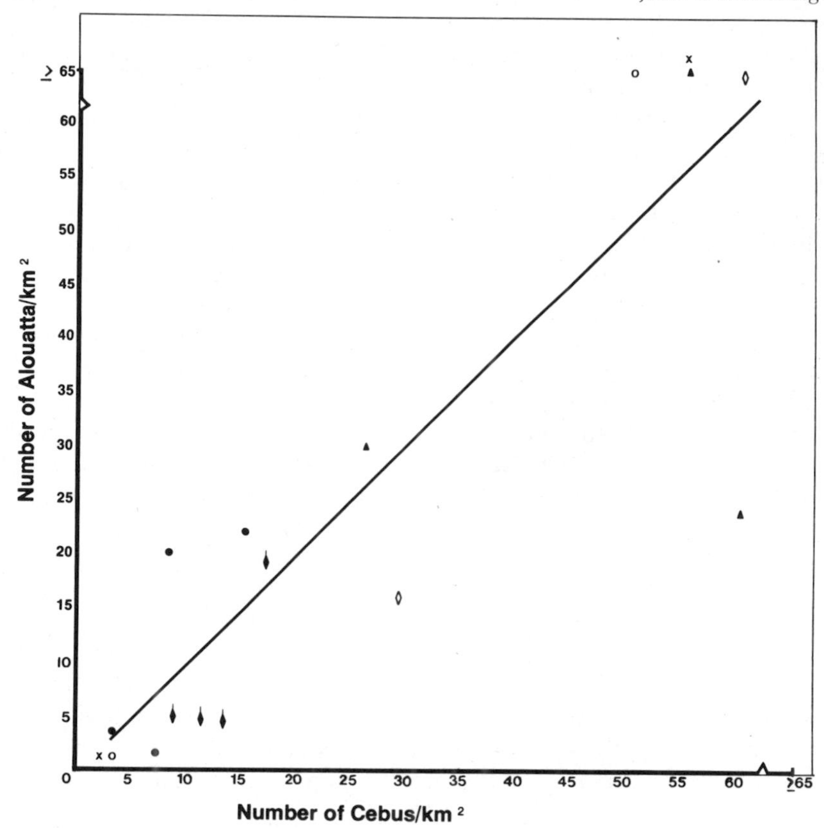

Figure 9. Density of *Alouatta* plotted against the density for sympatric *Cebus* from 15 areas. Although higher densities correlate for both genera, *Cebus* may be favored over *Alouatta* in some habitats (From Table 2). Symbols as in Figure 5.

Enders (1935). A comparison of Knight's maps and aerial photos with the distribution of *Alouatta* troops as determined by Carpenter in the early thirties (Carpenter, 1934) indicates that *Alouatta* had not colonized the then 20-year-old secondary forest. With the exception of the decline of *Alouatta* numbers in the late forties (Collias and Southwick, 1952), the howler numbers on B.C.I. have increased as the forest has matured. It also appears that *Cebus* has declined since the mid-sixties. The suggestion may be made that as the forest has approached a mature state on Barro Colorado the numbers of *Cebus* have declined, thus favoring *Alouatta*. *Saguinus geoffroyi* is also not favored in areas of continuous mature growth (see Table 5).

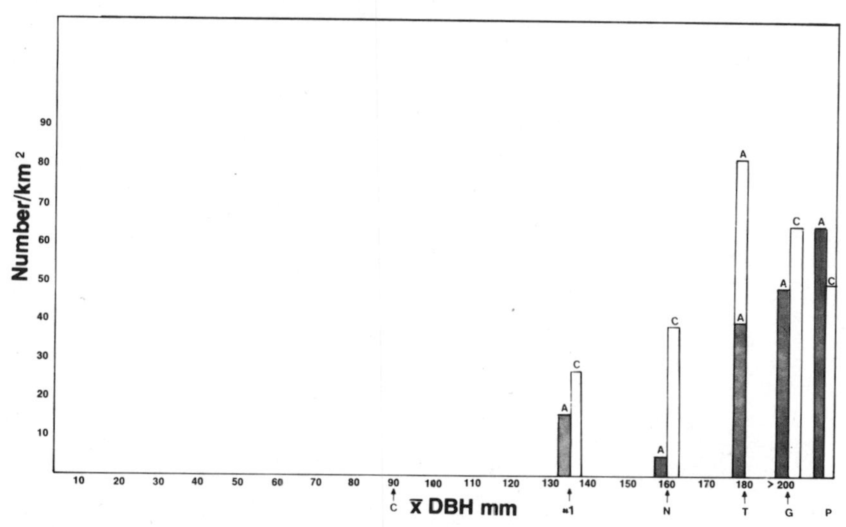

Figure 10. Numbers of *Alouatta* and *Cebus* per km² as a function of average DBH (diameter at breast height) for trees greater than 4 m in height. All densities are uncorrected for habitat quality except the values for 180 DBH where *Alouatta* is the only species present. The two values of 85/km² and 40/km² reflect "ecological" and "crude" densities respectively. A = Alouatta, C = *Cebus*; C = *Canaima*; 1 and N = two localities in Parque Nacional Guatopo; T and G = two localities at Hato Masaguaral—all in Venezuela. P = values for Barro Colorado, Panama.

The work of Izawa (1976), Durham (1975), and Freese (1975) suggests that *Lagothrix* and *Ateles* may negatively influence one another. Although these two genera exist in sympatry they appear to reach their maximum densities when their home ranges do not overlap considerably. In fact, *Ateles* may even reach high densities at the expense of *Alouatta* when it exists in areas unoccupied by *Lagothrix* (see Klein and Klein, 1976; and Green, 1978). Bernstein *et al.* (1976) found *Ateles belzebuth* and *Lagothrix lagotricha* in microsympatry in northern Colombia. In their study troop home ranges were adjacent, and occasionally troops were seen in proximity without hostility. This contrasts strongly with Durham's report (1975) of interspecific antagonism at several Peruvian localities.

All of the foregoing examples suggest that the feeding niches of the larger cebid primates overlap to some extent and a considerable reduction in niche breadth can occur in areas of high sympatry. Competitive exclusion is strongly indicated for *Pithecia* and the genus *Cacajao*.

Table 5. Crude Densities for Three Primate Species on Barro Colorado Island.

Year	Alouatta		Cebus		Saguinus	
	Number/km²	*Authority*	*Number/km²*	*Authority*	*Number/km²*	*Authority*
1933	32	Carpenter, 1965	Abundant	Carpenter, 1934	Abundant	Enders, 1935
1951	15	Collias and Southwick, 1952	?	——	?	——
1959	52	Carpenter, 1965	?	——	?	——
1964	67	Eisenberg, unpubl.	58	Eisenberg, unpubl.	3.6	Eisenberg, unpubl.
1970	85	Eisenberg, unpubl.	33	Eisenberg, unpubl.	5.6	Eisenberg, unpubl.
1974	95	Eisenberg, unpubl.	<20	Eisenberg, unpubl.	<5.0	Eisenberg, unpubl.

Polyspecific Associations

In spite of the foregoing remarks on competition, polyspecific associations among neotropical primates do occur. Some associations seem to occur over a wide range of habitats. The association between *Cebus* and *Saimiri* is well known for Colombia (Klein and Klein, 1973), Venezuela (Eisenberg and Rudran, unpublished) and Guyana (Muckenhirn *et al.* 1976). Associations between *Alouatta* and *Ateles* in Colombia may be characteristic of only certain regions (Klein and Klein, 1973), while the associations between *Cebus* and *Ateles* on Barro Colorado Island (Eisenberg and Kuehn, 1966) may have been unique to that particular situation where *Ateles* was introduced. Table 6 summarizes those cases of polyspecific association which are reasonably well documented.

The Social Systems of the Ceboid Primates

Uniquely expressed in callitrichid primates is the phenomenon of obligate monogamy throughout the rearing cycle (Kleiman, 1977). This social system derives from the fact that the male actively participates in carrying the neonates for varying periods of time after their birth. During the rearing cycle the core social group will consist of a mated pair, their newborn young and several subadults or juveniles which may be offspring of the founder pair. The stability of this core unit may be variable depending on the species (Neyman, 1977), but it is the modal social system for all of the Callitrichidae studied to date

Table 6. Polyspecific Associations in Neotropical Primates.

Association	*Characteristics*	*Authority*
1. *Cebus apella* and *Saimiri sciureus*	Nonantagonistic; active	Klein & Klein, 1973
2. *Cebus nigrivittatus* and *Saimiri sciureus*	Nonantagonistic; active	Eisenberg & Rudran, unpubl.
3. *Cebus capucinus* and *Ateles geoffroyi*	Friendly to hostile	Eisenberg & Kuehn, 1966
4. *Alouatta seniculus* and *Ateles belzebuth*	Passive association	Klein & Klein, 1973
5. *Cebus apella* and *Saimiri sciureus*		Muckenhirn, *et al.*, 1976
6. *Saguinus imperator* and *S. fuscicollis*	Nonantagonistic; active	Terborgh & Janson (personal communication)

Table 7. Grouping Tendencies for Some Neotropical Primates

Species	Troop Range	Troop Size (Average or mode)	Temporary Aggregations	Frequent Subgroup Size
Callithrix jacchus	6–13	7	––	––
Cebuella pymaea	4–5	4	20?	––
Saguinus oedipus geoffroyi	4–8	4.5	16	––
S. o. oedipus	4–13	4.5	––	––
S. midas	2–6	4.8	––	––
S. fuscicollis	2–11	6	30–40	1
S. leucopus	3–10	4.8	––	––
Aotus trivirgatus	3–6	4	15–20	1
Callicebus torquatus	2–4	3.5	––	––
Callicebus moloch	2–6	3	––	––
Pithecia monachus	2–6	2	––	1
Pithecia pithecia	2–5	3.1	10	1
Chiropotes satanus	4–18	10.3	––	––
Cacajao melanocephalus	?–30	?	?	?
Saimiri sciureus	8– >43	31	200?	4–10
Cebus capucinus	5–30	11	––	2
C. nigrivittatus	5–42	14.8	––	3
C. albifrons	10–30	16	––	1–5
C. apella	6–30	12	––	3
Alouatta seniculus	2–17	8.5	––	1–3
A. palliata	3–44	14	––	1–2
Lagothrix lagotricha	16–43	24	––	––
Ateles geoffroyi	14– >30	18–20	––	1–5
A. belzebuth	10–30	19.5	––	1–7
A. paniscus	8–25	18.5	––	1–3

including the genera *Cebuella, Callithrix, Saguinus, Leontopithecus* and *Callimico* (Hoage, 1977; Kleiman, 1977). This tendency for shared parental care duties is also characteristic of several genera of cebid primates but the details of the rearing cycle are less well understood. The species which have been investigated and show male carrying are: *Aotus trivirgatus* (Moynihan, 1964) and *Callicebus moloch* (Moynihan, 1976). *Pithecia pithecia* and *P. monachus* grouping tendencies suggest a similar system. The distribution of this behavioral trait in both the Cebidae and Callitrichidae strongly suggests that it is a plesiomorph character of great phylogenetic age (Eisenberg, 1977), and may be

Subgroup Composition			
	Females plus		
Males	*Infants*	*Mixed*	*Authority*
——	——	——	Stevenson, in press
——	——	——	Moynihan, 1976
——	——	——	Dawson, 1977; Moynihan, 1970; Muckenhirn, 1967
——	——	——	Neyman, 1977
——	——	——	Muckenhirn, *et al.*, 1976
——	——	——	Freese, 1975
——	——	——	Green, 1978
——	——	——	Durham, 1975; Moynihan, 1964
——	——	——	Freese, 1975; Izawa, 1976; Kinzey, *et al.*, 1977
——	——	——	Freese, 1975; Robinson, 1977; Mason, 1966
——	——	——	Freese, 1975; Moynihan, 1976; Izawa, 1976
——	——	——	Muckenhirn, *et al.*, 1976; Buchanan (in press)
——	——	——	Muckenhirn, *et al.*, 1976
——	——	——	Hernandez-Camacho & Cooper, 1976
Yes	Yes	Yes	Baldwin & Baldwin, 1972; Klein & Klein, 1976; Izawa, 1976
——	——	——	Oppenheimer, 1968; Eisenberg, unpubl.
Yes	——	——	Eisenberg, unpubl.
——	——	——	Green, 1978; Freese, 1975
?	——	——	Izawa, 1976; Freese, 1975
Yes	——	——	Eisenberg, unpubl.; Neville, 1972; *see* Table 4
Yes	——	——	Various, *see* Table 4
——	——	——	Izawa, 1976; Bernstein *et al.*, 1976
Yes	Yes	Yes	Carpenter, 1935; Freese, 1976
Yes	Yes	Yes	Klein, 1972; Eisenberg, unpubl.; Izawa, 1976
Yes	Yes	——	Muckenhirn, *et al.*, 1976; Durham, 1975

ultimately understood as an adaptation for rearing a litter of considerable neonatal mass (relative to the mother's body weight) when the foraging strategy demands both mobility and that the young be transported continuously from the day of birth (Kleiman, 1977). It should be noted that this parental care strategy has limited the effective group size during the rearing cycle, hence troop size generally falls at the low range for monogamous neotropical primates (see Table 7).

In those species which show strong pair bonds, the pair has a form of duet calling which is reminiscent of duetting in the Old World, monogamous genera *Hylobates* and *Symphalangus* (Lamprecht, 1970; Apfelbach, 1972), as well as the Mentawi langur *Presbytis* (Tilson and Tenaza, 1976). The neotropical monogamous genera for which duetting has been established include *Leontopithecus* (McClanahan and Green, 1977) and *Callicebus* (Robinson, 1977). Duetting in monogamous primates would seem to be a convergence with vocal duetting described for monogamous birds. To quote Armstrong: "Although countersinging and duetting may bear some resemblance acoustically the circumstances . . . are quite different. Countersinging is more or less alternate territorial song; duetting is mutual or reciprocal song by paired birds." (Armstrong, 1973.) Countersinging in primates has been documented for adjacent territorial female gibbons by Tenaza (1976). Alternate long calling by *Alouatta* troops may also be considered countersinging.

From a probable monogamous ancestor the other genera of cebids have developed a variety of social systems some of which show variation according to discrete habitat adaptation (Eisenberg, Muckenhirn, and Rudran, 1972). Central to the evolution of larger social groupings has been the reduction of the natural competitive tendencies that exist between unrelated adults of the same sex class as well as the reduction of competition between the sex classes. (*See* Dittus, 1974, for an analysis of the problem in the cercopithecine *Macaca sinica*.)

As a starting point in addressing the problem of the adaptive nature of social organization, Eisenberg, Muckenhirn, and Rudran (1972) suggested that one might focus on the male-male competitive axis and classify primate social groupings as monogamous, unimale, multimale or age-graded male. The argument implied in this classification is that in the monogamous system an adult male and female bond during a rearing cycle, while in a unimale system one adult male reproductively monopolizes several females. In the age-graded male troop one has the illusion of a multimale troop, but in fact one male does most of the breeding and the younger (or satellite males) are related to the older, founder male. In a true miltimale troop several adult males share reproductive roles; these males may or may not be related. If this classification has validity then one would seek to interpret the selective forces which promote intermale tolerance and cooperation.

From a survey of the primate data one gains the impression that the modal social structure for *Alouatta seniculus*, *Cebus capucinus*, and *Ateles geoffroyi* is either a unimale or age-graded male troop (Neville, 1972; Oppenheimer, 1968; Eisenberg and Kuehn, 1966). *Saimiri* and *Lagothrix* show a multimale troop

(Izawa, 1976; Baldwin and Baldwin, 1972). The genera *Cacajao* and *Chiropotes* may also have multimale troops. Several other factors should be considered in describing these social structures: (1) There is a strong tendency in *Saimiri* during the nonbreeding season for the adult sexes to be somewhat spatially separated as the troop moves about. Thus, unisexual subgrouping is strongly developed in this genus (Baldwin, 1971; Baldwin and Baldwin, 1972). (2) The troop in both *Ateles belzebuth* and *A. geoffroyi* tends to be fractionate into small foraging units, which often tend to be either all male or a female and her semidependent and dependent offspring (Klein, 1972; Carpenter, 1935). Klein argued that this subgrouping tendency reflected an adaptation to foraging on extremely patchy resources. No such long-term fractionation of troops is characteristic of *Lagothrix lagotricha* (Izawa, 1976; Kavanaugh and Dresdale, 1975) (*see also* Table 7).

As a word of caution, however, one should be wary of generalizing from studies of a single species to the attributes of genus or for that matter from a single population to an entire species. For example, we are only beginning to understand the ecology and behavior of the genus *Alouatta,* which is the best studied New World primate. Let us consider diet, troop size, and social structure. Coelho (this volume, Chapter 8) described the diet of *Alouatta villosa pigra* in Guatemala as 90% derived from the *Brosimum* fruit—an observation confirmed by Schlichte (1978). Yet for *A. palliata* (= *villosa*) in Costa Rica and Panama, the studies of Glander (1975), Hladik and Hladik (1969), Carpenter (1934), and Milton (1978) indicate from 40–60% of the diet is leaves. *A. seniculus* shows a similar level of leaf feeding in Colombia and Venezuela (Klein and Klein, 1976; Neville, 1972; Rudran, unpublished).

Troop size in *Alouatta* seems to differ when *A. palliata* and *A. seniculus* are compared. At high densities the troops of *palliata* are consistently larger than the troops of *seniculus.* Since the data represent a tremendous range of habitat types we must conclude that different adaptive strategies are manifested by the two species (*see* Table 8). It is also clear that *A. palliata* shows a range of social structure. Unimale, age-graded, and perhaps multimale troops have been described but these may represent stages in the growth and ultimate fractionation of troops. A clearer case is demonstrable with *Alouatta seniculus* where the unimale condition is modal (Klein, 1972) and the age-graded male troop is intermediate (Eisenberg, unpublished) (*see* Table 8). The unimale condition varies slightly depending on the

Table 8. Troop Size, Composition, and Density for Alouatta.

Species	Locality	$\overline{X}$ Troop Size	Average Number for Each Sex and Age Class						Percent of One-Male Troops	Density (No./km²)	Authority
			Adult Males	Adult Females	Subadult Males	Subadult Females	Juveniles	Infants			
Alouatta palliata	B. C. I., 1932	17.3	2.7	7.4	——	——	3.9	3.1	17%	26.5	Carpenter, 1965
	B. C. I., 1951	7.9	1.2	4,5	——	——	1.0	1.2	80%	15.9	Collias & Southwick, 1952
	B. C. I., 1959	18.5	3.3	9.1	——	——	3.1	2.9	13%	54.2	Carpenter, 1965
	Costa Rica	8.1	1.6	3.6	——	——	1.9	.9	63%	21.5	Heltne & Thorington, 1976
A. seniculus	Colombia	5.0	1.4	2.0	——	——	1.1	.5	75%	20.5	Heltne & Thorington, 1976
	Venezuela	8.4	1.6	2.5	——	——	2.9	1.3	54%	86	Heltne & Thorington, 1976
	Colombia	5.4	1.4	2.2	——	——	1.7	.3	——	<4.0	Izawa, 1976
	Venezuela	8.8	1.7	2.3	1.90	.67	1.2	1.0	——	85	Eisenberg, unpubl.
	Venezuela	8.4	1.6	2.5	.84	.28	1.8	1.4	——	108	Neville, 1972

relative density of *A. palliata*. It would appear that the unimale condition is more widespread in a population of *A. seniculus* but this may be more evident in smaller troops. Part of the problem results from the fact that not all investigators age their males into both subadult and adult classes. When this is not done, the so-called adult male age class may be inflated with the inclusion of younger males. More uniform techniques of aging must be adopted by several investigators before this problem can be resolved (*see* Table 8).

If the situation is unclear for the genus *Alouatta*, surely more data are needed for many other species before a final synthesis of behavioral evolution can be made for the ceboid primates. We badly need field data on *Chiropotes*, *Cacajao*, *Pithecia*, and *Lagothrix* although steps have been taken by D. Buchanan (unpublished) and J. Cassidy for the latter two genera, respectively.

Communication Mechanisms

The New World primates probably employ chemical communication to a greater extent than do most Old World primates (Eisenberg, 1977). Discrete glandular areas and scent-marking movements have been identified for all New World genera (Epple and Lorenz, 1967). The Callitrichidae appear to show a modal tendency to strongly integrate chemical signals into their affect repertoire (Eisenberg, 1977). Visual signals involve coloration, crests, body movements, and facial expressions. Facial mobility in the marmosets, tamarins, and saki monkeys is quite limited, but they have evolved a rich variety of color patterns, crests, and facial hair patterns. Intraspecific selection has played an important role in the evolution of these patterns tending to produce monomorphism in the callitrichids. *Pithecia pithecia* shows a pronounced sexual dimorphism in the facial hair which is snow white in the male. Sexual selection resulting in size dimorphism is not pronounced within the ceboidae, but *Alouatta*, *Lagothrix*, and *Ateles* generally have heavier males. Size dimorphism is most pronounced in *Alouatta* with one species, *A. caraya*, in addition showing color dimorphism; the females are brown and the males black (Krieg, 1948).

Conspicuous differences in the morphology of the external genitalia which may serve as visual signals have been highly developed in *Ateles* where the female possesses a pendulous clitoris (*see* Eisenberg, 1976). This is the most exaggerated case of morphological divergence in the form of the female external genitalia although a similar condition is found in *Lagothrix*. *Cebus*, *Ateles*, and *Lagothrix* have a very rich repertoire of facial expressions.

The vocalizations of the New World primates have been studied in some detail and certain aspects of their adaptive nature have been discussed elsewhere (*see* Eisenberg, 1976; and Oppenheimer, 1977, for a review). One interesting aspect appears to stand out. Most New World primate species have evolved a series of "long calls" which usually have several variations. These long calls may serve (1) to promote spacing among adjacent groups (for example, *Alouatta*, Chivers, 1969; Baldwin and Baldwin, 1976b; *Callicebus*, Robinson, 1977); (2) cohesion among separated subgroups (for example, *Ateles*, Klein, 1972; Eisenberg, 1976); or (3) both (for example, *Ateles*, Eisenberg, 1976; *Cebus*, Oppenheimer, 1968, 1977). A long call involving duetting is characteristic of the monogamous *Leontopithecus* (McClanahan and Green, 1977), and *Callicebus* (Moynihan, 1966; Robinson, 1977). The study of long calls in terms of function and propagation characteristics promises to be very rewarding in the future (*see* Eisenberg, 1976). Of special interest will be the study of variation between populations of the same species. Baldwin and Baldwin (1976b) have prepared the first thorough analysis of vocalizations in *A. palliata* which offers a useful base for comparisons with *A. seniculus* (Eisenberg and Morton, in preparation).

Trends in the evolution of auditory signal systems in the ceboid primates parallel those trends outlined for the Old World cercopithecoids (Oppenheimer, 1977). The importance of vocal signals in the life of forest dwelling primates must not be underestimated, but as yet no firm correlations can be made between the form of the social organization and the types of communication pattern other than the correlation between monogamy and duetting. Chorusing by troop members such as that shown by *Alouatta* may aid in keeping adjacent troops separated. Indeed chorusing may approximate individual countersinging when two troops call at an area of range overlap. Much more experimental research involving playback experiments will be necessary before the various functions of even the well-studied *Alouatta* vocalizations can be understood.

Summary and Conclusions

The larger ceboid primates would seem to consume a considerable amount of young leaves in their diet, at least during part of the annual cycle. *Alouatta* averages the highest leaf consumption of any of the ceboid primates. All of the larger ceboids (*Ateles, Lagothrix,* and *Alouatta*) contribute significant percentages to total primate biomasses for any given area provided the population has not been selectively hunted. Over a wide range of habitats the genus *Alouatta* tends to average a high percentage contribution to the biomass for a given primate community. However, the range of contribution by *Alouatta* is variable. It would appear that the ability of *Alouatta* to attain high biomasses in part results from its ability to utilize leaves as a significant portion of its diet during certain seasons of the year.

High primate diversity with respect to the number of sympatric species at any given area appears to occur when there is good reason to believe that the carrying capacity for primates is high. This generalization does not pertain at the limits of primate distribution when the diversity will be low for zoogeographical reason. Forest succession, however, has a great deal of influence on both primate diversity and carrying capacity. When a forest has been cut and begins to regenerate to a mature growth form, one can discern in the northern neotropics a progression of species which are able to use the forest as it becomes more mature. *Saguinus oedipus* appears to be favored in early forest succession, followed by *Cebus capucinus, C. nigrivittatus,* or *Cebus albifrons*. Mid to early mature succession favors *Alouatta palliata* and *Ateles geoffroyi*. Forests associated with major river systems in the northern neotropics may favor the squirrel monkey, *Saimiri*. Such formations, when surrounded by open savanna, are more appropriately termed gallery forests and may support considerable numbers of both *Saimiri* and *Cebus*.

White sand soils and their drainage systems tend to prove nutrient poor. Nutrient-poor drainage systems (more popularily called black water systems) appear to show low primate carrying capacities. This correlation, however, does not hold on downstream black water systems which drain mosaic habitats. *Alouatta* and *Cebus* appear to be very adaptable genera which have species that have adapted to the extremes of primate distributions in the neotropics. It would appear that the recent establishment of tropical rain forests in the lowlands of the Amazon basin and the Orinoco River system have led to recent faunal exchanges. Niche separation in some communities of neotropical primates may be imperfectly established and interspecific competition may be of a very frequent occurrence in some areas. Certain species dyads appear to show properties of a competitive system at moderate to low carrying capacities. *Cebus apella* appears to compete with *Cebus nigrivittatus* and *Saimiri sciureus* with *Saguinus midas*. In some areas *Ateles geoffroyi* may adversely affect the distribution of *Cebus capucinus*. However, at higher carrying capacities intertaxa associations can be formed between species to create mixed troops. Most notable are troops consisting of both *Saimiri* and *Cebus*.

The social systems of neotropical primates show some unique adaptive features. Monogamy as a rearing strategy appears to be a phylogenetically ancient trait within the Ceboidea. The adoption of this mating and parental care-system, with the exception of the Hylobatidae, is unique within the haplorhinni. Pair bonding and monogamy appears to be the modal rearing strategy of the Callithricidae as well as for the cebid genera *Aotus* and *Callicebus*. The modal social structures of other primate species are difficult to discern. We badly need further long-term studies since the social structure of a primate troop may show a longitudinal progression in its form, passing through a unimale phase to a multimale phase with an intervening age-graded male composition.

During the evolution of the ceboid primates olfactory communication seems to have declined in those genera showing the most advanced morphological features. Auditory communication retains a dominant role throughout all genera of the Ceboidea. Duetting is a special correlate of monogamous bonding and appears to have evolved in a form convergent to duetting shown in the monogamous Old World primates, such as *Hylobates* and *Symphalangus*.

Appendix

When primate densities are expressed in mean numbers per km^2 based on walking transects with a standard detection width, the standard deviation is usually quite high, yet transect census data usually positively covary with the real density and thus such data are useful as indicators of relative abundance. Census data derived from transect walking and actual troop counts are compared in Table I. Given the problems of overestimation and underestimation, transect census data are useful in the analysis of the relative abundance of each species making up a primate community at a given study site. Data from Peru, Bolivia, and Guyana are presented in Table II as back up material for the discussions in the text. The density estimates should be taken as indicators of abundance.

One alternative way to treat transect data is offered here from our results, which may sidestep the problem of high standard deviations when transect data are expressed as average densities. The following discussion derives in part from Caughley (1977:20–23). Set each transect length such that for a known detection width one will cover an area equal to 0.5 the average home range of the species in question. Each transect will then be treated as a single plot. Express the result of each transect as either the presence or absence of a troop or subgroup. Run transects until at least ten samples are available. Given that the transects are randomly placed in the habitat, one can use a Poisson distribution for estimating real density of troops or subgroups per unit area from the percent frequency of troops or subgroups per sampling unit.

The proportion of plots with no troops will be $1\text{-}f = e^{-\bar{x}}$ and the proportion of plots containing one or more groups will be $f = 1\text{-}e^{-\bar{x}}$. Using a table of exponentials the value of $-\bar{x}$ can be found. Table III includes a transformation of some of our previous data utilizing this method. Troops per unit area can be converted to individuals per unit area by multiplying by mean troop size.

Table I. Some Replicated Density Estimates for Neotropical Primates

Location and Species	Method of Estimation							Authority
	Actual Troop Count & Mapping				Transect			
	Ecological		Crude			Detection	No./km²	
	Base area	No./km²	Base area	No./km²	$\overline{X}$ Length[a]	Width	$\overline{X}$ (sd)	
I. Venezuela								
A. Hato Masaguaral								
1. *Alouatta seniculus*								
(a) West side of highway								
1969–70 East & Middle Forests	97 ha	108	——	——	——	——	——	Neville, 1972
1972 West, East & Middle Forests	160 ha	100	——	——	——	——	——	Neville, 1976
1975 East & Middle Forests	97 ha	87	——	——	——	——	——	Eisenberg & Kleiman, unpubl.
1975 East & Middle Forests	113 ha	75	200 ha	42	(2) 1.4 km	150 m	88 (±3.0)	Eisenberg, unpubl.
(b) Guarico Forest								
1975 North End	200 ha	48[b]	——	——	(2) ?	?	70[b]	Green, unpubl.
1976 South End	80 ha	8	——	——	——	——	——	Eisenberg, unpubl.
1976 Total Gallery Estimate	6 km²	40[b]	11 km²	20	——	——	——	Eisenberg, unpubl.
2. *Cebus nigrivittatus*								
(a) Guarico Forest								
1975 North End	200 ha	65[b]	——	——	——	——	——	Green, unpubl.
1976 South End	100 ha	23[b]	——	——	——	——	——	Eisenberg, unpubl.
1976 Total (Estimate)	6 km²	44[b]	11 km²	25	——	——	——	
B. Guatopo National Park								
1. *Alouatta seniculus*								
(a) North End	400 ha	20[b]	——	——	(4) 8 km	25 m	16	Eisenberg, unpubl.
North End	——	——	——	——	(2) ?	?	9	Green, unpubl.
(b) South End	——	——	100 ha	6	——	——	——	Eisenberg, unpubl.
2. *Cebus nigrivittatus*								
(a) North End	400 ha	22.5[b]	——	——	(4) 8 km	25 m	27	Eisenberg, unpubl.
North End	——	——	——	——	?	?	39[b]	Green, unpubl.

(b) South End	100 ha	35	——	——	——	——	——	Eisenberg, unpubl.
3. *Ateles belzebuth*								
(a) North End	——	——	400 ha	9[b]	(4) 8 km	25 m	5.6	Eisenberg, unpubl.
II. Panama								
A. B. C. I.								
1. *Alouatta palliata*								
1959	——	——	15 km²	52	——	——	——	Carpenter, 1965
1964	——	——	——	——	(11) 8.1 km	50 m	67	Eisenberg, unpubl.
1967	3.8 km²	70	——	——	——	——	——	Chivers, 1969
1970	——	——	——	——	(7) 9.7 km	50 m	85 (± 35)	Eisenberg, unpubl.
1974	——	——	——	——	(6) 8.0 km	50 m	95	Eisenberg, unpubl.
2. *Cebus capucinus*								
1964	——	——	——	——	(11) 8.1 km	50 m	58	Eisenberg, unpubl.
1967	90 ha	17	15 km²	12	——	——	——	Oppenheimer, 1968
1970	——	——	——	——	(7) 9.7 km	50 m	33.3 (± 20.3)	Eisenberg, unpubl.
1974	——	——	——	——	(6) 8.0 km	50 m	16.7	Eisenberg, unpubl.
3. *Ateles geoffroyi*								
1964	100 ha	10[c]	15 km²	0.3	——	——	——	Eisenberg & Kuehn, 1966
1970	100 ha	12	15 km²	0.9	——	——	——	Eisenberg, unpubl.
1972	110 ha	10	——	——	——	——	——	Dare, 1974
1974	150 ha	13	15 km²	1.1	——	——	——	Eisenberg, unpubl.
4. *Saguinus oedipus*								
1964	——	——	——	——	(11) 8.1 km	50 m	3.6	Eisenberg, unpubl.
1965	——	——	2.5 km²	5.6	(6) 7.1 km	——	——	Muckenhirn, 1967
1970	——	——	——	——	(6) 8.0 km	50 m	4.7	Eisenberg, unpubl.

[a]( ) refer to number of replicates.
[b]Extrapolation from known troop average size.
[c]Actually only five animals occupied about 45 ha; value is nevertheless recorded in theoretical number/km².

Table II. Density Estimates and Biomass Conversions for Guyana,

Area	Alouatta		Cebus		Ateles		Aotus	
	No./km²	kg/km² @ 5.5	No./km²	kg/km² @ 2.6	No./km²	kg/km² @ 5.0	No./km²	kg/km² @ 0.8
Guyana[b]								
Berbice River	4.7	26	13.7	36	4.9	25	——	——
24-Mile Reserve	5	28	11	29	——	——	——	——
Mahdia	9.6	52.8	——	——	——	——	——	——
Issano Rd.	——	——	——	——	——	——	——	——
HMPS	4.9	27	8.2	21	——	——	——	——
Moraballi	——	——	——	——	——	——	——	——
Apoteri Area	3.6	20	10.6	28	——	——	——	——
Mathews Ridge	1.6	9	9	23	——	——	——	——
Pakani Area	19.5	107	21.4	56	8.1	40.5	——	——

Area	Alouatta		Lagothrix		Ateles		Cebus apella		Cebus albifrons	
	No./km²	kg/km²	No./km²	kg/km²	No./km²	kg/km²	No./km²	kg/km²	No./km²	kg/km²
Peru[c]										
Saimiria	29.5	162	7	38.5	——	——	24	62.4	2	5.2
Cocha cashu	24	132	——	——	22.4	112	36	93.6	24	62
Orosa	——	——	——	——	——	——	17	44.2	——	——
Bolivia[d]										
El Triunfo	120	660	——	——	2	10	55	143	——	——
Rio Acre	——	——	——	——	——	——	7	18.2	——	——
Ixiamas	8	44	——	——	8.5	42.5	——	——	——	——
San Jose	0.8	4.4	——	——	——	——	2	5.2		

[a]A complete analysis of the Peru and Bolivia data is included in: C. H. Freese, 1977. *Population Densities and Niche Separation in Some Amazonian Monkey Communities*. Ph.D. Dissertation, Johns Hopkins University, Baltimore, Md.
[b]Muckenhirn, *et al.*, 1976.

Table III. Comparison of Density Estimates.

Locality and Species	*Transect Density (Number/km²)*	*Plot Density Estimate (Number/km²)*
1. Panama, 1970		
Alouatta	95	88
Cebus	33	30
2. Venezuela		
Alouatta	16	16
Cebus	27	34
Ateles	5.6	5.7

Bolivia, and Peru.[a]

Chiropotes		Pithecia		Saimiri		Saguinus		Total	Total
No./km²	kg/km² @1.2	No./km²	kg/km² @1.1	No./km²	kg/km² @0.3	No./km²	kg/km²	Individuals /km²	Biomass kg/km²
7.6	9	——	——	——	——	2.3	1.4	33	97
——	——	——	——	——	——	——	——	16	56[e]
——	——	——	——	——	——	——	——	9.3	53[e]
——	——	6.5	7.2	——	——	——	——	6.5	7.2[e]
——	——	12	13	41.5	33	——	——	67	94
——	——	——	——	33.8	27	5.7	3.4	40	30[e]
——	——	5.9	6.5	20.8	17	13.9	8.3	55	79
——	——	2.6	3	——	——	——	——	13.2	35[e]
12.7	15.2	——	——	——	——	3.7	2.2	65.4	221

Callicebus 0.9		Pithecia		Saimiri		Saguinus fuscicollis		Saguinus imperator		Total	Total
No./km²	kg/km²	No./km²	kg/km²	No./km²	kg/km²	No./km²	kg/km²	No./km²	kg/km²	Individuals /km²	Biomass kg/km²
——	——	9	10	72	58	15	9	——	——	158.5	345
2.1	1.9	——	——	84	67	10.8	6.5	5.4	3.2	208	479
13	11.7	4.5	4.9	36	32.4	9	5.4	——	——	80	98.6
——	——	——	——	100	90	——	——	*S. labiatus*		287	903
14.7	13.2	6.6	7.2	24	21.6	27.6	16.6	27.6	16.6	107	93.4
——	——	——	——	——	——	3.6	2.2	*C. argentata*		20.1	88.7
——	——	——	——	——	——	——	——	5.5	3.3	8.3	12.9

[c] Freese, 1975.
[d] Heltne, *et al.*, 1976.
[e] Black water areas.

Acknowledgments

Obviously this paper could not have been written without the input and data of the numerous individuals who are cited throughout the text. I owe a great intellectual debt to a number of colleagues who have sharpened the questions I put to the data. My wife, Devra G. Kleiman, has been the guiding force in our marmoset research, and her symposium in August 1975 was decisive in formulating some of the ideas concerning monogamy in the Callitrichidae. Margaret Ann O'Connell first pointed out to me the relevance of James H. Brown's work to my thinking. Nancy Muckenhirn, Martin Moynihan, and Richard Thorington have exchanged ideas so often with me that I cannot disentangle my own from theirs. Ken Green and Lewis Klein have done their best to acquaint me with primatology in Colombia. Bob Cooper and Jorge Hernandez-Comacho helped me immensely in Colombia as did R. Gonzala. Edgardo Mondolfi, Pedro Trebbau, Juan Gomez-Nunez, and Tomas Blohm made our work possible in Venezuela. Thanks are warmly rendered to Rafael Garcia and the staff of the Parque Nacional Guatopo. R. Rudran and Wolfgang Dittus converted our field primatology efforts from the descriptive to the advanced quantitative stage. Peter Marler and Tom Struhsaker have had their influence through discussion and example, and to them I offer my deepest gratitude. This research was made possible by a grant from the International Environmental Sciences Program of the Smithsonian Institution and Grant Number MH 28840 from the National Institutes of Mental Health. The author gratefully acknowledges the aid and courtesy shown by the Wenner-Gren Foundation in supporting the symposium which led to the preparation of this paper.

Notes

1. As many as 14 have been reported in macrosympatry, but are unconfirmed in microsympatry (Terborgh, personal communication).
2. Drought is defined according to the climate diagram developed by Walter (1973). In the tropics a period exceeding 4 weeks with rainfall less than 30 mm/month approaches drought conditions.
3. Certain forms of gallery forest may provide partial exceptions to this rule.
4. *Cebus albifrons, C. nigrivittatus; C. capucinus.*

References

Altmann, S. A. 1959. Field observations on a howling monkey society. *J. Mammal.* 40:317–330.

Alvarez, T. 1963. The recent mammals of Tamaulipus, Mexico. *Univ. Kansas Publ. Mus. Nat. Hist.* 14(15):363–473.

Apfelbach, R. 1972. Electrically elicited vocalizations in the gibbon *Hylobates lar* and their behavioral significance. *Z. Tierpsychol* 30: 420–430.

Armstrong, E. A. 1973. *A study of bird song.* 325 pp. New York: Dover.

Baldwin, J. D. 1971. The social organization of a semifree-ranging troop of squirrel monkeys (*Saimiri sciureus*). *Folia Primat.* 14:23–50.

Baldwin, J. D., and Baldwin, J. I. 1972. The ecology and behavior of squirrel monkeys (*Saimiri oerstedi*) in a natural forest in western Panama. *Folia Primat.* 18:161–184.

Baldwin, J. D., and Baldwin, J. I. 1976a. Primate populations in Chiriqui, Panama. Pages 20–31 in *Neotropical primates: Field studies and conservation.* Eds. R. W. Thorington, Jr. and P. G. Heltne. Washington, D.C.: National Academy of Sciences.

Baldwin, J. D., and Baldwin, J. I. 1976b. Vocalizations of howler monkeys (*Alouatta palliata*) in southwestern Panama. *Folia Primat.* 26:81–108.

Bernstein, I. S., Balcaen, P., Dresdale, L., Gouzoules, H., Kavanagh, M., Patterson, T., Neyman-Warner, P. 1976. Differential effects of forest degradation on primate populations. *Primates* 17:401–411.

Buchanan, D. In press. The genus *Pithecia.* In *Introduction to new world primatology.* Eds. R. A. Mittermeier and A. F. Coimbra-Filho. Washington, D.C.: Smithsonian Inst. Press.

Carpenter, C. R. 1934. A field study of the behavior and social relations of howling monkeys. *Comp. Psychol. Mon.* 10(2):1–168.

Carpenter, C. R. 1935. Behavior of red spider monkeys in Panama. *J. Mammal.* 16:171–180.

Carpenter, C. R. 1965. The howlers of Barro Colorado Island. Pages 250–291 in *Primate behavior: Field studies of monkeys and apes.* Ed. I. DeVore. New York: Holt, Rinehart and Winston.

Castro, R., and Soini, P. 1977. Field studies on *Saguinus mystax* and other callitrichids in Amazonian Peru. Pages 73–78 in *The biology and conservation of the Callitrichidae.* Ed. D. G. Kleiman. Washington, D.C.: Smithsonian Inst. Press.

Caughley, G. 1977. *Analysis of vertebrate populations.* New York: Wiley.

Chivers, D. J. 1969. On the daily behaviour and spacing of howling monkey groups. *Folia Primat.* 10:48–103.

Clutton-Brock, T., and Harvey, P. 1977. Primate ecology and social organization. *J. Zool.* 183:1–39.

Coelho, A. 1974. Energy budget of the Guatemalan howler and spider monkey: A socio-bioenergetic analysis. Ph.D. dissertation, University of Texas, Austin.

Collias, N., and Southwick, C. 1952. A field study of population density and social organization in howling monkeys. *Proc. Amer. Phil. Soc.* 96:143–156.

Dare, R. 1974. The social behavior and ecology of spider monkeys *Ateles geoffroyi* on Barro Colorado Island. Ph.D. dissertation, University of Oregon.

Dawson, G. A. 1977. Composition and stability of social groups of the tamarin, *Saguinus oedipus geoffroyi* in Panama. Pages 23–38 in *The biology and conservation of the Callitrichidae*. Ed. D. G. Kleiman. Washington, D.C.: Smithsonian Inst. Press.

Dittus, W. J . P. 1974. The ecology and behavior of the toque monkey, *Macaca sinica*. Ph.D. dissertation, University of Maryland.

Durham, N. M . 1971. Effects of altitude differences on group organization of wild black spider monkeys (*Alteles paniscus*). *Proc. 3rd. Int. Congr. Primatol.* 3:32–40.

Durham, N. M. 1975. Some ecological, distributional, and group behavioral features of Atelinae in Southern Peru. Pages 87–101 in *Socioecology and psychology of primates*. Ed. R. H. Tuttle. The Hague: Mouton.

Eisenberg, J. F. 1976. Communication mechanisms and social integration in the black spider monkey, *Ateles fusciceps robustus*, and related species. *Smithson. Contrib. Zool.* 213:1–108.

Eisenberg, J. F. 1977. Comparative ecology and reproduction in New World primates. Pages 13–22 in *The biology and conservation of Callitrichidae*. Ed. D. G. Kleiman. Washington, D.C.: Smithsonian Inst. Press.

Eisenberg, J . F. 1978. The evolution of arboreal herbivores in the class Mammalia. Pages 135–152 in *The ecology of arboreal folivores*. Ed. G. G. Montgomery. Washington, D.C.: Smithsonian Inst. Press.

Eisenberg, J . F., and Kuehn, R. E. 1966. The behavior of *Ateles geoffroyi* and related species. *Smithson. Misc. Collect.* 151(8):1–63.

Eisenberg, J . F., Muckenhirn, N. A., and Rudran, R. 1972. The relationship between ecology and social structure in primates. *Science* 176: 863–874.

Eisenberg, J. F., and Thorington, R. W. 1973. A preliminary analysis of a neotropical mammal fauna. *Biotropica* 5:150–161.

Enders, R. K. 1935. Mammalian life histories from Barro Colorado Island, Panama. *Bull. Mus. Comp. Zool.* Harv. Univ. 78:385–502.

Epple, G., and Lorenz, R. 1967. Vorkommen, Morphologie und Funktion der Sternaldrüse bei den Platyrrhini. *Folia Primat.* 7:98–126.

Fittkau, E. J., Irmler, U., Junk, W. J., Reiss, R., and Schmidt, G. W. 1975. Productivity, biomass, and population dynamics in Amazonian water bodies. Pages 289–311 in *Ecological studies, vol. 11, Tropical ecological systems*. Eds. F. B. Golley and E. Medina. Berlin: Springer Verlag.

Freese, C. 1975. A census of non-human primates in Peru. Pages 17–42 in *Primate censusing studies in Peru and Colombia*. Washington, D.C.: Pan American Health Organization.

Freese, C. 1976. Censusing *Alouatta palliata*, *Ateles geoffroyi*, and *Cebus capucinus* in the Costa Rican dry forest. Pages 4–9 in *Neotropical primates: Field studies and conservation.* Eds. R. W. Thorington, Jr. and P. G. Heltne. Washington, D.C.: National Academy of Sciences.

Galindo, P., and Srihongse, S. 1967. Evidence of recent jungle yellow-fever activity in eastern Panama. *Bull. W. H. O.* 36:151–161.

Glander, K. E. 1975. Habitat and resource utilization: An ecological view of social organization in mantled howling monkeys. Ph.D. dissertation, University of Chicago.

Green, K. M. 1978. Neotropical primate censusing in northern Colombia. *Primates* 19:537–550.

Handley, C. O., Jr. 1966. A checklist of the mammals of Panama. Pages 753–795 in *Ectoparasites of Panama.* Eds. R. L. Wenzel and V. J . Tipton. Chicago: Field Museum of Natural History.

Handley, C. O., Jr. 1976. Mammals of the Smithsonian Venezuelan project. *Brigham Young Univ. Sci. Bull. Biol. Ser.* 20(5):1–91.

Heltne, P. G., Freese, C., and Whitesides, G. 1976. *A field survey of nonhuman primate populations in Bolivia.* Washington, D.C.: Pan American Health Organization.

Heltne, P. G., and Thorington, R. W., Jr. 1976. Problems and potentials for primate biology and conservation in the new world. Pages 110–124 in *Neotropical primates: Field studies and conservation.* Eds. R. W. Thorington, Jr. and P. G. Heltne. Washington, D.C.: National Academy of Sciences.

Hernandez-Camacho, J., and Cooper, R. W. 1976. The nonhuman primates of Colombia. Pages 35–69 in *Neotropical primates: Field studies and conservation.* Eds. R. W. Thorington, Jr. and P. G. Heltne. Washington, D.C.: National Academy of Sciences.

Hladik, A., and Hladik, C. M. 1969. Rapports trophiques entre végétation et primates dans la forêt de Barro Colorado (Panama). *La Terre et la Vie* 116:25–117.

Hoage, R. J . 1977. Parental care in *Leontopithecus r. rosalia.* Pages 293–306 in *The biology and conservation of the Callitrichidae.* Ed. D. G. Kleiman. Washington, D.C.: Smithsonian Inst. Press.

Izawa, K. 1976. Group sizes and compositions of monkeys in the upper Amazon basin. *Primates* 17:367–399.

Janzen, D. 1974. Tropical black water rivers, animals and mast fruiting by the Dipterocarpaceae. *Biotropica* 6(2):69–103.

Kavanaugh, M., and Dresdale, L. 1975. Observations on the woolly monkey *Lagothrix lagotricha* in northern Colombia. *Primates* 16:285–294.

Kinzey, W. G. 1977. Diet and feeding behavior in *Callicebus torquatus.* Pages 127–152 in *Primate ecology.* Ed. T. H . Clutton-Brock. New York: Academic Press.

Kinzey, W. G., Rosenberger, A. L., Heisler, P. S., Prowse, D. L., and

Trilling, J. S. 1977. A preliminary field investigation of the yellow-handed titi monkey *Callicebus torquatus torquatus* in northern Peru. *Primates* 18:159–181.

Kleiman, D. G. 1977. Monogamy in mammals. *Quart. Rev. Biol.* 52: 39–69.

Klein, L. L. 1972. The ecology and social organization of the spider monkey, *Ateles belzebuth.* Ph.D. dissertation, University of California, Berkeley.

Klein, L. L., and Klein, D. J . 1973. Observations on two types of neotropical primate intertaxa associations. *Am. J. Phys. Anthrop.* 38: 649–653.

Klein, L. L., and Klein, D. J. 1975. Social and ecological contrasts between four taxa of neotropical primates. Pages 59–85 in *Socioecology and psychology of primates.* Ed. R. H. Tuttle. The Hague: Mouton.

Klein, L. L., and Klein, D. J. 1976. Neotropical primates: Aspects of habitat usage, population density, and regional distribution in La Macarena, Colombia. Pages 70–78 in *Neotropical primates: Field studies and conservation.* Eds. R. W. Thorington, Jr. and P. G. Heltne. Washington, D.C.: National Academy of Sciences.

Knight, D. H . 1975. A phytosociological analysis of species-rich tropical forest on Barro Colorado Island, Panama. *Ecol. Mono.* 45:259–284.

Krieg, H. 1948. *Zwischen Anden und Atlantik.* Munich: Hanser.

Lamprecht, J. 1970. Duetgesang beim Siamang. Z. *Tierpsychol.* 27:186–204.

Mason, W. 1966. Social organization of the South American monkey, *Callicebus molloch. Tulane Stud. in Zool.* 13:23–28.

Mason, W. 1971. Field and laboratory studies of social organization in *Saimiri* and *Callicebus.* Pages 107–137 in *Primate behavior: Developments in field and laboratory research. vol. II.* Ed. L. A. Rosenblum. New York: Academic Press.

McClanahan, E. B., and Green, K. M. 1977. The voccal repertoire and an analysis of the contexts of vocalizations in *Leontopithecus rosalia.* Pages 251–270 in *The biology and conservation of Callitrichidae.* Ed. D. G. Kleiman. Washington, D.C.: Smithsonian Inst. Press.

McKey, D. 1978. Soils, vegetation and seed-eating by black colobus monkey. In *The ecology of arboreal folivores.* Ed. G. G. Montgomery. Washington, D.C.: Smithsonian Inst. Press.

Milton, K. 1978. Behavioral adaptations to leaf eating by the mantled howler monkey (*Alouatta palliata*). In *The ecology of arboreal folivores.* Ed. G. G. Montgomery. Washington, D.C.: Smithsonian Inst. Press.

Moynihan, M. 1964. Some behavior patterns of platyrrhine monkeys. I. The night monkey (*Aotus trivirgatus*). *Smithson. Misc. Collect.* 146(5):1–84.

Moynihan, M. 1966. Communication in the titi monkey, *Callicebus. J. Zool. Soc. Lond.* 150:77–127.

Moynihan, M. 1970. Some behavior patterns of platyrrhine monkeys, II: *Saguinus geoffroyi* and some other tamarins. *Smithson. Contrib. Zool.* 28:1–77.

Moynihan, M. 1976. *The new world primates.* Princeton, N.J.: Princeton University Press.

Muckenhirn, N. A. 1967. The behavior and vocal repertoire of *Saguinus oedipus* (Hershkovitz, 1966) (Callithricidae, Primates). M.S. thesis, University of Maryland.

Muckenhirn, N. A., Mortensen, B., Vessey, S., Fraser, C. E., and Singh, B. 1976. *Report on a primate survey in Guyana.* Washington, D.C.: Pan American Health Organization.

Neville, M. K. 1972. The population structure of red howler monkeys (*Alouatta seniculus*) in Trinidad and Venezuela. *Folia Primat.* 17: 56–86.

Neville, M. K. 1975. Census of primates in Peru. Pages 3–16 in *Primate censusing studies in Peru and Colombia.* Washington, D.C.: Pan American Health Organization.

Neville, M. K. 1976. The population and conservation of howler monkeys in Venezuela and Trinidad. Pages 101–109 in *Neotropical primates: Field studies and conservation.* Eds. R. W. Thorington, Jr. and P. G. Heltne. Washington, D.C.: National Academy of Sciences.

Neyman, P. 1977. Aspects of the ecology of free-ranging cotton-top tamarins (*Saguinus o. oedipus*). Pages 39–72 in *The biology and conservation of the Callitrichidae.* Ed. D. G. Kleiman. Washington, D.C.: Smithsonian Inst. Press.

Oppenheimer, J. R. 1968. Behavior and ecology of the white-faced monkey, *Cebus capucinus*, on Barro Colorado Island, Canal Zone. Ph.D. dissertation, University of Illinois, Urbana.

Oppenheimer, J. R. 1977. Communication in new world monkeys. Pages 851–889 in *How animals communicate.* Ed. T. Sebeok. Bloomington: University of Indiana Press.

Pope, B. L. 1966. Population characteristics of howler monkeys (*Alouatta caraya*) in northern Argentina. *Amer. J. Phys. Anthrop.* 24: 361–370.

Robinson, J. G. 1977. The use of vocalizations to regulate the spatial relationships both within and between groups in a territorial, monogamous primate the titi monkey *Callicebus moloch.* Ph.D. dissertation, University of North Carolina, Chapel Hill.

Schlichte, H. J . 1978. A preliminary report on the habitat utilization of a group of howler monkeys (*Alouatta villosa pigra*) in the national park of Tikal, Guatemala. Pages 551–559 in *The ecology of arboreal folivores.* Ed. G. G. Montgomery. Washington, D.C.: Smithsonian Inst. Press.

Simpson, V. B. 1971. Pleistocene changes in the fauna and flora of South America. *Science* 173:771–780.

Sioli, H. 1975. Tropical rivers as expressions of their terrestrial environments. Pages 275–288 in *Ecological studies, vol. 11, tropical ecological systems.* Eds. F. B. Golley and E. Medina. Berlin: Springer Verlag.

Smith, C. C. 1977. Feeding behaviour and social organization in howling monkeys. Pages 97–126 in *Primate ecology.* Ed. T. Clutton-Brock. New York: Academic Press.

Stevenson, M. F. In press. The behaviour and ecology of the common marmoset, *Callithrix jacchus jacchus* in its natural environment. In *The biology and conservation of the Callitrichidae II.* Ed. H. Rothe.

Struhsaker, T., Glander, K., Chirivi, H., and Scott, N. J. 1975. A survey of primates and their habitats in northern Colombia. Pages 43–78 in *Primate censusing studies in Peru and Colombia.* Washington, D.C.: Pan American Health Organization.

Tenaza, R. 1976. Songs, choruses and countersinging of Kloss' gibbons (*Hylobates klossii*) in Siberut Island, Indonesia. *Z. Tierpsychol.* 40:37–52.

Thorington, R. W., Jr., and Heltne, P. G. (Eds.) 1976. *Neotropical primates: Field studies and conservation.* Washington, D.C.: National Academy of Sciences. 135 pp.

Tilson, R., and Tenaza, R. 1976. Monogamy and duetting in an old world monkey. *Nature* 263:320–321.

Trapido, H., and Galindo, P. 1956. The epidemiology of yellow fever in middle America. Exp. Parasitol. 5:285–323.

Walter, H. 1973. *Vegetation of the earth in relation to climate and the eco-physiological conditions.* London: English Universities Press, Ltd. and New York: Springer Verlag.

World Weather Records, 1951–1960, Vol. 3. 1951–1960. Washington, D.C.: U.S. Department of Commerce.

Chapter 11

Biological Parameters and Pleistocene Hominid Life-Ways

C. Loring Brace

Having examined a number of environmental factors which influence social organization and behavior in a variety of nonhuman primate taxa, we now return to look at our own species, but this time from an evolutionary perspective. In this chapter, Dr. Brace will take us through three million years of our evolutionary past in considering the extent our environment has shaped our social organization and behavior. Based primarily on the available dental material and the theoretical implications relative to diet, Dr. Brace offers some tentative conclusions about some of the significant factors in our evolution.

The first major event in our evolutionary past, which would separate us from the rest of our primate relatives, was acquisition of bipedal locomotion. From the fossil evidence, we know that our locomotory patterns have been essentially unchanged for the last three million years. The interesting question is what advantage did bipedalism confer in the early evolutionary stages. Hominid bipedalism has no locomotor advantage when compared with the rest of the locomotory modes available, but did result in the freeing of the hands for nonlocomotor functions. With the freeing of the hands, we see the evolution of the manufacture of stone tools about two million years ago. The next major event, systematic hunting, does not appear in the fossil record until much later. It is with the acquisition of this hunting life-style that we see the transition from the australopithecines to members of our own genus. Human hunting relies more on guile than physical prowess and, therefore places a premium on learning ability and efficient communication. As a consequence, the acquisition of systematic hunting is hypothesized to have resulted in profound influences on the brain and neural development, increased learning abilities, and so on.

Brace now invites us to test his hypotheses against the available data and begins by correlating each of these significant evolutionary events with changes in dentition. He concludes that careful study of hominid dental remains can go a long way in elucidating the interaction between our environment and the evolution of our social behavior.

"Man the hunter" is a made-over ape, and many of the specific differences between pongids and hominids may derive from this retooling job even though much evidence suggests it was an incomplete one. The expanded brain, the capacity for language, the vastly increased importance of learned behavior, the hairless, sweaty skin, and the adjustment of our digestive physiology are all major departures from the presumed ancestral condition. Note that I have not mentioned our mode of locomotion and our compulsive and habitual dependence on tools. In both of these aspects we also depart from our closest pongid relations, but from a reading of the fossil record, it would appear that the latter of these had become incorporated in the ancestral hominid line long before there is any evidence for predatory behavior.

The skeletal remains from the Afar region of Ethiopia discovered as a result of the work of D. C. Johanson (Taieb *et al.*, 1974, 1975; Johanson and Taieb, 1976) display evident hominid features—bipedalism and nonprojecting canine teeth—at a time fully a million years before the oldest known stone tools, and two million years before we have evidence that deliberate hunting of large prey had become an important hominid activity (Isaac, 1976).

The material is really most interesting and deserves all the attention it has received. I mention it in conjunction with a caveat, however, since there have been consistent attempts to portray it as well as other material from East Africa (and elsewhere) in terms of a paradigm that at bottom is reluctant to accept Darwinian evolution as applicable to the human line. The oft-repeated argument that the European Neanderthals could not have been the ancestors of Modern Europeans because populations of modern form already existed, has been projected back to

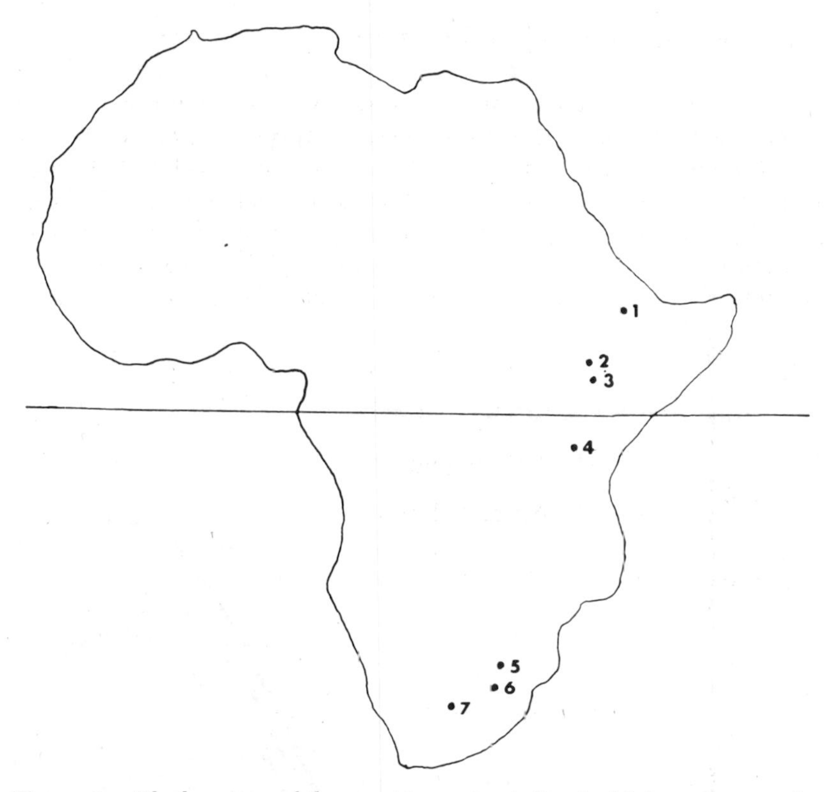

Figure 1. The location of the most important sites in Africa where early hominids have been discovered. (1) Hadar in the Afar Depression, Ethiopia. (2) The Omo Valley of southern Ethiopia. (3) The East Rudolf (Turkana) area of northern Kenya. (4) Olduvai Gorge in northern Tanzania. On a map of this scale, Laetolil is in practically the same spot. (5) The Makapan Valley in the northern Transvaal of South Africa. (6) Swartkrans, Sterkfontein and Kromdraai in the central Transvaal of South Africa. (7) Taung, the site of the original *Australopithecus* discovery.

the late Pliocene/early Pleistocene. Australopithecines, according to this interpretation, could not have been ancestral to late hominids because "true" *Homo* already existed. The reader should be warned, however, that both the advocates and the opponents of this view tend to owe their interpretive perferences more to theoretical expectations than to a careful comparative analysis of the available data. But if it is still too early to make a definitive choice between competing interpretations, it is not inappropriate to indicate something concerning what our expectations ought to be.

The Tool Dependency Revolution

The Afar hominids give hints of a pongid ancestry not evident later on. Arm length appears to be relatively long in comparison to leg length. The canine teeth show a cervical-occlusal elongation that has an echo in one or two specimens from Sterkfontein (Sts 50 and maybe 51) but not in more recent hominids. As can be seen in Figure 2, even the cross-sectional area of the canine of the earliest East African specimens is larger in an absolute sense

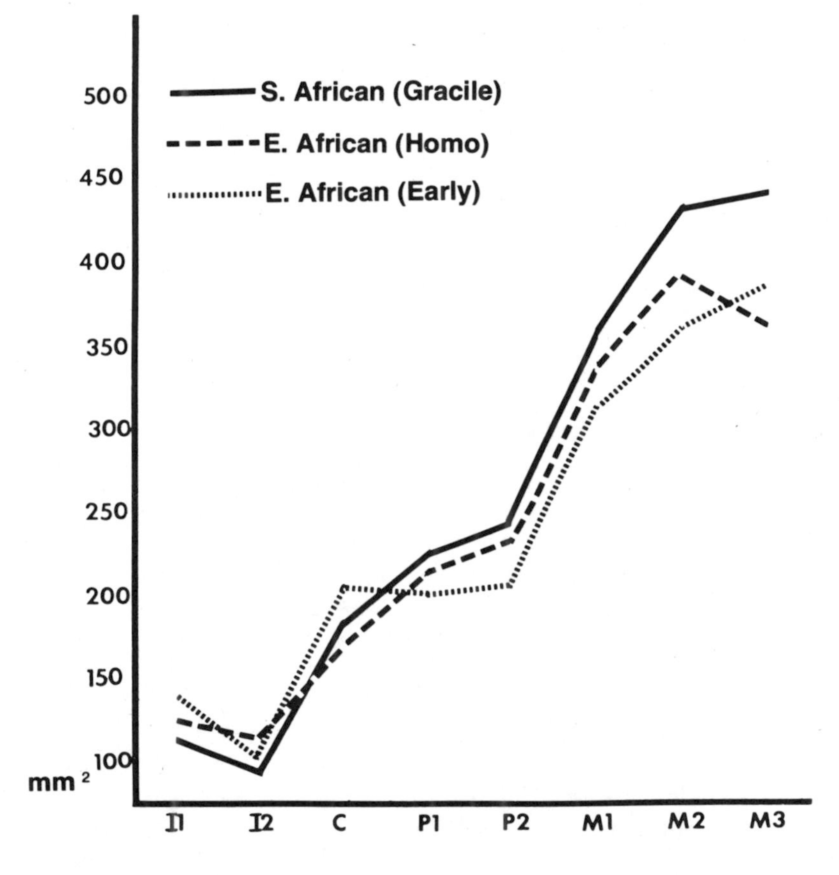

Figure 2. Composite tooth-size profile of cross-sectional areas of (mesial-distal x buccal-lingual = MD x BL) where each point represents the sum of both the upper and lower tooth areas ($I1 = I^1 + I_1$; $I2 = I^2 + I_2$; ... $M3 = M^3 + M_3$). South African graciles include all the material from Sterkfontein and Makapan; East African "Homo" includes all the published material from Olduvai, Omo, and East Turkana tentatively given that designation; and East African Early includes the recent material from Afar and Laetolil. The figures on which this graph is based are listed in Table 1.

(201 mm as opposed to 183 mm for South African graciles, 180 mm for what is being claimed as early *Homo* in East Africa, and 156 and 152 mm for South and East African robusts) than for any other hominid in the fossil record (the late third interglacial Neanderthals from Krapina in Jugoslavija at 173 mm, *see* Table 1).

In addition, the lower first premolar shows a degree of mesial-distal elongation and a mesio-buccal to disto-lingual orientation of the long axis that approaches the form of the sectorial tooth of primates in which projecting canines are characteristic. Finally there is a thickening visible low down on the lingual face of the mandibular symphysis which, if not homologous to a simian shelf, certainly appears to be its functional analogue. Indeed, if we were to judge on the basis of the jaws and teeth alone, we might reasonably conclude that the material was pongid, even if not developed to the extent that we think of as typical for modern representatives. This assessment is based upon an examination of descriptions and illustrations

Table 1. The Composite (Upper and Lower) Cross-sectional Area (MD x LL) Figures on which Figures 2, 3, and 5 Are Based.

	I1	I2	C	P1	P2	M1	M2	M3
E. African Early[a]	139.3	100.9	201.0	205.3	201.1	310.0	368.1	381.0
E. African "Homo"[b]	128.6	104.8	169.2	217.3	225.8	346.7	285.2	367.5
S. African Gracile[c]	111.2	93.6	183.2	224.2	235.7	360.2	439.3	447.9
S. African Robust[d]	105.5	93.8	156.2	253.3	306.2	387.9	456.4	490.9
E. African Robust[e]	99.8	88.0	151.6	307.1	365.5	484.1	570.7	578.1
Gorilla[f]	202.9	181.4	509.4	369.9	309.7	431.8	518.2	475.4
Pan[f]	180.1	160.1	278.8	168.0	145.1	225.7	236.6	208.4
Papio[g]	155.2	113.3	191.4	138.6	117.2	191.4	264.1	314.4

[a]The East African early group is composed of specimens from Afar and Laetolil and the measurements are reported in Johanson and Coppens (1976), Johanson and Taieb (1976), and White (1977).

[b]The East African "Homo" group consists of specimens from Olduvai measured by this author and specimens from East Turkana reported in Day and Leakey (1973), and Leakey and Wood (1973, 1974).

[c]The South African gracile group is composed of the specimens from Makapan and Sterkfontein.

[d]The South African robusts are from Kromdraai and Swartkrans, and the measurements were taken by this author.

[e]The East African robust measurements are reported in Day, *et al.* (1976), Hurford (1974), Leakey, Mungai, and Walker (1971, 1972), and Leakey and Wood (1974).

[f]The gorilla (*Gorilla gorilla*) and chimpanzee (*Pan troglodytes*) measurements are from Mahler (1973).

[g]The baboon (*Papio cynocephalus*) measurements are from Swindler (1976).

For the compilation the author is indebted to Alan S. Ryan.

published by Johanson and Taieb (1976) and White (1977) and also from a perusal of the available casts, but I should stress that the appraisal is my own and not derived from the interpretive comments of the original describers.

The available pelvis, however, is completely hominid in form. Even though, from the fragmentary evidence, it would appear that the arms were relatively long, yet it would seem that arboreal clambering or terrestrial quadrupedalism had been given up some time since in favor of a two-footed orthograde mode of locomotion. Whatever other things that may mean, speed of foot is not one of them.

Once upon a time anthropologists wrote some fatuous nonsense in praise of hominid bipedalism. Actually their concern was principally to contrast the supposed incompletely upright posture of the European Neanderthals with the tall "straight-limbed" hunters of the Upper Palaeolithic. According to this scenario, their greater speed and agility gave them a significant degree of superiority over the crude and clumsy Neanderthals, hastening the demise of the latter. Popular illustrated accounts still depict Pleistocene hunters sprinting across the plains in hot pursuit of this or that kind of ungulate—a harmless fantasy, perhaps, but one that tells us more about contemporary ignorance than about the life ways of the past.

We know now that the Neanderthals were just as well adapted to effective bipedal locomotion as their Upper Paleolithic successors, and that, in fact, hominid locomotion has not changed in any significant way for perhaps three million years. For the first half of that time span no evidence has yet been found to indicate that the hunting and butchering of large-sized game animals constituted a major part of hominid activities. The evidence for subsequent hominid hunting activities seems clearly demonstrated by the rise in the ratio of artifacts to bones, the increase in the degree of bone break-up, and the increase in the representation of large animals (Isaac, 1976), but this does not appear until one and one-half to two million years after the oldest known hominid bipeds.

Clearly, bipedal locomotion did not develop in response to the selective pressures engendered by a hunting mode of existence. A little reflection will make us realize that hominid bipedalism has no locomotor advantage when compared with the modes employed by the rest of the animal world. A human being in full sprint cannot catch anything larger than a house cat; in fact, it cannot even catch a house cat. Nor can it escape the threat of a potential predator by rapid flight. Consider the case of the cat again, only turn the chase around the other way and expand the cat to the size of a leopard, and what we have is a recipe for rapid extinction.

Even in the absence of archeological confirmation, then, bipedalism should indicate to us that tools had become an essential element in the behavioral repertoire. Stone tools are of less than two million years in antiquity but evidence for bipedalism goes back more than three million years. If indeed bipedalism could not have been viable without the use of hand-held weapons and a whole complex of learned behavior governing their manufacture and employment, then we should be able to take the evidence for bipedalism alone as indicative of the presence of culture as a necessary adaptive strategy. Then, despite the equivocal form of its jaws and teeth, such a creature would qualify for the definition of hominid.

If we accept this as indicative of the presence of culture even in the absence of any other tangible clues, then we must infer that the elements of which that culture was composed were entirely of a perishable nature. At any time during hominid evolution, of course, the vast bulk of culture was composed of aspects of behavior which by their very nature leave no enduring traces. Even the tangible foci of the behavioral world, tools, have been predominantly constructed of perishable materials.

The suggestion has often been made that the addition of a simple digging stick to the food-getting strategy of a savanna baboon would effectively double its food supply. The use of such a tool by the earliest hominids would have given them access to underexploited food resources and enabled them to vie effectively with baboons for the occupancy of the tropical savanna lands.

It has also been observed that the digging stick redirected is an even better defensive implement than the formidable canines that form the usual terrestrial primate armamentarium. Canines, after all, become functional only when their employer literally comes to grips with an opponent, and this can be a precarious position when the situation involves a hungry leopard on the one hand and frightened baboon on the other. The separation of would-be predator and potential prey by the length of five feet of stout pointed stick, however uncomfortable that distance may be, is nonetheless just a little less desperate.

The wood of such a hypothetical dual-purpose implement of course, is perishable, and, however reasonable our guesses, there is simply no direct evidence to substantiate them. Obviously, at this point, we are well beyond anything for which we can ever have direct evidence. To have gotten this far, I have had to use a degree of analogy and inference that are unacceptable to the laboratory scientist. Having gone this far, however, the same process can be used to take one further step.

Clearly there are no surviving counterparts of the first bipedal hominids. If we are to get any insight from modern nonhuman primates, we have to consider both those that are phylogenetically close to the human line and those that are adapted to the ecological conditions that prevailed where the earliest hominids lived. It is generally agreed that the closest living pongid relatives are gorillas and chimpanzees, and that the best ecological counterparts are the savanna baboons. And for the pongids, if it is recognized that chimpanzees are closest to humans in a phylogenetic sense, it is also felt that gorillas, as predominantly terrestrial animals, should display some of the same kinds of responses to the problems of terrestrial existence as did the earliest hominids.

Although no nonhominid primate is bipedal, nevertheless, with the exception of the specialized patas monkey (*Erythrocebus*), their quadrupedalism simply does not allow them to run as fast as the predators with whom they have to contend. A slow-footed terrestrial primate, whether it be quadrupedal or bipedal, can be expected to have similar survival problems, and one would expect some sort of similarity in the adaptive response. In the case of the quadrupeds, one clear aspect of that response is the development of enlarged canine teeth that can serve as defensive weapons.

If terrestrial bipedalism is even less efficient as a means of predator avoidance, it is evidently compensated for by the more efficient defensive strategy embodied in the employment of hand-held weapons. The canines, however, being released from their role in defensive activities, should undergo the kind of reduction predicted by the Probable Mutation Effect (Brace, 1963). From the composite tooth-size profile based on the early Omo, Afar, and Laetolil material, it is clear that the canines are reduced when compared to pongids or baboons (Figure 3). It is also clear that the postcanine-size profile pattern resembles that of baboons to a greater extent than that of the closest pongid relatives, chimpanzees. A consideration of available tooth size for the earliest hominids, then, does not contradict our expectations that these savanna-dwelling bipeds survived on a baboon-like diet and defended themselves with hand-held weapons.

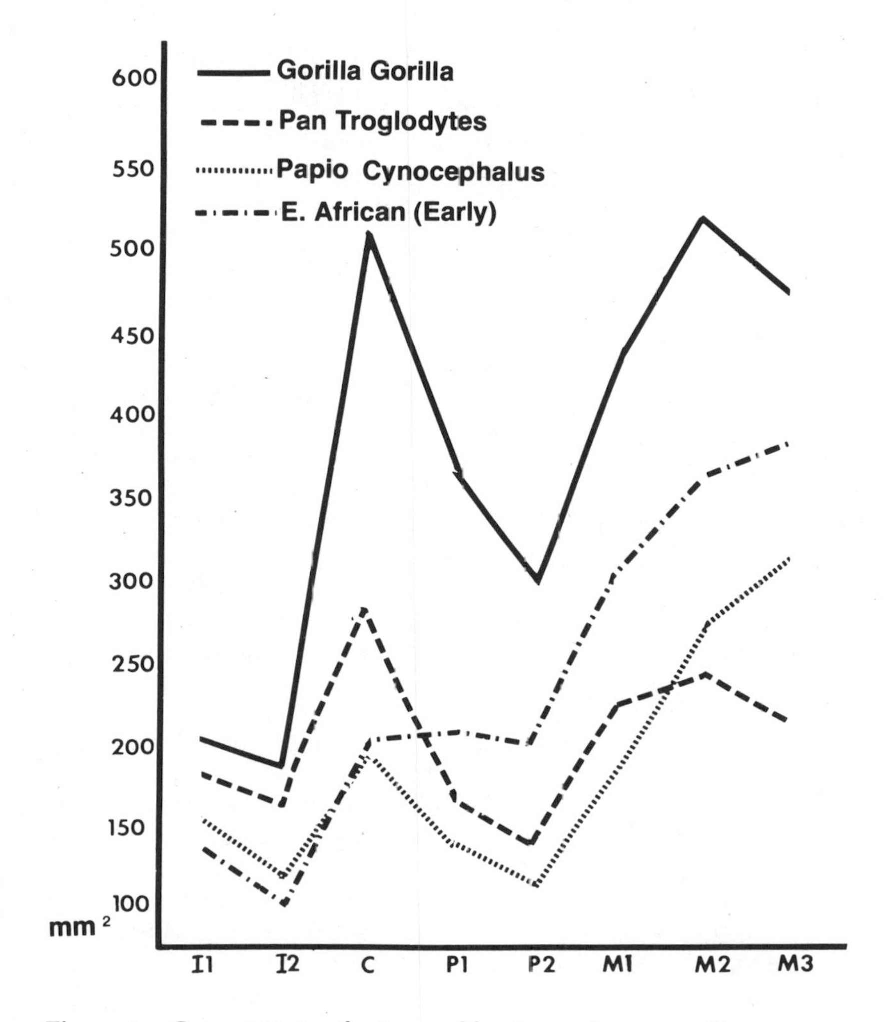

Figure 3. Composite tooth-size profiles (sum of upper and lower mean cross-sectional areas) for gorillas, chimpanzees, baboons, and early East African hominids from Afar and Laetolil. The figures from which this graph was constructed are given in Table 1.

Finally, there is one more aspect that the analogy of contemporary primates would lead us to expect. In modern primates, the enlargement of canines for defensive purposes is not equally a property of all adults, often being developed to its maximum extent in males only. Accompanying this development and clearly related to its employment is the development of a substantially greater bulk and strength in males as compared to females. In gorillas and baboons, males are literally double the bulk of females and they are endowed with vastly greater bodily strength. Maintenance of group integrity, including defense against predation, is largely the responsibility of adult males, a situation principally imposed by the primate reproductive strategy where much care is lavished on a few offspring. The developmental basis for a creature that employs learned rather than instinctive behavior is provided in part by a prolonged period of infant dependency during which the youngster is carried as well as nourished by its mother. If, during that time, the mother actively engages in group defense, the dependent infant becomes especially vulnerable to injury and death. Keeping infants out of danger by putting the burden of defense on the unencumbered males has obvious survival value for a group that has adopted such a reproductive strategy—the *K* strategy of McArthur and others (Nichols *et al.*, 1976).

It has long been realized that humans have carried this strategy of prolonging infant dependency to something of an extreme; and although we can only guess at the timing of development among the earliest hominids, surely this was not less than that observable in other terrestrial primates. We should expect then, that even in the absence of defensively enlarged canines, the burden of group maintenance fell upon weapon-wielding males especially.

Furthermore, at least in the initial stages of defensive weapon-wielding when the tools were of less than reliable and regular form, their effectiveness as deterrents must have been closely tied to the size and strength of the wielders. A 50-pound baboon can be quite a handful, but even equipped with a vigorously employed hand-held tool, it would easily be knocked off its feet by a swipe or two from the paws of a leopard four times that size. But a defensively positioned hominid in excess of 100 pounds cannot be so easily batted around. With the butt of its pike planted on the ground and the tip positioned towards the impending threat, the hominid has a chance of surviving even a concerted attack. Should the predator charge, it would literally impale itself.

In any case, the survival chances of the defending hominid are considerably improved if it is of chimpanzeelike rather than baboonlike bulk, and a good case can be made for the expectation that populations of the earliest hominids displayed a pronounced degree of sexual dimorphism in this regard. Unfortunately, this expectation is not an easy one to test, at least on what we have to date. Evidence is fragmentary and incomplete;

and, at present, it is difficult to tell much about the nature of within-population variation among the Plio-Pleistocene hominids. Part of the problem has been in the typological expectations on the part of many paleoanthropologists. Each differing fragment tended to receive special taxonomic recognition without taking normal within-population ranges of variation into account, and until recently, the possibility of marked sexual dimorphism simply was not considered.

But even if we are alert to such possibilities, it is still hard to tell whether a given degree of difference reflects normal individual variation, male-female distinction, or separation at the species or genus level. Complete long bones are rare; the recovery of both arm and leg bones from the same individual is rarer still; and a sampling of comparable and measureable portions of the skeleton for a series of individuals from the same population is almost nonexistent. We do have teeth, but the majority of them are isolated instances. There are just enough in implanted arrays so that we can recognize the distinction between robust and gracile hominids in the early Pleistocene of East Africa, but, to my mind, they are still insufficient to test my expectation that there were pronounced male-female distinctions within each such group.

But if the evidence is not adequate either to prove or to disprove the case, the expectation remains, and with it some still further implications for early hominid behavior and social organization. Given a situation where males are double female size and are also subject to the occasionally mortal consequences of defensive activity, we can envision groups in which the large males are outnumbered by the smaller females. In effect, what we could see would be bipedal stick-wielding pongids with a multi-male social organization. Whether or not we would call this a state of "primitive promiscuity" or "group marriage," one can suspect that the long-term pair-bonding of the known forms of human social organization had yet to develop. The circumstances that developed males to cope with occasions of vigorous sexual activity, and others of desperate physical activity, have had their echoes throughout the remaining course of human evolution.

You may complain, as you have every right to do, that this is an extraordinary amount of interpretation to be based on a mere hatful of jaws and teeth and a single pelvis from back in the Pliocene. However, this evidence is all that we possess, and we are obliged to squeeze everything out of it that we can. That I have tried to do. For the moment, I would argue that we can do little more, and I know that there are more than a few who would maintain with some conviction that in fact we should be doing a great deal less. But it is *I* who am crawling out on my limb to afford *you* the pleasure of chopping at it from a position of skeptical safety. Whatever the outcome, I hope that it shall have been worth the effort.

But what are these Pliocene hominids? Where do they fit, and precisely what are their affinities? Hominids they are, presumably dependent upon culture for their ability to survive, and the very earliest such for which we have evidence. This should give them claim of place as ancestors of all subsequent hominids including ourselves. I think, however, that if some were alive today, we would be even more hard put than was the judge considering the tropis in Vercors' engaging novel, and I doubt that we would recognize them as being truly "human" (Vercors, 1953).

Given my reconstruction of their mode of living, I should have to accord them generic distinction from *Homo*. The other genus that has attained general recognition within the family Hominidae is *Australopithecus*. Members of this genus come in large (robust) and small (gracile) forms which surely deserve separation at the specific level. The first of the genus was the Taung child, christened *A. africanus* in 1925 (Dart, 1925). It is awkward, to say the least, when the type specimen for a new taxon is an immature individual, but most scholars have felt that the group to which it belonged is exemplified by material subsequently found at the Transvaal sites of Sterkfontein and the Makapan Limeworks Dump.

To be sure, Tobias has recently (1974a, b) tried to claim that the Taung specimen belonged to a different and larger species of the same genus, which would leave us in the still more awkward position of having no type specimen at all for a fundamental taxon. This can be countered by noting that the one measurable adult feature of the Taung individual is an almost perfect representative of the same feature in the Sterkfontein/Makapan material and differs by an appreciable, although not significant amount from the Kromdraai/Swartkrans grouping to which he has attributed it. I refer to the permanent first molar. The sum of the cross-sectional area of the upper and the lower permanent first molars of the Taung specimen is 363 mm^2. This is only 3 mm^2 over the 360 mm^2 mean for Sterkfontein/Makapan grouping, and some 25 mm^2 smaller than the 388 mm^2 mean for the

Kromdraai/Swartkrans grouping. I see no reason not to regard the Taung specimen as a member of the group represented at Sterkfontein and Makapan, which can correctly be regarded as *A. africanus.*

As the first and most widely recognized australopithecine group, it should be instructive to use it as a standard against which to compare the hominids from the Pliocene of East Africa. For reasons previously enumerated, this is most easily done by using tooth measurements. Although the East African canine is somewhat larger and molars slightly smaller than those of *A. africanus,* the composite tooth-size profile of the two groups is remarkably similar. The summary tooth size figure for *A. africanus,* at 2,095, is 188 mm larger than that for the East African group at 1,907. (For a discussion of summary tooth size, see Brace, in press.)

Elsewhere I have spoken of the difficulties in computing a variance for summary tooth size, noting, however, that an average difference of 100 mm probably has some basic biological meaning. This being the case, surely the nearly 200 mm difference between the East and South African groups being compared has some significance. To put some perspective on this, however, the difference between the summary tooth size figures of the Australian aborigines from the northern part of the Cape York peninsula, and those at Broadbeach in southeast Queensland at 204 mm is even greater. The point is not that a difference of 188 mm has no meaning because it is exceeded by differences between summary tooth size figures for modern populations. On the contrary, I have argued that the Australian figure I cite does indeed have interpretable significance. I use this to demonstrate, however, that not only do differences of this magnitude occur within modern *Homo sapiens,* but they occur between populations of what has been regarded as a single "race"—furthermore one that some reputable students of morphology (for example, Abbie, 1968; Simmons, 1973) have claimed, although I think mistakenly, to be especially homogeneous.

The most disparate populations of the genus *Homo* differ by more than 500 mm^2. Even without elaborate tests, then, the similarity of the tooth-size profiles in Figure 2 and the summary tooth sizes in Figure 4 should warrant the inclusion of both the E. African "Homo" and the E. African Early groups in the same genus, namely *Australopithecus.* And if it is premature to argue about the specific identity of the Early East African representatives, it would certainly be premature to suggest that they are specifically distinct from their South African congeners. For the

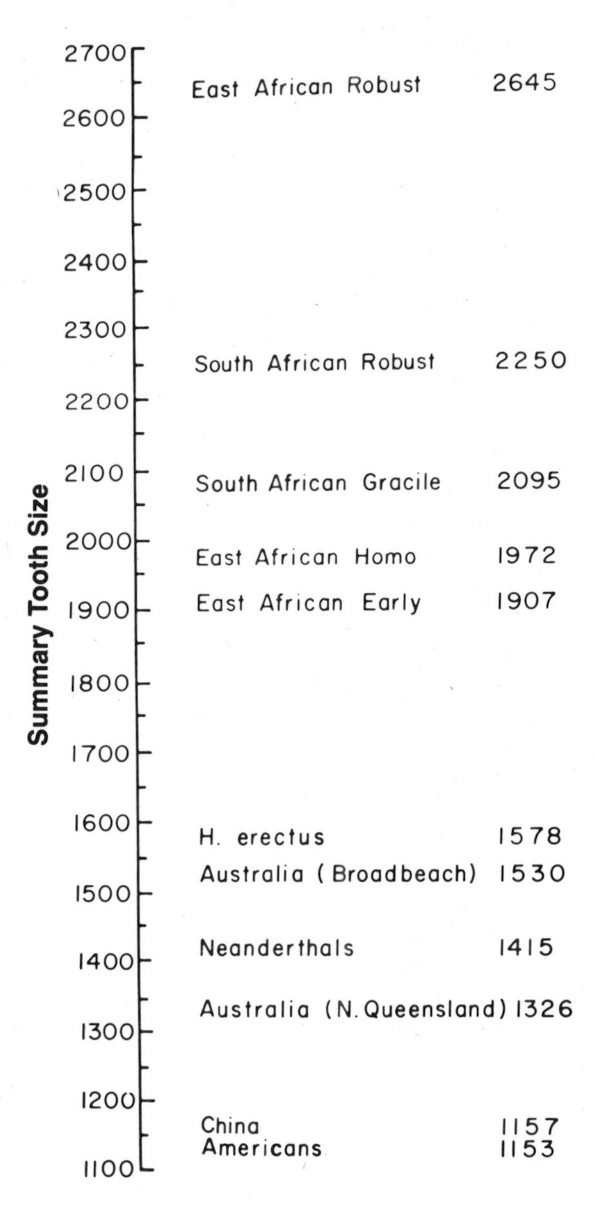

Figure 4. The summary tooth-size figure is simply the summed total of the mean cross-sectional area for each tooth category, upper and lower, for a given population (described in greater detail in Brace, *in press*). The South African figures were calculated from measurements I made in Johannesburg and Pretoria in 1968; the East African figures were

moment we are surely justified in referring colloquially to both groups as australopithecines.

Although absolute dates of the South African specimens are not possible, careful work on the datable dimensions of suid evolution has allowed the establishment of broadly secure time categories for the Transvaal sites. Evidently the East African Pliocene sites are considerably older. The larger size of their canine teeth then would reflect their closer proximity to the pongid ancestor in which the canine played a valid defensive role. The slightly smaller postcanine dentition may simply be a reflection of smaller gross body size, or it may indicate that the diet-related expansion in the latter australopithecines had yet to develop. At the moment we simply do not have enough evidence available to frame a useful discussion. Suffice it to say the Pliocene australopithecines of East Africa form an excellent model for the ancestor from which all subsequent hominids evolved.

Now if the Pliocene hominids from East Africa do not show enough dental differences from the original South African taxon to warrant formal separate status, the same can also be said for the other South African group that has been formally recognized. The robust australopithecines from Kromdraai and Swartkrans are slightly larger than the gracile ones from Sterkfontein and Makapan. Again, quantitative comparison is most easily done with tooth measurements. As Figure 4 shows, the summary tooth-size figure for the South African robust specimens, at 2,250, is only 155 mm^2 higher than that for the gracile specimens. This is not enough to warrant specific let alone generic distinction from either of the groups previously discussed although, again, the difference probably does have some "basic biological meaning" which may well involve the application in incipient form of the "dietary hypothesis" offered by Robinson more than 20 years ago (Robinson, 1956, 1961, 1962).

---

calculated from the measurements reported in Table 1. The Australian measurements were made by myself on specimens in the Queensland Museum and in the Department of Anatomy at the University of Queensland in 1974; the Chinese measurements were made by myself on specimens in the Department of Anatomy at the Queen Mary Hospital in Hong Kong; the American measurements were from extractions collected at the University of Michigan Dental School in Ann Arbor; the *erectus* figures were calculated from Weidenreich (1937) and the European Neanderthals were from Wolpoff (1971).

But if the original robusts from South Africa cannot be formally distinguished from the other australopithecines discussed so far, the same cannot be said for the robust specimens from East Africa, heralded initially by the famous "Zinjanthropus" *boisei* of L. S. B. Leakey in 1959. The summary tooth size for the East African robusts at 2,645, is very nearly 400 mm larger than that for the South African robusts and a good 550 mm larger than the South African reference group from Sterkfontein and Makapan. The composite tooth-size profile is graphic evidence of the enormous postcanine expansion, and it is evident that average anterior tooth size is absolutely as well as relatively smaller (Figures 4, 5). In fact, the East African evidence appears to exemplify the description that Broom offered, but which study has trouble confirming, for the South African robust material so long ago (Broom, 1939, 1950). If some of the writings of the next few years reverberate to the echoes of a ghostly chuckle, it would not be the first time that Robert Broom has gotten the last laugh. The old reprobate deserves to enjoy a posthumous "I told you so."

If Broom's description applies to the East African robust material, then some of Robinson's logic must do so as well. And finally, if the phenomenon so described does indeed warrant taxonomic recognition, as the dental data seem to indicate, then the name should be that given to the first demonstrable specimen In this case, ironically, it was discovered after the description had been offered and developed. The specimen is Leakey's Zinj (Leakey, 1959a, b, c), and if the generic designation is unacceptable, the specific one must stand. As a biped, it was surely a hominid. As a biped with an ape-sized brain and lacking projecting canines, it would be hard to deny it status in genus *Australopithecus*. But, although the brain was no larger than that of other australopithecines, the postcanine teeth were an order of magnitude greater and indicate a significantly different dietary adaptation. Together, it certainly qualifies to be known as *Australopithecus boisei*.

Before we listen for the echo of another posthumous chuckle, however, we have to realize that Leakey would have found a measure of satisfaction but little amusement in this appraisal. Initially he heralded Zinj as the maker of the Oldowan tools, disparaged its australopithecine affinities, and emphasized the supposed presence of incipiently modern features. Actually this approach was a good example of paleoanthropological obfuscation, although he was hardly the first nor, as it would seem, the last to employ it. It involved focusing on aspects of his material

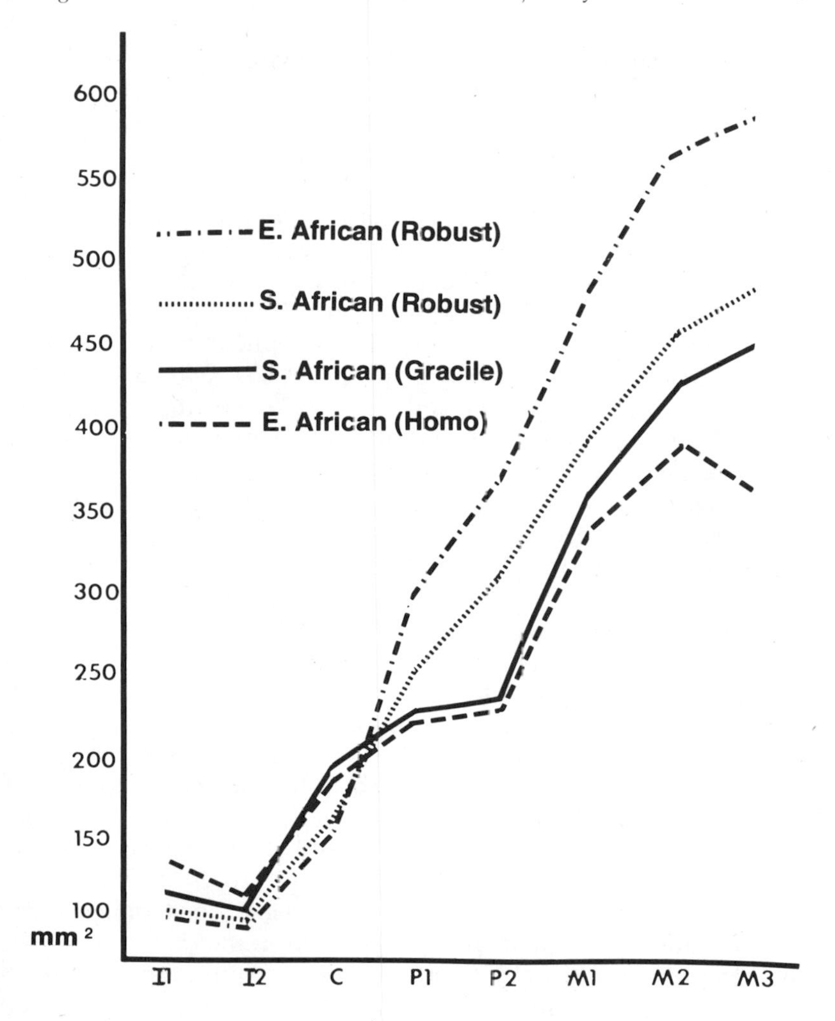

Figure 5. Composite tooth-size profiles of cross-sectional areas for robust and gracile hominids from South and East Africa, based on the data in Table 1.

wherein it differed from typical australopithecine form, and implying it was therefore on the line of evolution towards modern form without first determining whether or not the difference was actually in a modern direction. (As it turned out in the case of Zinj, many such differences in fact were *farther* from the direction of modern morphology.) The obverse of this technique was the identification of features that did indeed resemble the modern condition as evidence for the connection of his fossil with the main line of hominid evolution, as opposed to that of the supposedly aberrant australopithecines, when in reality the australopithecine manifestation of the feature was unknown.

After it became obvious that Zinj was not only an australopithecine but an especially robust one at that, this, probably Leakey's most significant discovery, was downgraded to exemplify what he was *not* looking for. Since he had spent a third of a century looking for the origin of "true man," and since Zinj seemed to have been convincingly disqualified, the tendency was to identify an early hominid fragment that was "not-Zinj" as "Homo" by default. That it might yet be australopithecine has rarely been considered because, after all, if Zinj was an australopithecine and the piece in question was non-Zinj, then, of course, in syllogistic fashion, it could not be an australopithecine. As a consequence the hominids in the early Pleistocene of East Africa have generally been identified as either robust australopithecines or as a form of Homo, but nothing comparable to *A. africanus* has achieved general recognition.

Just as an exercise, then, I have utilized the comparative technique employed earlier in this paper and focused exclusively on tooth measurements. I have assembled all of the early Pleistocene teeth that the East African workers have provisionally referred to "Homo" and used them to construct a composite tooth-size profile (Figures 2, 4, 5). Actually, I did this twice—once with everything so labelled, and once omitting OH 13 which, by my own previous work (Brace, 1973), I managed to convince myself was better regarded as a genuine *Homo erectus* in the middle of Bed II at Olduvai. The removal changed the profile only trivially and does not materially alter the conclusion suggested by the bulk of the earlier Pleistocene material called "Homo."

The summary tooth-size figure at 1,972 mm^2 (it was 1,945 with OH 13 included) falls right in between the Pliocene australopithecines (1,907) and the Sterkfontein/Makapan group from South Africa (2,095). The shape of the composite profiles of the three groups is also quite similar and, given the small sample sizes involved, there would seem to be little reason to distinguish the early Pleistocene East African teeth from either of the other two groups. Based on tooth size alone, the material from East Africa that has been labelled "Homo" with self-declared caution would seem to be indistinguishable from the well-established taxon, *A. africanus*.

The Hunting Revolution

But if the Pliocene hominids, especially as exemplified in the Afar and Laetolil material, could be regarded as primitive forms of *Australopithecus* that continued with minor modifications into the early Pleistocene where they can be seen both in South and East Africa, what happened next is of the greatest interest to the student of human evolution. This was no less than the final transformation of the bipedal, weapon-wielding, behavioral baboon into a genuine if primitive human being. Again, as with the transformation of a pongid into a hominid, but maybe even more so, the visible changes in the skeleton were not particularly dramatic, but they provide positive evidence for a major adaptive shift that had profound implications for the shaping of human physiology and behavior. Also for the first time we have slight but encouraging confirmation from the archaeological record for the inferences that we can make from our study of the anatomical evidence.

The adaptive shift I have mentioned was the adoption of hunting as a major component of hominid subsistence activities. Unlike my earlier expectations in this regard, the weight of the evidence suggests that this was initially spurred by a form of niche divergence that led to the increasing differentiation of two forms of early Pleistocene hominids. As one australopithecine increasingly focused its traditions and activities on the pursuit and capture of animal food sources, it would appear that another took the tack of concentrating on nourishment from the plant world.

The most tangible if enigmatic evidence for this adaptive shift is the appearance of undoubted stone tools at Omo, East Rudolf, and Olduvai at roughly two million years ago. Similar tools of apparently comparable antiquity also appear in the Transvaal sites in South Africa. The tools are tangible and their refinement and diversification as time goes on are clear, but the precise, behavioral implications have been the subject of considerable debate. The earliest and crudest are recognized as the products of hominid agency only because they occur in places where they could not have gotten through the normal consequences of geological action. As time goes on, their refinement of form bespeaks deliberate shaping that we take to be the product of hominid activity, but the question remains: What were they used for?

One standard artist's rendition shows this or that hominid in defensive or attacking posture, clutching variously shaped pieces of rock firmly in hand. This review regards the stone implements as weapons. Another unique interpretation regards the bifacially flaked objects as projectiles, although it is difficult to see, in the absence of an as yet unsuspected Paleolithic cannon, how they could have been hurled with sufficient force so that their supposed aerodynamic shape was of any significance, or that they could have constituted anything more than a minor annoyance to the elephants and hippopotami among whose bones they are found. Going up through the stratigraphic sequence at Olduvai from Bed I and Bed II and from the KBS tuff through the Karari Industry of the Upper Member in the Koobi Fora Formation at East Rudolf (Turkana), there is a rise in the ratio of artifacts to bones, an increase in the representation of large animals, a more evident degree of bone break-up, and a proliferation of differentiated tool forms including an increasing frequency of bifaces (Isaac, 1976:493, 496).

The association of stone tools of hominid manufacture with bones of large sized mammals in increasing numbers is evidence that the utilization of animals for food was becoming an increasingly important part of the lifeway of the early Pleistocene hominids. And if the tools were not weapons, surely they played a role in the processing of the animals after they were dead. The most defensible interpretation is that they were used for butchering. The addition of sharp pieces of stone to the less effective teeth and fingers must have made quite a difference to the early hominid faced with the problem of trying to get through the skin of a defunct hippopotamus.

The early association of hominid tools and animal bones suggests a focus on slow game, infants, and the immature of larger animals, and the occasional fortuitous use of trapped or already dead representations of the large-sized mammalian fauna. But by the middle of Bed II times at Olduvai and the equivalents elsewhere, it seems clear that large Pleistocene mammals are being regularly and successfully hunted. The archaeological evidence provides us with an interpretable picture of the development of hominid hunting activities between two and one million years ago.

The hominid skeletal remains, rare and fragmentary though they may be, also give us a clear indication of the changes that occurred in the hominids that were associated with this developing hunting tradition. Simply put, brain size doubles, going from about 500 to about 1,000 cm^3, and total tooth size drops by a third, going from around 2,000 to just over 1,500 mm^2—entirely the result of a decrease in the crushing surface area of the postcanine dentition.

This is the tangible evidence. The implications, if only in the realm of inference, are nonetheless profound. The most obvious and probably the most significant change was the increase in brain size. Even the australopithecines ranked at the top end of the primate scale of encephalization, and the ensuing cerebral expansion, coming as it did without a significant increase in body size, would have to indicate an adaptive development of major import in the course of primate evolution. The brain size sequence from Sts 5 to ER 1,470, through ER 3,773 and OH 9 marks the conversion from chimpanzee to human size. It is hard to avoid the conclusion that this must be integrally related to the development of hunting as a major part of the emerging way of life.

The evidence shows that by the Middle Pleistocene, hominids had joined the ranks of the major predators. And yet they were just as slow of foot as before. Possessing neither speed nor anatomical weaponry, we are forced to conclude that the key to their success was guile. It must be significant, then, that the organ by which guile is generated—the brain—literally doubles in size.

One hominid lineage surviving to the threshold of the Middle Pleistocene did not show an increase in brain size. In this, the robust *A. boisei,* it is just the molars that increase by about a third, at the same time the molars in the lineage with the increasing brain show a decrease of about a third. Since the primary function of molars is in the processing of ingested food, it is legitimate to suggest that the two lineages were concentrating on different diets—just as Robinson claimed 20 years ago. The large-toothed lineage can be thought of as concentrating on plant foods, and the small-toothed lineage utilized quantities of animal protein which, however tough it may have been in its uncooked condition, simply required less chewing to be rendered digestible. Interestingly enough, a Middle Pleistocene level of dental development was maintained virtually without alteration in some hunting and gathering populations right up to the twentieth century. A picture of the course of hominid evolution seen solely from the point of view of tooth size is presented in Figure 6. In a sense, each point on the graph represents the total dental crushing surface available to the average member of the group named.

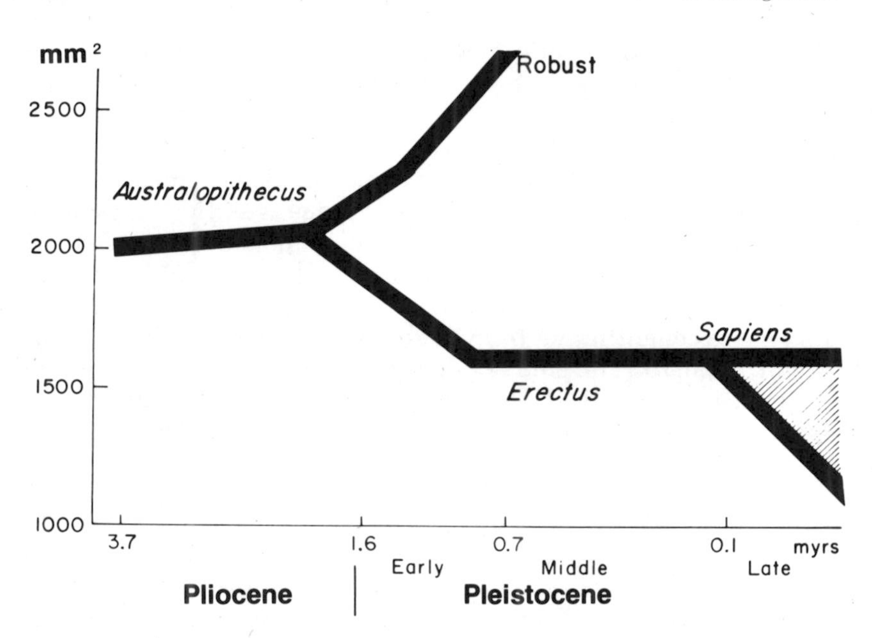

Figure 6. Hominid evolution depicted solely from the point of view of tooth size, based on the data in Figure 4. The date for the Plio/Pleistocene boundary is taken from Haq *et al.*, 1977.

Brain cases and teeth are tangible phenomena, and their changes are easy to tie to the archeological evidence for the development of hunting. Although we shall never have a comparable direct way of testing it, I cannot resist the temptation of suggesting that the adjustment to a hunting way of life also involved the development of many if not all of those additional features by which humanity is sharply separated from the rest of the primate world: the hairless, sweaty skin; the reorganization of digestive physiology featuring a daily rather than an hourly periodicity of bowel movement; the elimination of a breeding season; and the still further prolongation of infant and juvenile dependence, allowing a focus on the accumulation of learned behavior that culminates in the development of language.

The last of these points involved the greatest extrapolation and consequently is the most controversial and difficult to sustain. But, if it really was the adoption of a hunting way of life that created a human out of a precocious ape, then we should suspect that it was involved in the shaping of just those particular features—excepting the ones previously conditioned by the adoption of tool-using—wherein humanity departs from the expected forms of its primate heritage. And further, if it was a focus on learning that allowed a creature that, at best, was to be regarded as anatomically unpromising, to develop into a major predator, then surely this was the necessary background for the most remarkable of all human achievements, the development of a linguistic dimension.

Anatomical determinists at the turn of the century tried to claim that the capacity to speak was determined by the shape of the chin or the development of genial tubercles, and a version of this has recently been revived, but, as speech therapists are well aware, the most extraordinary alterations in the form of the oral cavity, including the complete destruction of the jaw, have no effect on the capacity for speech. The key is evidently in the mind, and we have no right to examine the jaws and facial skeletons of various fossils for evidence of their linguistic capabilities.

The determinants were in the forces that shaped the neurological bases for control, and it seems most logical to seek these in the circumstances that conditioned the major behavioral features wherein humans are distinguishable from their closest primate relatives. I suggest that these circumstances were encountered when the early Pleistocene tool-wielding biped adopted hunting as a major part of its way of life. Language allows the coordination of disparate activities for the benefit of all members of a group, whether these be related to a sexual division of labor or cooperative hunting activities. Without it, the hunting capabilities of the anatomically ill-equipped early hominids are no more believable than those of baboons or chimpanzees—sufficient to capture occasional infants and small game, but not adequate to acquire large game animals on a regular basis.

The hunting of large Pleistocene mammals, the reutilized camp sites, and the manufacture of stone tools to a pattern that itself can almost be regarded as the embodiment of symbolic behavior all suggest that the hominid lifeway had become completely culture-conditioned prior to the beginning of the Middle Pleistocene 700,000 years ago. The remaking of human digestive physiology and epidermal appearance can be comfortably associated with this altered mode of life, and so too can the elimination of periodic estrus as a normal part of reproduction physiology.

One of the notable physiological distinctions between nonhuman and human primates is in the prolonged nonseasonal sexuality of the latter, a condition that might well be termed an extension of estrus. The question of why there should be such a difference has often been pondered, but the direction in which a solution has been sought may tell us more about human self-satisfaction than evolutionary biology. In many aspects of the human condition, especially those concerning sex, we tend to think that things are the way they are because that is "desirable." Sometimes, however, I suspect that we could benefit from looking at matters from the other end. It is just possible that we consider things to be desirable because they are there. Modern scientists have occupied themselves for a century on the adaptive reason for such features as breasts, buttocks, beards, and pubic hair, and the best they have been able to come up with is the lame explanation that they exist because they are desirable and somehow must have been selected for. But I think a less rigid appreciation was expressed by the eighteenth-century novelist Laurence Sterne whose libidinous protagonist, Uncle Toby, in reflecting on the female form divine, remarked that he really preferred breasts as paired phenomena, but that, if the good Lord had chosen to create woman with a single one in the middle, then no doubt men would savor the nuances of its shape with the same degree of lecherous pleasure as they now devote to the appreciation of two at a time.

It is worth considering the possibility that the lack of marked seasonality in human reproductive behavior, as in those other primates where such is also the case, may not be the positive phenomenon it has been portrayed as being. A breeding season is only advantageous when it can be geared to seasonally abundant resources. For most mammals, the breeding season is timed so that births take place when food resources begin to occur in quantities sufficient to sustain the demands first of nursing mothers, then of rapidly growing young. Both of these extra demands have been reduced by the onset of the season of scarcity, and the whole process is set to start over again.

Human reproductive involvement, however, entails such a prolonged responsibility that it way exceeds a periodicity that can be articulated with anything so short as a yearly recurrence. Pregnancy alone takes nine months and, until recently, the infant human being depended totally on mother's milk for sustenance for at least a year following birth. Even after it begins to add solid food to its diet, it continues to get the same amount of milk that had formerly been its sole intake.

In the world which shaped human physiology, pregnancy and lactation represented a constant and not a periodic drain for the human female. The nutritive requirements for both men and women through their reproductive years were both constant and equal. Since seasonal breeding behavior represented no adaptive advantage, it ceased being selected for and disappeared through normal accumulation of disruptive mutations. Enough residual interest in sex was maintained to ensure that, following weaning, abortion, or loss of an infant, pregnancy would ensue as rapidly as possible and a new generation would be coming along as the old one started to fail.

By the beginning of the Middle Pleistocene, I suggest that all the major biobehavioral transformations necessary to produce humanity had already taken place conditioned literally by a milieu of its own manufacture—the cultural ecological niche. The creator and consequence of this happening we recognize in a formal sense as a member of the genus *Homo*, specifically designated *erectus*. As our own lineal forbear, only one major change was required before *Homo erectus* became *Homo sapiens*, and this was a further 50% expansion of the brain. If this was not quite so fundamental as the full 100% increase that accompanied the transformation of *Australopithecus* into *Homo*, it was still enough to warrant specific recognition, and it surely must represent a major expansion of behavioral capacities. These we can suspect were related to those aspects of behavior principally involved in the cultural realm. I suggest that it was no accident that the era when cultural growth and differentiation seems to have acquired dynamics all of its own only occurs after human brain size had reached its modern level and ceased to expand.

References

Abbie, A. A. 1968. The homogeneity of Australian aborigines. *Archaeol. Phys. Anthropol. Oceania* 3:323–231.

Brace, C. L. 1963. Structural reduction in evolution. *Amer. Nat.* 97: 39–49.

Brace, C. L. 1973. Sexual dimorphism in human evolution. *Yrbk. Phys. Anthrop.* 16:31–49.

Brace, C. L . In press. Australian tooth size clines and the death of a stereotype. *Current Anthropology..*

Broom, R. 1939. The dentition of the Transvaal Pleistocene anthropoids, *Plesianthropus* and *Paranthropus*. *Ann. Transvaal Mus.* 19(3): 303–314.

Broom, R. 1950. The genera and species of the South African fossil ape-man. *Amer. J. of Phys. Anthrop.* 8:1–13.

Dart, R. A. 1925. *Australopithecus africanus:* The man-ape of South Africa. *Nature* 115:195–199.

Day, M. H ., and Leakey, R. E. F. 1973. New evidence for the genus *Homo* from East Rudolf, Kenya (1). *Amer. J. of Phys. Anthrop.* 39:341–354.

Day, M . H., Leakey, R. E. F., Walker, A. C., and Wood, B. A. 1976. New hominids from East Turkana, Kenya. *Amer. J. of Phys. Anthrop.* 45:369–436.

Haq, B. U., Berggren, W. A., and van Couvering, J . A. 1977. Corrected age of the Pliocene/Pleistocene boundary. *Nature* 269:483–488.

Hurford, A. J . 1974. Fission track dating of a vitric tuff from East Rudolf, Kenya. *Nature* 249:236–237.

Isaac, G. Ll. 1976. The activities of early African hominids: A review of archaeological evidence from the time span two and a half to one million years ago. Pages 483–514 in *Human origins: Louis Leakey and the East African evidence.* Eds. G. Ll. Isaac and E. R. McCown. Menlo Park, Calif.: Benjamin.

Johanson, D. C., and Coppens, Y. 1976. A preliminary anatomical diagnosis of the first Plio/Pleistocene hominid discoveries in the central Afar, Ethiopia. *Amer. J. Phys. Anthrop.* 45:217–234.

Johanson, D. C., and Taieb, M. 1976. Plio-Pleistocene discoveries in Hadar, Ethiopia. *Nature* 260:293–297.

Leakey, L. S. B. 1959a. A new fossil skull from Olduvai. *Nature* 184: 491–493.

Leakey, L. S. B. 1959b. A "stupendous discovery": The fossil skull from Olduvia which represents "the oldest well-established stone toolmaker ever found." *Illustrated London News,* Sept. 12, pp. 217–219.

Leakey, L. S. B. 1959c. The newly discovered skull from Olduvai: First photographs of the complete skull. *Illustrated London News,* Sept. 19, pp. 288–289.

Leakey, R. E. F., Mungai, J. M., and Walker, A. C. 1971. New australpithecines from East Rudolf, Kenya (I). *Amer. J. Phys. Anthrop.* 35:175–186.

Leakey, R. E. F., Mungai, J . M., and Walker, A. C. 1972. New Australopithecines from East Rudolf, Kenya (II). *Amer. J. Phys. Anthrop.* 36:235–251.

Leakey, R. E. F., and Wood, B. A. 1973. New evidence for the genus *Homo* from East Rudolf, Kenya (II). *Amer. J. Phys. Anthrop.* 39:355–368.

Leakey, R. E. F., and Wood, B. A. 1974. A hominid mandible from East Rudolf, Kenya. *Amer. J. Phys. Anthrop.* 41:245–249.

Mahler, P. E. 1973. *Metric variation in the pongid dentition.* Ph.D. dissertation (anthropology). University of Michigan.

Nichols, J . D., Conley, W., Batt, B., and Tipton, A. R. 1976. Temporally dynamic reproductive strategies and the concept of *r*- and *K*-selection. *Amer. Natur.* 110:995–1005.

Robinson, J. T. 1956. *The dentition of the Australopithecinae.* Transvaal Museum Memoir No. 9. Pretoria: Transvaal Museum.

Robinson, J. T. 1961. The australopithecines and their bearing on the origin of man and of stone tool making. *S. Afr. J . of Sci* 57:3–13.

Robinson, J. T. 1962. The origin and adaptive radiation of the australopithecines. Pages 120–140 in *Evolution and hominization.* Ed. G. Kurth. Stuttgart: Gustav Fischer.

Simmons, R. T. 1973. Blood group genetic studies in the Cape York area. Pages 13–24 in *The human biology of aborigines in Cape York.* Ed. R. L. Kirk. Australian Aborigine Studies 44, Australian Institute for Aboriginal Studies, Canberra.

Swindler, D. R. 1976. *Dentition of living primates.* New York: Academic Press.

Taieb, M., Johanson, D. C., Coppens, Y., Bonnefille, R., and Kalb, J. 1974. Decouverte d'hominides dans les series Plio-Pleistocenes d'Hadar (Bassin de l'Awash: Afar, Ethiopie). *C. R. Seances Acad. Sci., Paris* Serie D, 279:735–738.

Taieb, M., Johanson, D. C., Coppens, Y., and Aronson, J. L. 1976. Geological and palaeontological background of Hadar hominid site, Afar, Ethiopia. *Nature* 260:289–293.

Tobias, P. V. 1974a. The Taung skull revisited. *Natural History* 83: 38–43.

Tobias, P. V. 1974b. Taxonomic implications of the transfer of the Taung skull to the robust australopithecine lineage. *Nature* 252:85–86.

Vercors (Jean Bruller). 1953. *You shall know them.* Trans. Rita Barisse, Boston: Little Brown. 249 pp.

Weidenreich, F. 1937. The dentition of *Sinanthropus pekinensis. Palaeontologia Sinica,* New Series D., no. 1; Whole Series no. 101:1–180.

White, T. D. 1977. New fossil hominids from Laetolil, Tanzania. *Amer. J. Phys. Anthrop.* 46:197–230.

Wolpoff, M. H . 1971. *Metric trends in hominid dental evolution.* Case Western Reserve Studies in Anthropology, no. 2. Cleveland: The Press of Case Western Reserve University.

Chapter 12

Ecological Factors and Social Organization in Human Evolution

Bernard Campbell

In further pursuit of the major adaptive shifts which occured in human evolution, Campbell considers how shifts to different ecological niches both became possible due to the acquisition of behavioral adaptations, and resulted in the development of further behavioral modifications. He sees the shift from the forest to the savanna as being the first major shift in hominid phylogeny, and the shift from tropical to temperate environments as being the second major shift resulting from, and in, major changes in hominid life-styles.

Movement into the temperate zone required considerable control over the environment and the evolution of a host of support technologies far beyond simple stone tools. The winter season in these northerly climes would require considerable technological sophistication in the form of the use of fire, manufacture of clothing, and erection of shelter. With the acquisition of these characteristics, hominids have made a significant behavioral adjustment to ecological constraints and have come a step closer toward environmental control through learned behavior. Hominids, thus, continue to respond to their ecology, but do so by acquired behavioral patterns. Such behavior is, nonetheless, in response to ecological constraints. As seen in Birdsell's chapter, traditions and genes (memes and genes of Dawkins, 1976) share many attributes in common, and both represent ways in which ecology may influence social organization. They are, perhaps, two similar mechanisms, but both respond in analogous fashions.

Reference

Dawkins, R. 1976. *The selfish gene.* New York: Oxford University Press.

The main purpose of this chapter will be to discuss some of the chief environmental factors which have impinged upon and constrained the behavioral repertoire of the Hominidae, and to review the evidence we can use to indicate the nature of early hominid social organization and adaptation.

It seems clear that during the approximately 15 million years of hominid evolution man's lineage must have undergone two successive major socioecological adjustments. These were (1) a change in trophic level of consumption due to the development of scavenging, hunting, and a fully omnivorous diet; (2) a major change in environment as hominids adapted to cooler regions. These will be briefly considered in turn. Following this, we shall review and speculate about some of the accompanying social changes which have been postulated and the biosocial background in which it seems most reasonable to view them.

The Evolution of Omnivory

The environment of chimpanzees described by Suzuki (this volume, Chapter 7), Nishida (1968), Sugiyama (1968), and Goodall (1968) would appear to approximate quite closely that in which we can postulate the emergence of the first Hominidae. Present evidence suggests that the earliest candidates for the hominid lineage (14–12 myr) were associated with a forest fauna at Fort Ternan (Andrews and van Couvering, 1975), at Pasalar (Andrews and Tobien, 1977), and at Rudabanya (Kretzoi, 1975), while later (12–9 myr) forms appear in a woodland savanna habitat with a decreasing proportion of arboreal species (for example, Athens: Simons, 1978; Candir: Tekkaya, 1974; and Potwar: Pilbeam *et al.*, 1977). By the time of *Australopithecus* (4–2 myr) we find a fully bipedal hominid adapted to savanna woodland, evidence of which is commonly found in sites along lake shores and stream beds. The difference between the adaptation of the earliest forest-bound *Ramapithecus* forms and later

Australopithecus populations is not so much due to a complete change of environment (though the earliest *Dryopithecus* ancestors were undoubtedly forest-dwelling), but rather a slow expansion from the forest fringe into more open country which, the fossil record makes clear, was accompanied by a change in diet from a predominantly vegetable to a mixed omnivorous diet.

Almost all primates eat at least some animal food, including tree frogs, bird's eggs, and insects. Chimpanzees and baboons in woodland and savanna environments are well known to hunt and kill various mammals smaller than themselves (Harding, 1975) and have been described as omnivores (Teleki, 1975). Some cooperation has been observed and meat is clearly enjoyed as a supplement to a fruit and vegetable diet. On the assumption that dryopithecine ancestors also carried these behavior patterns, there is, as Teleki points out (1975) no novelty in the early hominid adaptation. If early dryopithecines were adapted to dense forest, then chimpanzee-style predation possibly *would* have been a novel behavior pattern, but one which did not evolve solely in the hominid line, but among monkeys and apes as well.

Goodall (1968) and Teleki (1973) have described how the male chimpanzees act as primary predators at Gombe and often kill other primates and small mammals. They have been observed to operate as a group of 2–5 individuals with what looks like some degree of premeditation, but they certainly show no evidence of setting out on safari with the express purpose of hunting. Their prey is rarely more than one-fifth of their own weight. Thus, for the subsistence of the earliest hominids, we do not need to postulate anything very different from that described for chimpanzees.

It follows from these considerations that the most significant developments in the earlier stage of hominid evolution were not so much predation, but scavenging, big game hunting, and collecting. The appearance of these patterns of behavior is the point at which the primate analogy breaks down and has to be replaced by a strictly hominid pattern of subsistence and social organization. This organization is associated with new technology and carnivore hunting behavior, characterized by cooperation and food sharing (Thompson, 1976).

These events are not *necessarily* associated with a move from forest to a more open habitat, but the fossil record makes it clear that such a move did occur between 14 and 9 million years ago. Woodland and woodland savanna is undoubtedly a far richer environment than forest for a predator. The evidence therefore suggests the development of meat-eating at this time.

Hominid Hunting and Collecting

According to some authors, the first evidence which suggests the existence of big-game hunting comes from Olduvai Gorge, Bed I, site FLK N (Leakey, 1971), a butchery site dated approximately 1.7 myr B.P. However, the existence of an elephant here cannot be taken of itself as evidence of organized big-game hunting. While the elephant may have been killed at the lake shore, it may equally well have died there and been butchered on the spot. It clearly would not have been removed complete to that place after being killed elsewhere. For unquestioned evidence of big-game hunting, we need to find bones of a number of large animals accumulated at a single site, such as we find at Olduvai, Bed II, site BK (Leakey, 1971), Olorgesailie (Isaac, 1966), and Torralba/ Ambrona (Howell, 1966). It appears that soon after 1 million years B.P. we can be fairly certain that cooperative big-game hunting was established in a number of places. From this time onwards, the hunting of large mammals became an increasingly important factor in evolving hominid subsistence and social behavior. There are some important and quite obvious distinctions which separate this evolved hominid style of hunting from the predation seen in chimpanzees and baboons:

1. In the absence of canine fangs, weapons must have been used, if not at first, certainly in the later stages when large animals were regularly killed. [There is however no archaeological evidence of weapons until the Middle Pleistocene (Hoxnian interglacial) when the first wood spear points are found, for example, at Clacton; Oakley *et al.*, 1977.] It is hard to imagine how horses, elephants, and other powerful game could have been regularly killed without spears.
2. Regular and sustained cooperation between hunters would have been required.
3. The carrying of surplus meat would also be essential for this adaptation, and was made possible by a fully evolved bipedalism which certainly predates it.
4. Food sharing is equally implied, and on the basis of our knowledge of humans and chimpanzees, would be associated with a high status being assigned to successful hunters. Meat is much appreciated, when available, by monkeys, apes, and men. Meat sharing with kin would be adaptive (Hamilton, 1964a, b).

5. Meat would have come to play a far more important part in the diet, rising from something around 1% in monkeys and apes to 30% or more (Teleki, 1975).
6. With infants becoming increasingly helpless in the course of hominid evolution, due to a slower rate of development, mothers would have been unable to join hunting parties which accordingly became a male prerogative.

Thus, the hunting of big game, rather than chimpanzee-style predation, involves the appearance of these most important hominid characteristics, all of which have far-reaching social implications. This point has been made in the past by Campbell (1966), Washburn and Lancaster (1968), and Isaac (1978), among others.

Judging by the nature and importance of food gathering techniques, it seems clear that at some point early hominids began gathering vegetable and other small food items which could easily be collected (for example, nuts, fruit, roots, snails, eggs, lizards, and so on). This involved behavior entirely different from that seen in any other primate (compare Tanner and Zihlman, 1976):

1. Gathering implies the collection of more food than the animal's immediate needs, for later consumption.
2. It implies the use of containers for the food gathered, and eventually required the employment of sticks for digging and knocking down food out of reach.
3. The consistent sharing of the gathered food, especially with small children, is probable.
4. It is also likely that an exchange would have occurred with the hunters who brought surplus meat into camp.

Thus it appears that food gathering is a more novel and original behavior in hominid history than predation or small-game hunting. It requires some technology as well as a sense of purpose, which is not so clearly required for small-game hunting.

The combination of both evolved big-game hunting and food gathering has further implications:

1. A division of labor appears, based to some extent on age and sex differences.
2. Extensive food sharing and exchange, especially between the sexes.

3. A resulting interdependence between the sexes.
4. A more varied diet, which means a more reliable food resource base.

Thus, the successful exploitation of big game and the efficient collection of small portable items of food took hominid subsistence and behavior into a new realm which is not paralleled either by living primates or living carnivores. The combination of the traits of premeditation, hunting and collecting technology, altruistic cooperation, exchange and sharing, together with an increasingly deep division of labor (ultimately due to the helplessness of hominid infants), distinguish the hominid adaptation and bring it closer to that seen in modern hunter gatherers than to that seen in any living animal.

Adaptation to North Temperate Zones

This second environmental change which not all, but many, hominids underwent, follows clear evidence of competence in big-game hunting. This ability, which opened up vast food resources, was an essential preliminary for entry into areas of the world in which cold winters reduced the amount of vegetable foods available. As Bartholomew and Birdsell pointed out in their well-known papers (1953; Birdsell, 1968) the nature and extent of food resources would directly control the size and density of populations of early African protohominids. In northern biomes the dependence of population size and density on food resources is much less direct and is complex: it has much deeper cultural implications.

We are concerned here with the *Homo erectus* populations from north temperate zones and *Homo sapiens* populations which succeeded them and eventually occupied arctic zones in Neanderthal times. (People of these periods were of course also occupying tropical zones.) In northern biomes, food resources were subject to far more profound variation than those in tropical biomes: winter and summer replaced the wet and dry seasons of the tropics, and winter brought with it not only a scarcity of vegetation, but low temperatures which themselves presented further energy problems to hominids.

Liebig's Law of the Minimum is of particular relevance in trying to understand hominid adaptations to seasonal temperate and arctic environments of low equability. This law states that a chemical reaction at any level is controlled not by factors present in excess, but by that essential factor which is present in minimal quantity. In ecological terms it follows that the season which yields minimum resources determines the size of the population, and this implies that adaptations directed to increasing resources during such a season are extremely beneficial to the species. It follows that adaptations to the exploitation of summer food resources were of less value to the hominids (and therefore of less interest to ourselves). Adaptations to the winter are, in contrast, of primary interest.

Bearing this in mind, we can postulate three kinds of cultural adaptation to winter.

1. Adaptations to heat energy conservation
   a. Clothing
   b. Shelter
2. Adaptations to heat energy production: the control of fire
3. Adaptations to the increased extraction of resource energy
   a. Increase in species exploited
   b. Improved methods of collection and hunting
   c. Improved resource utilization
   d. Resource storage and conservation

Generally speaking, these adaptations fall outside the realm of tool-making as such (though hunting and butchery required tools), but involve the development of *facilities*, defined by the geographer Wagner (1960) as "devices which restrict or prevent motion or energy exchanges." We are concerned with the extraction, conservation, and production of energy.

Obvious examples of facilities fall in category a. Clothing made from skins, tied with sinews or vegetable fibers, requires no great innovations and allows an enormous saving in energy used to maintain body heat which implies a saving in food resources. Clothing has some serious drawbacks if continuously worn: it encourages ectoparasites and reduces ultraviolet light absorption by the skin. The effects of these two factors must have become apparent by the time clothing was worn continuously in northern regions to keep warm. Some evidence of rickets has been recognized in the skeletal remains of Neanderthal man (Mayr and Campbell, 1971). Alternative sources of vitamin D from fish oils were evidently not generally available at this time. Fish was not yet a stable food resource.

Shelters have a long history. There is evidence that some

kind of shelter, a windbreak, was constructed as early as 1.7 myr B.P. in Bed I times at Olduvai (site DKI) (Leakey, 1971), so there is therefore no difficulty in postulating shelter construction at later times in northern zones. Caves acted as ready made shelters (for example, Choukoutien) and at Lazaret in southern France, we have cave dwellings of Mindel-Riss age (about 300,000 B.P.) containing artificial windbreaks for extra protection (de Lumley, 1969). Primitive methods of building structures and eventually sewing clothing are necessary for life in the north temperate region.

The control and utilization of fire is perhaps the most obvious and important adaptation in the realm of energetics which comes to mind. It is not only a direct source of heat energy, but may even allow physical survival under conditions which would otherwise be too extreme for tropical hominids, for we are still tropically adapted creatures.[1] No doubt once fire was captured, there would be a very high selection pressure to control and preserve it. In turn its successful control would allow the exploitation of cooler regions, opening up huge new areas and resources for exploitation.

But fire does more than this. Besides acting as an effective protective weapon against large carnivores, and its value in tool construction, it improves resource utilization by making certain food more nutritious and more palatable. Heat breaks down complex chemicals and frees amino acids normally locked into indigestible forms. The control of fire (its manufacture certainly followed much later) was without doubt a very important milestone in human evolution. It was the first extra-bodily source of heat energy beside the sun, and was in due course to put an immense impact into the development of technology. Without the mastery of fire, the conquest of the northern biomes and the development of metal technology would never have occurred. Its great importance has rarely been recognized in studies of man's cultural evolution and cannot be overstressed.

In the archaeological record (*see* Perles, 1977) we see questionable evidence of fire in Europe at L'Escale cave (France) which could be as early as Mindel I, about 600,000 B.P. Definite hearths appear around 400,000 B.P. at Vértesszöllös (Hungary) and continuously maintained hearths occur at Choukoutien (about 350,000 B.P.). The earliest fire known from Africa is from the Kalambo Falls site and probably carries a date of about 60,000 yrs B.P. It seems clear that fire is not really vital to survival in tropical biomes, though it may be very valuable.

When we come to consider our third category, the extraction of resource energy, we find a very different picture from that painted by Bartholomew and Birdsell for the protohominids. It is well documented that high latitude north temperate and arctic environments carry a lower diversity of species and a correlated less stable community, while low latitude tropical environments carry an immense species diversity accompanied by great community stability (Odum, 1969). This fact has considerable implications for early hominids in northern areas. In order to maintain a regular food supply in winter the hominids must have either been able to specialize in the development of hunting techniques and technology, or alternatively been prepared and able to kill and consume a broader range of species than their tropical ancestors. But, at the same time, the diversity of herbivorous mammals was greatly reduced. This paradoxical situation implies that almost any animals that could be caught would probably be a welcome source of food—at least at some period of the year. There is some evidence of this at Choukoutien, because although the bones around the hearths in the cave suggest a diet which was 70% deer, it also includes not only sheep, zebra, pig, buffalo, and rhinoceros, but also carnivores including wolves, foxes, badgers, leopards, and other cats. This evidence suggests that at least some of the latter must have been the food remains of the inhabitants: the total number of food species at the site is in the region of 45, which represents a diverse larder. The presence of carnivores seems to suggest dependence on the next highest trophic level above the herbivores, which usually shows slightly greater stability in a community and so more reliability as a resource, though population size and density were very much lower. The fact that this wide range of food species was hunted from a reduced community diversity implies improved hunting techniques and technology.

Because of the lack of community stability, hominid arctic adaptations in particular must have been accompanied by a very flexible approach to hunting techniques and patterns. Waning populations of game and irregular migrations could have been fatal to small hunting bands. Flexibility in making plans and a readiness for a number of hunting possibilities must have characterized these peoples.

The Barren Ground Caribou of northern Canada supply the well-known modern example of the extensive annual migrations which we associate with temperate mammals. They move between their summer home, feeding on the lichens and mosses of the tundra, and their winter home, feeding in the vast spruce forests further south. A number of more temperate North American species such as the elk, mule deer, and dall sheep, still manage an annual migration, and no doubt many Eurasian species once showed similar patterns before men disturbed and decimated them. Hominids dependent on species with an annual migratory cycle of this kind must study their behavior in detail to intercept the migratory herds. The low diversity of arctic forms would imply even greater dependence on carnivores of the next trophic level such as bears, fish, and marine mammals (which themselves are often migratory). We now have evidence of Neanderthal settlements on the Pechora river in northern Russia with remains of polar bear present in their shelters.

One of the most obvious ways to defeat Liebig's Law is to store food—a behavior only necessary in Africa with the coming of agriculture. Any hominid bands which succeed in maintaining a small food store for the winter will be at an inestimable advantage. In the warmer north temperate zones, meat can be stored by drying in strips in the sun, or smoking in a fire. Fruit can also be sundried for winter storage. Salt would facilitate the storage of both meat and vegetables. Further north, meat and fruit could be frozen in the winter and stored at low temperatures in underground pits (with permafrost at less than 2 ft deep, this would present few problems). We have evidence of such pits from Molodova, USSR (Klein, 1973). The development of containers and storage pits for food (both are facilities) was of paramount importance in the evolution of *Homo*.

In these few paragraphs it has become clear that the development of certain facilities—clothes, shelter, fire, and containers—were essential for survival in the north. But their development must have been accompanied by a new flexibility in hominid behavior which implies a more intelligent assessment of survival strategies.

Social Adaptations

King has recently discussed the question of early hominid social organization (1975, 1976). He has pointed out the similarities to be found in the social organization of human hunter-gatherers, monkeys, and apes which parallels the view of their subsistence taken by Teleki. As Suzuki and others make clear, the chimpanzee community emerges as a stable social unit with the capacity to vary between compact and fragmented states (compare van Lawick-Goodall, 1973, and Sugiyama, 1968). King points out that three species of social carnivores with broadly similar social organizations (lions, hyenas, and cape hunting dogs) are all capable of killing animals larger than themselves which may be as much as 100 times their own weight. Thus the carnivore analogy seems to confirm that no new major social adaptation need be considered as a *necessary* stage in the evolution of big-game hunting.

Attempts to understand hominid behavioral evolution have, therefore, been based on analogies with animals phylogenetically related such as chimpanzees, and ecologically and behaviorally related such as baboons and cape hunting dogs. It now appears that phylogenetically related forms are most valuable as models at the earliest phases of protohominid evolution (*Ramapithecus–Australopithecus*), but become less relevant with the appearance of big-game hunting and food collection (*Homo habilis–Home erectus*). Although social carnivores are successful big-game hunters, man's reliance on weapons and cutting tools, and his ability to subsist on a large proportion of vegetable food renders any analogy with them of limited value. It is indeed rather surprising that some behavioral parallels do exist.

In the evolution of *Homo*, these animal analogies are left far behind both in terms of behavior and subsistence, and man's dependence on new social and technological adaptations becomes the primary factor in his survival. It is therefore clear that the best model for early man is modern man—in the form of the remaining tribes of hunter-gatherers. However, these groups, though they follow an archaic adaptation, have certain rather basic cultural advantages over early man in the form of bows and arrows, and boomerangs—both relatively recent inventions. (Many also use iron today for various purposes including the replacement of flint arrowheads and spearheads.) It is also noteworthy that today there are no surviving hunter-gatherers in temperate regions: those which we can still study live for the most part in what we and early man would consider marginal areas.

We will now consider some aspects of early man's social adaptations in this framework.

Troops, Bands, and Food Sharing

It is now clear that since such close similarities in social organization exist between monkeys, apes, men, and carnivores, we do not need to postulate any major development in social organization during early hominid evolution. A recognizable troop or community of about 50 animals can be fragmented into subgroups of different kinds, including all-male bands, under different circumstances. Among chimpanzees, the environment constrains and permits certain behaviors so that in the open savanna woodland we see a different organization from that seen in forest. This organizational plasticity is part of chimpanzee behavioral potential, no doubt shared by hominids.

Perhaps the feature of early hominid social evolution that we can most safely predict is the regular fragmentation of the group into hunting bands (mainly males) and gathering bands (mainly mothers and children). We can also predict some kind of home base; and although we have no direct evidence of banding, we do have evidence of such a base. While some are short-lived butchery sites, many of the occupation levels in Bed I Olduvai Gorge are relatively long-term, probably seasonal, camp sites, along streams or lake shores (Isaac, 1978). In view of this we can probably postulate a third subgroup in our hominid group, which stayed at home, consisting of infants and old people. The implications for food exchange and reciprocity are clear: the closer the kin and the more interdependence within groups, the more such reciprocity can be expected.

Social obligations brought on by such "altruism" (which is always selfish) have become a central factor in all tightly knit human societies. Koestler (1960) has brilliantly described the web of *gimus* and *giris*, that is, obligations, to repay *ons*, that is, debts, which grips Japanese society. The passion for exchanging gifts, he writes, "is governed by elaborate rules for the reciprocation of the gift, or the service rendered, by precisely the right gift or service after the right interval of time—neither too short, because it would convey keenness to get rid of the *on*, nor too long, lest the burden of the *on* become unbearable." This is a sophisticated approach to the reciprocity which has surely pervaded all human society for perhaps a million years.

It is worth noting that kin selection (Hamilton, 1964a, b) could account for the appearance of such reciprocity in the form of food sharing, if human groups consisted only of kin and their mates. If, however, the social groups contain individuals not genetically related or recognized as mates, we would need to invoke reciprocal altruism (Trivers, 1971) to explain the adaptive value of such sharing between unrelated members of the group. There is no reason why reciprocal altruism should not reinforce the reciprocity due to kin selection.

Thompson, Vertinsky, and Krebs (1974) have pointed out that where food is patchy and the patches widely dispersed, centralized food sharing could theoretically become an essential adaptation to a social animal. These conditions do seem to be met among the evolving hominids who came to rely on an omnivorous and varied diet with patchy and mobile food resources, in a woodland savanna context. It follows that such food sharing would maximize resource utilization from such an environment: even if it was not essential for survival, it would be a most valuable adaptation.

Population Equilibrium and Control

Animal populations maintain a dynamic equilibrium at a level below the carrying capacity of their environments in such a way that individuals are not found to be seriously suffering from malnutrition under normal circumstances. Physiological control mechanisms operate and they have been well reviewed by Stott (1962), both for animals and humans. The inadequacy of such mechanisms among humans, however, is clear when we realize the huge extent of malnutrition seen today in many parts of the world (the medical profession fights to destroy the often painful operation of such mechanisms: an obvious example is the life-support given to premature babies). It is clear that for a long time both behavioral and cultural mechanisms have also operated in human society to check population growth and maintain its equilibrium.

Simple behavioral responses to the stress of overpopulation have been observed in animals, but they are not known to include infanticide. Infanticide is very widespread among human hunter-gatherers and agriculturists and was common in the West until quite recently. It appears to have been a worldwide phenomenon and *the* primary cultural means of population control. (In societies subject to western influence it was first proscribed and now, when the result of this has proved disastrous, it is becoming replaced by the more humane abortion.)

It is very hard to understand how natural selection could have brought about or permitted such a means of population control, especially since, among human beings, it is often carried out by the mother or grandmother. However, it would clearly be advantageous for a mother to kill a second nursing infant to maximize the chances of survival of the older individual which carries the greater parental investment. In a strongly interdependent group, in which the successful reproduction of any female depended absolutely on cooperative food collection and sharing, we can also envisage a situation in which it might be advantageous for females to kill their infants in order to retain the cooperation of band members. It is an extraordinary fact that among some hunter-gatherers, mothers will even kill their *first* infants (Birdsell, personal communication), such is the power of social pressures, which can be seen as due to the operation of reciprocal altruism. In the Ellice Islands of the Pacific, infanticide was ordered by law; only two children were allowed to a family, as "they were afraid of the scarcity of food" (Turner, 1884). In this way the society came to control the most biologically vital components of human behavior in contradiction to what might be expected from the theories of natural and kin selection.

Another worldwide method of birth control which can be accounted for by reciprocal altruism is the appearance of a whole range of taboos on sexual intercourse. These can develop so broadly that there can be very few occasions left when a female can have intercourse with a man. For example, the Abipones of Paraguay were described by Dobrizhoffer (1822) as follows:

> The mothers suckle their young for three years during which time they have no conjugal relations with their husbands, who, tired of the delay, often marry another wife. The women therefore kill their babes through fear of repudiation, sometimes getting rid of them by violent arts, without waiting for their birth.

This is one of hundreds of examples which might be quoted. Often, an extension of the period of lactation can delay a further pregnancy for some years. Although there are many kinds of sexual taboo operating in different societies, perhaps the most important is the taboo against getting pregnant outside marriage. Marriage has been described by Miller (1931) as a social device "to check and regulate promiscuous behavior in the interest of the human economic schemes." The restriction of childbearing to married females is bound to have the effect of limiting the birthrate, and in any case puts the reproductive processes under full social control. The minimum age of marriage may be raised by law, or, more commonly, a bride-price is required, which, like a dowry, will raise the age at which people marry and cause extra delays during hard times when the husband, or his family, find it more difficult to accumulate the necessary gifts. In this way reproduction comes to be dependent on a certain minimal surplus of resources.

These and many other ingenious and complex devices do seem to have been effective until recent times, at least among agricultural and hunting peoples. The proscription of many of these practices by western invaders and missionaries has had a near disastrous effect on the equilibrum which was once to be found in most human societies (*see*, for example, Wilkinson, 1973). Western man has increased his food production vastly beyond that available to nonwestern peoples, and then exported his new, modified, social behavior patterns without the appropriate accompanying technical knowledge to increase resorces. The result has been overpopulation and widespread malnutrition.

Language, the Hominid Adaptation

The most important characteristic of man's culture, which distinguishes it from the protoculture of some animals,[2] is the symbolic dimension which it carries as a result of the evolution of language. While technology is in a real sense the "leading edge" of culture, which comes into direct contact with the environment and makes possible the most effective extraction of resources, language allows the complementary development of social aspects of man's cultural adaptation, which are a central part of human nature. Language is not essential for the development of social hunting (compare dogs) nor for a simple technology (compare chimpanzees); rather it is concerned with the social framework in which both of these might further evolve.

The nature and origin of language, and its peculiar character, have been discussed elsewhere (for example, Hockett and Ascher, 1964; Altmann, 1967; Thorpe, 1972; Campbell, 1974). Here we shall attempt to review briefly some of the adaptive aspects of language which made it such a basic and essential feature of hominid evolution. The adaptive strength of language can be summarized as follows:

1. Language constitutes a method of *environmental reference* through symbols. It allows communication about objects out in the environment, rather than about the inner state of individuals.
2. Symbols allow the *storage* of far more information. Thus the inner record of the environment (Craik, 1943) can be immensely enriched, and further enlarged through the increasing development of exploratory behavior, which has clearly occurred during human evolution.

3. The symbolic nature of a language permits *classification* of such referents, and indeed of all memory data. A classified data bank allows far more efficient retrieval than an unclassified temporal sequence of memory images. Classification of data also allows storage of cultural information and norms; it strengthens the "decider sub-system" (Nagel, this volume, Chapter 13) and provides better based criteria leading to more adaptive decisions. The extended data bank is especially valuable where there is environmental unpredictability with rare events, as occur in semiarid and cold regions.
4. Language permits the *transmission* of such symbolic data. Thus it comes about that each individual carries the pooled experience of the group, past and present, as knowledge is passed on and accumulated. Language also allows the transmission of information between bands or groups which may be intermittently extended through space and time: it maintains information sharing through the division of labor. It is an essential factor for successful cooperation among extended groups; it permits broader dispersion and wider interdependence.
5. Language therefore implies *displacement*, so that reference can be made not only to the present, but to the past and future: this permits the appearance of instruction as well as mere observational or social learning.
6. Language permits *discussion*, bargaining, planning, and democratic processes, generally. Individuals could, for the first time, share their thoughts as well as their feelings. It permits more flexibility in social behavior and more originality. It gives new grounds for both antagonism and cooperation.
7. The articulation of the demands of *reciprocity* was probably essential for the development of complex webs of social obligations in time and space, which bond extended social groups.
8. Language permits the appearance of a fully *hierarchical system* with positive control of executive subsystems (Nagel, this volume, Chapter 13). It allows prescription as well as proscription (which alone is well developed among nonhuman primates).
9. Language permits the appearance of more *complex social organizations* with different, named, individual roles.

Because of the neural organization and development required for the evolution of language, we can be sure that it was a long, slow process; language development was surely a necessary concomitant of the development of more complex technology and more complex social systems. Through all the aforementioned factors language made possible the development of more efficient resource extraction.

Male–Female Bonding

Studies of animals make it clear that all possible patterns of mating system occur in nature, from monogamy to full promiscuity, without cultural norms to control them. A glance at the higher primates makes it clear that the human mating system is closer to that of the polygynous chimpanzee than that of the monogamous (and faithful) gibbon.

The conditions in which polygyny (in which individual males frequently control or gain access to multiple females) is most likely to occur in animal populations have recently been reviewed and discussed by Emlen and Oring (1977). These conditions may be summarized as (1) a long phase of parental care with one sex freed from such duties, (2) asynchronous female receptivity, (3) food resources, clumped and patchy, but rich enough to allow members of harems to live in reasonable proximity.

Some 90% of bird species studied are monogamous, but monogamy is believed rare among mammals. The only examples among the Old World higher primates are found in the Hylobatids, the Mentawi Island langur, and man. Some 84% of human societies are, however, recorded as permitting polygyny, and it is more accurate to classify humans as polygynous with facultative monogamy under cultural sanctions. Those which live in societies which ban polygyny are often serially monogamous, or have extramarital bonds with women, in at least a temporary manner.

It has been previously suggested that by controlling his mating system, *Homo sapiens* has developed a new population control mechanism. This mechanism is probably more effective in monogamous than in polygynous societies. However, it should be noted that even in societies in which polygyny is permitted, it is normally only practiced by the wealthier men, and so it does appear to depend on resource availability. Polygyny appears to lead to greater social stability than does monogamy, because the family is a larger social group with more extended bonds of interdependence.

Without the restrictions imposed on both sexes by the social formalities and sanctions of the society, the human mating system appears little different from that found among the chimpanzee and gorilla. In a sense, monogamy is only a limiting instance of polygyny, and human behavior gives us no reason to suppose we exhibit as a general rule the kind of male-female bonding which characterizes gibbons. The formalization and power of the marriage contract finds its basis in the need to stabilize human societies so that extensive reciprocity can flourish. Reciprocity is the economic adaptation which specifically characterizes *Homo sapiens*.

Conclusions

This short chapter can in some sense be considered a postscript to the earlier Burg Wartenstein Conference, *The Social Life of Early Man* (Washburn, 1961). As a result of extensive new studies of the higher primates, and of modern surviving hunter-gatherers (for example, Lee and DeVore, 1968, 1976; Bicchieri, 1972) it seems possible today to draw a more direct behavioral continuum between modern man and his pongid ancestors. We also know more today about the diet and environment of early man.

It seems clear that the behavioral repertoire of evolving hominids was extensive and flexible, and was constrained differentially by a series of differing environments. In this short-term sense, environment can be seen as permissive, not causal (though in a very long-term sense it is of course also causal, through natural selection). Culture and technology constitute the means by which tropical man has positively adapted to so many different environments: culture gives man his behavioral and adaptive plasticity. This plasticity is perhaps less evident in his social patterns and mating systems, as Birdsell has indicated (this volume, Chapter 6), than in his technology.

Man's cultural adaptations have been the means by which he has come to differ so markedly from other primates. Probably the most striking and important biological adaptation is the appearance of linguistic ability. This has placed the power of symbols at our disposal and enhanced to an extraordinary degree both our intelligence and our consciousness. There is a real sense in which these cultural developments are products, not so much of our tropical, as of our temperate history. The icy north demanded thought and stimulated extraordinary advances in energy control and use. Today, with both symbols and energy at our disposal, we can rise above our immediate environment and look back on ourselves and our history. To survive in outer space is an achievement which finally frees man from the constraints of his terrestrial environment and gives him a divine view of his place in nature.

Notes

1. Recognized anatomical and physiological adaptations to temperate and arctic environments are limited in number. *Homo sapiens* follows the so-called rules of Bergmann and Allen (especially clear among hunter gatherer groups); certain facial features such as nose form appear to relate to absolute humidity (at its highest in the tropics); and arctic and subarctic people have a raised metabolic rate and some increase in extremity temperature (Weiner, 1971).
2. Animals with clearly recognized *traditions* can be said to show *protoculture*. Well-known instances are the washing of potatoes covered with sand among Japanese macaques (Kawai, 1965) and the tool-using and tool-making traditions among chimpanzees (van Lawick-Goodall, 1968). Equally fascinating is the tradition of pecking through milk-bottle tops to reach the cream which has been established since 1945 among the tits (*Parus sp.*) in those parts of western Europe where milk bottles carry aluminium foil tops and are daily placed on doorsteps (Hinde and Fisher, 1952). These traditions have in two instances originated in recent times and have been spread and transmitted through observational learning. All of them increase resource utilization.

References

Altmann, S. A. 1967. The structure of primate social communication. Pages 325–362 in *Social communication among primates*. Ed. S. A. Altmann. Chicago: University of Chicago Press.

Andrews, P., and van Couvering, J. A. H . 1975. Palaeoenvironments in the East African Miocene. *Contrib. Primatol.* 5:62–103.

Andrews, P., and Tobien, H. 1977. New Miocene locality in Turkey with evidence on the origin of *Ramapithecus* and *Sivapithecus*. *Nature* 268:699–701.

Bartholomew, G. A., and Birdsell, J . B. 1953. Ecology and the protohominids. *Amer. Anthrop.* 55:481–498.

Bicchieri, M. G. (Ed.) 1972. *Hunters and gatherers today*. New York: Holt, Rinehart and Winston.

Birdsell, J. B. 1968. Some predictions for the Pleistocene based on equilibrium systems among recent hunter-gatherers. Pages 229–240 in *Man the hunter*. Eds. R. B. Lee and I. DeVore. Chicago: Aldine.

Campbell, B. G. 1966. *Human evolution*. Chicago: Aldine.

Campbell, B. G. 1974. The physical basis of language use in primates. Pages 290–309 in *Frontiers of anthropology*. Ed. M . Leaf. New York: Van Nostrand.

Craik, K. J . W. 1943. *The nature of explanation*. Cambridge: Cambridge University Press.

Dobrizhoffer, M. 1822. *An account of the abipones and equestrian people of Paraguay*, vol. II. London: John Murray.

Emlen, S. T., and Oring, L. W. 1977. Ecology, sexual selection and the evolution of mating systems. *Science* 197:215–223.

Hamilton, W. D. 1964a. The genetical evolution of social behaviour I. *J. Theor. Biol.* 7:1–16.

Hamilton, W. D. 1964b. The genetical evolution of social behaviour II. *J. Theor. Biol.* 7:17–52.

Harding, R. S. O. 1975. Meat-eating and hunting in baboons. Pages 245–257 in *Socioecology and psychology of primates*. Ed. R. H. Tuttle. The Hague: Mouton.

Hinde, R. A., and Fisher, J. 1952. Further observations on the opening of milk bottles by birds. *Brit. Birds* 44:393–396.

Hockett, C. F., and Ascher, R. 1964. The human revolution. *Curr. Anthrop.* 5:135–168.

Howell, F. C. 1966. Observations on the earlier phases of the European Lower Palaeolithic. *Amer. Anthrop.* 68:88–201.

Isaac, G. L. 1966. Studies of early culture in East Africa. *World Archaeol.* 1:1–28.

Isaac, G. L. 1978. The archaeological evidence for the activities of early African hominids. 1978. Pages 219–254 in *African hominidae of the plio-pleistocene*. Ed. C. Jolly. London: Duckworth.

Kawai, M. 1965. Newly acquired pre-cultural behavior of the natural troop of Japanese monkeys on Koshima Islet. *Primates* 6:1–30.

King, G. E. 1975. Socioterritorial units among carnivores and early hominids. *J. Anthrop. Res.* 31:69–87.

King, G. E. 1976. Society and territory in human evolution. J . Hum. Evol. 5:323–332.

Klein, R. G. 1973. *Ice-age hunters of the Ukraine*. Chicago: University of Chicago Press.

Koestler, A. 1960. *The lotus and the robot*. London: Hutchinson.

Kretzoi, M. 1975. New ramapithecines and *Pliopithecus* from the Lower Pliocene of Rudabanya in N-E Hungary. *Nature* 257:578–581.

Leakey, M. D. 1971. *Olduvai Gorge, III*. Cambridge: Cambridge University Press.

Lee, R. B., and DeVore, I. (Eds.) 1968. *Man the hunter*. Chicago: Aldine.

Lee, R. B., and DeVore, I. 1976. *Kalahari hunter-gatherers*. Cambridge, Mass.: Harvard University Press.

Lumley, H. de. 1969. Une Cabane Acheuléene dans la grotte du Lazaret (Nice). *Mem. Soc. Prehist. Franc.* 7.

Mayr, E., and Campbell, B. G. 1971. Was Virchow right about Neandertal? *Nature* 229:253–254.

Miller, G. S. 1931. The primate basis of human sexual behavior. *Quart. Rev. Biol.* 6:379–410.

Nishida, T. 1968. The social group of wild chimpanzees in the Mahali Mountains. *Primates* 9:167–224.

Oakley, K. P., Andrews, P., Keeley, L. H., and Clark, J. D. 1977. A reappraisal of the Clacton spear point. *Proc. Prehist. Soc.* 43:13–30.

Odum, E. P. 1969. The strategy of ecosystem development. *Science* 164:262–270.

Perles, C. 1977. *Préhistoire de feu.* Paris: Masson.

Pilbeam, D. R., Meyer, G. E., Badgley, C., Pickford, M. H. L., Behrensmeyer, A. K., and Shah, S. M. I. 1977. New hominoid primates from the Siwaliks of Pakistan and their bearing on hominoid evolution. *Nature* 270:689–695.

Simons, E. L. 1978. Diversity among the early hominids: A vertebrate paleontologist's viewpoint. Pages 543–566 in *African hominidae of the Plio-Pleistocene.* Ed. C. Jolly. London: Duckworth.

Stott, D. H. 1962. Cultural and natural checks on population growth. Pages 355–376 in *Culture and the evolution of man.* Ed. H. F. A. Montagu. New York: Oxford University Press.

Sugiyama, Y. 1968. Social organization of chimpanzees in the Budongo forest, Uganda. *Primates* 9:225–258.

Tanner, N., and Zihlman, A. 1976. Women in evolution, I. *Signs* 1:585–608.

Tekkaya, I. 1974. A new species of Tortonian anthropoid from Anatolia. *Bull. Miner. Res. Explor. Inst. Turk.* 83:148–165.

Teleki, G. 1973. *The predatory behavior of wild chimpanzees.* Lewisburg, Pa.: Bucknell University Press.

Teleki, G. 1975. Primate subsistence patterns: collector-predators and gatherer-hunters. *J. Hum. Evol.* 4:125–184.

Thompson, P. R. 1976. A behavior model for *Australopithecus africanus. J. Hum. Evol.* 5:547–558.

Thompson, W. A., Vertinsky, I., and Krebs, J. R. 1974. The survival value of flocking in birds: A simulation model. *J. Anim. Ecol.* 43:785–820.

Thorpe, W. 1972. The comparison of vocal communication in animals and man. Pages 27–47 in *Non-verbal communication.* Ed. R. A. Hinde. Cambridge: Cambridge University Press.

Trivers, R. L. 1971. The evolution of reciprocal altruism. *Quart. Rev. Biol.* 46:35–57.

Turner, G. 1884. *Samoa.* London: Macmillan.

van Lawick-Goodall, J. 1968. The behaviour of free-living chimpanzees in the Gombe Stream Reserve. *Animal Behav. Mon.* 1:161–311.

van Lawick-Goodall, J. 1973. Cultural elements in a chimpanzee community. Pages 144–184 in *Symposia of the fourth international*

congress of primatology, vol. 1, precultural primate behavior. Ed. E. W. Menzel. Basel: Karger.
Wagner, P. 1960. *Human use of the earth.* Glencoe, Ill.: Free Press.
Washburn, S. L. 1961. *Social life of early man.* Chicago: Aldine.
Washburn, S. L., and Lancaster, C. S. 1968. The evolution of hunting. Pages 293–303 in *Man the hunter.* Eds. R. B. Lee and I. DeVore. Chicago: Aldine.
Weiner, J . S. 1971. *The natural history of man.* New York: Universe.
Wilkinson, R. G. 1973. *Poverty and progress.* New York: Praeger.

Chapter 13

On Describing Primate Groups as Systems: The Concept of Ecosocial Behavior

Ueli Nagel

After having considered factors other than ecology which influence social behavior, intervening variables which modify the effect of ecology upon social behavior, mechanisms by which ecology may influence social behavior and the available evidence that ecology does influence social behavior, we now turn to a theoretical model giving a functional interpretation of social behavior in terms of ecological influences on the individual. Ueli Nagel employs a systems approach identifying types of social interactions and relationships. He views the problems posed by the environment as ones which demand individual solution and then asks how an individual might cope with his ecology.

Behavior is one obvious technique that animals use to adapt to the demands and opportunities of their habitat. Primate behavior includes a considerable proportion of what we term social behavior, in that the individual interacts with social partners in addition to the physical environment. In this chapter, Nagel relates the individual's social behavior to behavior directed at solving the problems posed by ecology.

Nagel begins by considering first those responses of the individual which are directly related to utilization of the physical environment. We are then asked to note that some environmental problems are handled in a social manner, that is, by joint action of two or more individuals to produce a specific environmental outcome. This activity Nagel calls ecosocial behavior and offers this as a conceptual tool in understanding how social behavior may relate to the ecology. Furthermore, in order to insure the availability and future joint efforts of partners in behavioral responses to the environment, individuals engage in behavior which is designed solely to maintain social relationships. This results in social organizations which permit and insure the expression of ecosocial responses to environmental problems. Pressures upon the individual originating in the environment thus influence the development of social solutions and the development of social relationships which make possible such social solutions. The link between social organization and ecology is indirect, but nonetheless present.

Nagel's theoretical model thus allows us to understand the nature of the relationship between ecology and social organization. He goes further, however, in that he has developed a classification scheme which will permit the precise identification of various types of mutually dependent ecological behavior—ecosocial behavior. This system then focuses attention back on the individual. Social organization is seen to result from the ecological pressures exerted on individuals and it is the individual solution to such pressures, which involves joint action with others, that produces social organization.

This model, or rather approach to the classification of behavior, may be combined with other theoretical perspectives. The classification system may be used to objectively define "role" behavior. Individuals who interact with one another according to discrete classification categories may be viewed as exemplifying role relationships. What has been a subjective or haltingly quantitative definition in the past may now be hopefully more quantitatively and objectively achieved. The role concept may be related to the way in which individuals interact in responding to their environment. The expression of behavior which establishes and maintains the relationships which operate in joint action towards the environment, however, may be related to such factors as kinship and reciprocity. The evolution of role relationships may therefore be approached from this perspective.

Thus we conclude with a mechanism for understanding how social organization (the set of interrelated roles that define a society) operates to influence social behavior which acts upon the environment. Investigators may proceed from either direc-

tion, that is, from an analysis of the role behavior in a society, or from an analysis of the environmental factors which individuals respond to in social fashion. Regardless of which approach is used, we should all meet at the junction where social behavior acts upon the ecology. It will be here that we may hope to understand the significance of social organization to the basic adaptation of the primates to their ecology. It is perhaps appropriate to once again quote from Hans Kummer (1971:38), "Non social ecological techniques are poorly developed in primates. Their specialization must be sought in the way they act as groups."

Reference

Kummer, H. 1971. *Primate societies: Group techniques of ecological adaptation*, 160 pp. Chicago: Aldine-Atherton.

Traditionally, relationships between primate social organization and ecology are investigated by recording, for a given primate group, social parameters (for example, frequencies of dyadic interactions, spatial arrangements), ecological parameters (for example, ranging patterns, feeding behavior), and parameters of environmental quality (for example, amount of available food, resource distribution). Correlations among these parameters can be examined when an adequate sample of groups and species has been studied. Early generalizations, derived from correlations in a relatively small sample, indicated simple relationships between types of primate social organization and types of habitat and habitat use (for instance, Crook and Gartlan, 1966). Since then, with the rapid accumulation of field data, the picture has become more complicated. It is now becoming clear that environmental variables influence social organization in an indirect way only. In a recent comparison of 100 primate species, Clutton-Brock and Harvey (1977) found no simple overall correlations between social and environmental parameters. Rather, results from correlational studies suggest a direct influence of the amount and distribution of essential resources on population density and on group ranging patterns, which, in turn, are among the variables determining what types of social organization are adaptive (for example, Clutton-Brock, 1974; Hladik, 1975; Nagel, 1973). A better understanding of these latter processes, therefore, seems to me an important step in the development of primate socioecology. The question as to whether a specific social organization is adaptive in a given environment, can now be formulated more precisely: does a social organization, that is, a specific network of relationships within a social unit, allow an *optimal coordination of member activities* (including ranging patterns and group sizes) in a given environment?

Functional aspects of behavior coordination among primate group members are currently treated either in sociobiological terms—classifying social behavior as beneficent (or altruistic) and disruptive (or selfish) (Clutton-Brock and Harvey, 1976)—or in sociological terms—categorizing behavior according to roles (for example, Fedigan, 1976)—both operating with functional categories, which are based on (contrasting) preassumptions about mechanisms of selection (compare p. 320 and Nagel, 1978). In this paper, instead, an inductive approach to the same question is developed. It starts from a *descriptive* catalog of basic patterns of ecosocial interactions (compare p. 319) among "group system" members in several primate species, and explores some dimensions of variability of such interactions in different relationships. It appears to be too early, however, for the formulation and testing of *predictions* about the functional value of such interactions and relationships.

The Primate Group as a System

General definitions of "system" are given by Hall and Fagen as "a set of objects together with relationships between the objects and between their attributes" (1956:81) and by Bertalanffy as "a complex of interacting elements" (1968:55).[1] A primate group can certainly be described as a complex of interacting members. Primatologists, however, want to analyze how the interactions and relationships of primate group members are organized.

How do system-theoretically oriented workers analyze the nature of interactions and relationships of the members of a human group? A few psychiatrists started pragmatically by describing the homeostatic relationships in the families and living groups of their patients and the regulative behavior conserving such groups, without deducing generalized statements about living systems (Watzlawick *et al.*, 1967). However, general systems theorists apply their models, which were originally derived from the cell and from the organism (for example, Weiss, 1969), to living systems at all levels, including social groups (Bertalanffy, 1968; Miller, 1965a, 1975). This method carries with it the danger of introducing a new form of the old organism analogy into the theory of social groups. The relationships between members are (implicitly or explicitly) defined as being hierarchical, with a "decider subsystem" (Miller, 1965b) regulating and governing the input-output processes of the system. In Miller's (1971) study of human face-to-face groups a general scheme of critical subsystems is applied, which is evidently derived from the organ structure of the body (*see* Table 1).

In Miller's terms, then, the organs of the body, which carry out the essential processes for the survival of the body, are its critical subsystems; the same critical subsystems are assumed to operate in human groups. The survival of the system is the norm for the adaptive processes carried out by the subsystems; in biological terms: the system is the unit of selection, not its components.

When this model is applied to primate groups, one implies that group selection is prevailing in primates and that groups are organized hierarchically, with members carrying out particular

Table 1. The Critical Subsystems of Living Systems in General and Human Face-to-face Groups in Particular, According to Miller (1971, 1975).

Matter–Energy Processing Subsystems	*Subsystems which Process Both Matter–Energy and Information*	*Information Processing Subsystems*
	Reproducer	
	Boundary	
Ingestor		Input transducer
		Internal transducer
Distributor		Channel and net
Converter		Decoder
Producer		Associator
Matter–energy storage		Memory
		Decider
		Encoder
Extruder		Output transducer
Motor		
Supporter		

functions for the whole group under the control of a "decider subsystem." It is obvious to the observer, who is familiar with primate groups in action, that these two implications cannot be accepted as a priori assumptions: there is too much "selfishness" in the everyday behavior of primates. Besides the observational evidence, there are also theoretical reasons why we should not apply a systems model of this kind to primate groups. First, every member primate is able to reproduce, that is, to proliferate its individual genotype. Population genetical models show that differential reproductive success at the individual level (in terms of inclusive fitness) is the basic mechanism for selecting primate genotypes; only under particular conditions will it be supplemented, limited, or even counteracted by true group selection, that is, interdemic or interpopulation selection (Lewontin, 1970; Maynard Smith, 1976; Wilson, 1975). Second, in a hierarchically organized system, the decider subsystem must have a *program* or "memory" of supraindividual *group norms* to rely on in its decisions, and it must have *effective means to control* the behavior of the executive subsystems in order to have them conserve the group norms and keep essential variables in a homeostatic range. In nonhuman primates, however, the "group memory" and the means for member control are limited to nonverbal mediums of information transmission and storage. The ways in which this condition constrains the development of a control hierarchy will be analyzed later in this chapter. Some forms of regulation and control will be discussed, which can develop and successfully operate without the medium of language.

In short, this review has shown that we cannot profitably apply the living system model developed by Miller and other authors to the nonhuman group. If we are to profit from the systems approach, we have to start with a systematic description of all possible member interactions in different ecosocial contexts without imposing any preassumptions about a particular principle of organization of adaptive group processes. This will lead us to a better understanding of the dynamics of functional differentiation and, eventually, to a set of predictive models about the modificatory and evolutionary adaptations of social organizations in the primates. The conceptual framework of ultimate causation must come from a thorough understanding of the selective forces which act on group living primates; the systems approach can provide the descriptive tools for investigating the mechanisms of differentiation and regulation.

Basic Patterns of Ecosocial Behavior

At this point it is necessary to define a few terms and symbols which will be used on the following pages.Group members will be symbolized as $M_1, M_2, M_3 \ldots M_m$ while the environmental situations for this group will be symbolized as $Z_1, Z_2, Z_3 \ldots Z_n$ (with Z standing for the environment in general). When the primate group is regarded as a system emphasis is put on the fact that the members interact with reference to the system's environment and respond to this environment by interacting with each other: this mutually dependent ecological behavior, which will be called "ecosocial behavior," is the subject of this section. *Ecosocial behavior* is defined as every behavior of M_i towards Z, which is influenced by other group members, and which in turn influences their behavior towards Z. Looking at two animals—the simplest case—we define an ecosocial interaction to be an exchange of behaviors between M_i and M_j in regard to Z. For description of a group member's ecosocial behavior over a period of time, the term "ecosocial activity profile" is used; this is defined as the totality of all ecosocial interactions toward Z between M_i and M_j. (The problem of defining Z is discussed later.)

If these new terms are to be meaningful, we have to differentiate between an ecosocial interaction and a social interaction. The latter is a broader and more pragmatic category, defined as any observable exchange of behavior between two animals (compare Hinde, 1975). Examples are: M_1 presents to M_2, M_2 lipsmacks to the approaching M_1, M_1 and M_2 groom each other, and so on. The environmental context of this sequence of interactions, which is not considered in this description, could well be recorded in addition: M_1 and M_2 would still exchange social interactions. An ecosocial interaction, however, is clearly connected with a particular environmental situation and therefore Z must be included in the description. Examples are (in sequence): M_2 is alerted by a call of M_1, M_2 approaches M_1 quickly, M_1 cautiously approaches an unfamiliar object (Z_1), while observed by M_2; after exploration M_1 begins to play with Z_1 and is joined by M_2, the two get into a quarrel about Z_1, and so on.

Now, if we observe only two members of a group (for example, a mother with her baby), they could be regarded as a small system within the suprasystem of the whole group; other group members would then be part of their environment. The question arises, whether there is a real—not just conceptual—difference between the ecosocial interactions of M_i and M_j (in regard to the nonsocial environment) and their interactions with reference to the social environment, that is, the other group members. The latter are, in the majority, of the type of triadic interactions, which are defined as the exchange of behaviors of M_i and M_j with each other and with M_k.[2] "Polyadic interactions" (between more than three group members) are also observed; most of them, however, can be broken down into triadic and dyadic components (van Hooff and de Waal, 1975).

This question cannot be resolved theoretically; the answer depends mostly on how far the monkeys and apes are able to generalize their perceptions of similar but not identical triadic situations. Since there are not enough observations available to resolve this problem at the moment, it is safer for a first approach to treat the two situations as separate cases. The question will be discussed further in terms of selection on p. 333. In Figure 1, all three categories of interactions are presented schematically in a graphic comparison.

Finally, the concept of ecosocial behavior should be compared with the concept of role, which is being used by an increasing number of authors to describe functional differentiation in primate groups. The origins of the role concept and the limitations of its application to the study of primate groups are discussed elsewhere in the framework of the systems approach (Nagel, 1979).

The analysis of the literature on roles in nonhuman primates reveals that the concept is defined by two essential criteria.

1. Role activities are associated with a certain position in the group. When an animal changes position, its role can be taken over by other group members, which move into this position. This is an example of behavioral regulation within the group.
2. Role activities benefit other group members and often the whole group.[3] This is not always stated explicitly, but it is implicitly deduced from the fact that roles are regulated within the group. The word role is even sometimes used as a synonym for function. From the preceding definition it becomes clear that the activities associated with a role can be understood as a special case of ecosocial behavior, namely ecosocial behavior which benefits other group members. Thus, a monkey role is a portion of the monkey's ecosocial activity profile.

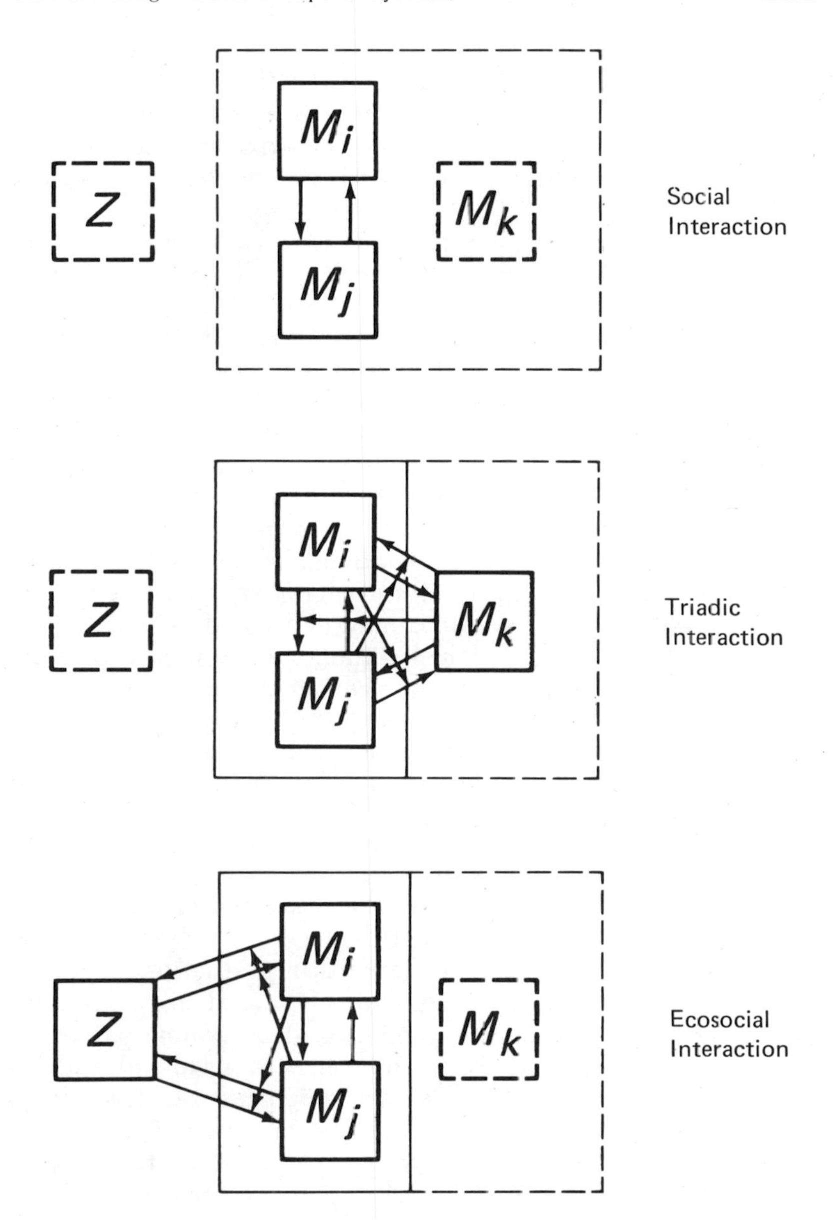

Figure 1. Schematic representation of different concepts for describing interactions among members of a primate group. M_i, M_j, M_k, are members of an idealized primate group. Z symbolizes the environment of their group.

By defining in this manner, we are able to conceptually separate the problems of describing behavioral differentiation among group members and of investigating the immediate causes of differentiation from the problem of explaining role differentiations in terms of ultimate causation, that is, by evolutionary theory.

Having localized the concept of ecosocial behavior in the landscape of current terms and concepts in primate sociology, we can now proceed to the task of compiling a systematic inventory of all possible ecosocial interactions within a particular primate group or species. This brings up yet another problem of definition: how should the environmental situation ($Z_1, Z_2, Z_3 \ldots Z_n$) be defined in order to get nonoverlapping categories of ecosocial interaction? The problem can be approached pragmatically by collecting a list of all observable and formally discernible ecosocial interactions between two members of a group, together with an array of Z-situations, which are potentially related to each of the recorded interactions. From this list, all cases in which an interaction cannot be related to one or a few definable Z-situations are then eliminated through further observation; similarly, all cases of clearly triadic interactions are eliminated, for separate consideration. In this approach it is assumed that the animals have the same problem of "defining" the Z-situations as the observer, and that their way of interacting is a result of their defining, that is, perceiving Z. With this method, the completeness of our list depends entirely on the number and quality of observations.

Another, more theoretical approach starts from the recognition that it makes no sense to define the Z-situation for *both* M_i and M_j at the same time: in the natural group situation in the field it is impossible to make sure that both M_i and M_j are responding to the same Z_a. Even in a standardized laboratory situation, where one controlled stimulus input can be given to both animals simultaneously, this stimulus may be *perceived* differently by each one. Thus, Z should be defined "subjectively," that is, separately for M_i and M_j according to their individual perceptions of Z; Z should therefore be regarded as consisting of Z_a and Z_b, which may be different or similar. In practice, such an individual definition of Z is seldom possible in the natural group situation. Under experimental conditions, however, it is possible to define Z operationally by repeatedly recording each single animal's response to a controlled Z-situation. From such tests and from existing descriptions and classifications of individual ecological behavior, Z-situations can be defined and classified. There is no need to present possible Z-classifications in detail in this paper; they are discussed elsewhere (Nagel, 1979). The range goes from a simple dichotomy into Z_a (being approached or searched) and Z_b (being withdrawn from and avoided) to an elaborated list based on the animal's repertoire of ecological behavior.

The next step is to derive, for a given Z-classification, the matrix of theoretically possible Z_a/Z_b-combinations, when M_i and M_j are put together. Each Z_a/Z_b-combination can then be related to particular categories of ecosocial interaction. Empirically, this can be done by testing M_i and M_j together in the same controlled Z-situations which are used in the single animal tests. It would be expected in most cases that Z-perceptions of M_i and M_j are mutually influenced and changed in the pair test: this is an important aspect of ecosocial interactions which can be documented by this procedure. However, each entry in the matrix of Z_a/Z_b-combination can be related with a whole catalog of basic patterns of ecosocial interaction. When tried for every Z_a/Z_b-combination, this catalog can show us many unexpected possible forms of ecosocial interaction; in addition, functional aspects of each basic pattern can be evaluated with this procedure. This approach, then, has heuristic values which the pragmatic approach lacks.

A catalog of basic patterns of ecosocial interactions is presented in Table 2. It was compiled by combining a theoretical and an empirical approach. Possible categories of mutual influencing of two acting components of a system were developed from theoretical considerations and comparisons at a general systems level. These were then subdivided and supplemented according to our empirical knowledge about the capacities of nonhuman primates to influence each other's behavior. The resulting list was checked for completeness against an (unsystematic) inventory of ecosocial interactions observed in three nonhuman primate species: baboons (*Papio hamadryas*, with some supplements on other baboon species), sykes monkeys (*Cercopithecus mitis albogularis*), and chimpanzees (*Pan troglodytes*) (Nagel, 1979). (The "chimp" inventory is the least complete, because it is entirely based on published materials. For the other two species I could partly rely on my own observations, partly on personal communicatins, as well as on published material.)

The catalog presented in Table 2 contains nonoverlapping

Table 2. Catalogue of Basic Patterns of Ecosocial Interactions (Explanation of Symbols and Criteria of Systematization in the Text, p. 319.)

Type 1 Situations: "Z_a similar to Z_b"
Z is perceived similarly by M_i and M_j who are therefore similar in motivation and "goal." Z may be a common food source, a common danger, and so on; Z may be dispersed in space so long as it still can be perceived by both M_i and M_j. (Further explanations in the text.)

Type 2 Situations: "Z_a not similar to Z_b"
Z is perceived differently (as Z_a and Z_b) by M_i and M_j; this may result from various causes, for example, from a difference in knowledge about Z. M_i and M_j are therefore dissimilar in motivations and "goals." Z_a and Z_b may be localized (in the same object) or dispersed.

Basic patterns in Type 1 Situations

1.1 Mutual stimulation of M_i and M_j in their similar activities towards Z, without mutual interference or coordination (see 1.2) of their activities. Increased efficiency may result functionally for both.

1.2 Simultaneous and coordinated, that is, noncompetitive, mutually tuned behavior of M_i and M_j, directed at the same Z (not dispersed in space). May be combined with pattern 1.1. Cooperation with mutual aid, increased efficiency and stabilization of the relationship may result functionally.

1.3 Mutual interference between M_i–Z and M_j–Z, without exclusion of one partner from Z. May be combined with pattern 1.1. Functionally, this pattern results in a situational compromise of Z-utilization.

1.4 M_i and M_j alternate with each other in their similar activities towards Z. Functionally, this alternating or time sharing may result in successive cooperation or in a compromise of Z-utilization. May lead to Type 2 situations.

1.5 The access of M_i (dominant partner) to Z is either respected by M_j without conflict or (re-)established by means of aggressive intervention of M_i against $M_j \rightarrow Z$. M_j's proximity often stimulates M_i in his Z-directed activity. Leads to Type 2 situations, when M_j does not wait until M_i leaves Z.

M_j-exclusion from Z may be diminished by the use of various behavioral strategies of the subordinate; this leads to forms which are transitional to patterns 1.3, 1.4, 1.7, and 1.8. A situational privilege for M_i results functionally. In a few special Z-situations, being excluded

Basic patterns in Type 2 situations

2.1 By his presence or his interaction with M_j, M_i intensifies M_j's explorative activities towards Z_b in a general manner. M_i's Z-directed activity is different from M_j's. Possible functions of this pattern are not apparent.

2.2 M_i and M_j mutually tolerate each other's different activities (this may result in their temporary separation). The different activities of M_i and M_j may have complementary functions, or functions with no relation to each other.

2.3 M_i conceals his additional information about Z_a from M_j by refraining from Z_a-directed activities, which might transmit such informations, without being forced by M_j. No apparent function of this pattern for M_i.

2.4 M_i and M_j's interactions result in a compromise activity, for example, with regular alternating of the two. This leads to Type 1 situations. Functionally, mutual inhibition or a compromise of Z-utilization may result.

2.5 In the presence of M_i (dominant partner), M_j either refrains from Z_b-directed activities or M_j's restraint towards Z_b is (re-)established by means of aggressive intervention of M_i against $M_j \rightarrow Z_b$. M_i does not necessarily show Z_a-directed activity. Functionally, a coordination of movements (from aggressive herding) according to M_i's intentions results. Refraining from Z_b-directed activities may be dangerous for M_j.

2.6 By his more intensive and persistent activity towards Z_a, M_i influences M_j to abandon his former activity and to follow or join $M_i \rightarrow Z$. Causally, this may start from a

Table 2 (continued)

Basic patterns in Type 1 Situations	*Basic patterns in Type 2 situations*
from Z may be advantageous for M_j.	motivational or an informational difference. This leads to pattern 2.8 and 1.8, if soliciting signals are used. Transition to pattern 2.4 is also possible. Functionally, a coordination of movements and activities may result, which combines individual needs (motivations) and environmental challenges.
1.6 The priority of M_i, as the first partner to interact with Z, is either respected by M_j or (re-)established by the more intensive and persistent $M_i \to Z$ activity and gradual inhibition of the less active M_j. M_j's activities towards M_i and Z are inhibited even if M_j is dominant over M_i. M_j's proximity often stimulates M_i in his Z-directed activity. Leads to Type 2 situations, when M_j does not wait until M_i leaves Z. M_j-exclusion from Z may be diminished in various ways, leading to forms which are transitional to patterns 1.3, 1.4, 1.7, and 1.8. Functionally, a situational privilege for M_i results. Activities of M_i are reinforced which lead to priority of interaction with Z. Such activities may have functions for M_j, too.	2.7 By initiating locomotion and/or Z_a-directed activity, M_i influences M_j to abandon his Z_b-directed activity and pay attention to $M_i \to Z_a$. The resulting attention structure and coordination may have functions for M_i and for M_j.
1.7 M_i observes M_j–Z interactions and directs his attention and activity at the resulting Z-situation. $M_j \to Z$ may be stimulated by M_i's presence, but is not solicited or inhibited by M_i. May alternate with pattern 1.5 or with pattern 1.2. While M_i may profit from M_j's efforts, no function is apparent for M_j.	2.8 M_i influences M_j to switch attention, motivation, and activities from Z_a towards Z_b by means of specific signals (alerting, soliciting signals, and so on). Often leads to pattern 1.8. (For possible functions, see pattern 1.8.)
1.8 M_i solicits and reinforces a specific $M_j \to Z$ activity, which differs from $M_i \to Z$. Different behavioral means may be employed for this; most frequently, specific signals are used, which elicit automatic responses. Partner-action soliciting is mostly displayed by a weaker animal with the function of enlisting the protection and aid of a stronger animal in regard to Z.	2.9 Z_a-directed activity of M_i transmits information to M_j which results in a persisting shift of Z_b-perception towards Z_a. Z_a and Z_b are different perceptions of the same object. Leads to Type 1 situations. More efficient M_j–Z interactions may result functionally.
1.9 Z-directed activity of M_i transmits information to M_j about possible interactions with Z, which have not yet been experienced by M_j, and M_j imitates M_i. More efficient M_j–Z interactions may result functionally.	
1.10 Nonaggressive interference of M_i with $M_j \to Z$ activity transmits information to M_j about how to change his interactions with Z. More efficient M_j–Z interactions may result functionally.	

categories, which could be discriminated clearly in form and/or mechanism; form differences are considered only at the level of partner influencing, not at the level of behavior elements. Interaction patterns are not related to many particular Z_a/Z_b-combinations, but to two basic types of Z_a/Z_b-combination: "Z_a similar to Z_b" versus "Z_a not similar to Z_b." "Z_a similar to Z_b" does not mean identity of Z-perceptions of M_i and M_j, but they should perceive the same Z-quality, orienting with the same functional type of behavior towards the same type of object in their common environment. For example, M_i and M_j both show by their behavior that they perceive an orange as food, although they may not be equally hungry. In practice it may not be possible in every case to distinguish between the two types of Z_a/Z_b-combination; the distinction in Table 2 is made for easier orientation. Two categories of partner influencing—namely, aggressive intervention against $M_j \rightarrow Z$ and soliciting and reinforcing of $M_j \rightarrow Z$ by means of specific signals—occur in both Type 1- and Type 2-situations (patterns 1.5 and 2.5 and patterns 1.8 and 2.8); they will be considered further in the next section. In other cases it is not clear whether the mechanisms involved are identical or only similar in the two types of Z_a/Z_b-combinations (patterns 1.4 and 2.4, patterns 1.7 and 2.7, and patterns 1.9 and 2.9). In all other cases no such parallels between Type 1- and Type 2-situations could be found.

Patterns are not defined functionally; thus, several patterns may have the same possible function and one pattern may have different possible functions. This is illustrated by short comments on possible functions of each of the listed patterns. In the following sections, functional aspects will be discussed and examples will be given for some basic patterns. However, there is not enough space to give examples for every pattern listed in Table 2.

How can this catalog be useful for empirical research? Understanding the interdependency of an optimal coordination of member activities (in a given environment) with the network of social relationship within the group is becoming a central research problem in primate socioecology. This problem can be investigated by making use of the catalog of ecosocial behavior patterns. Several research procedures are possible.

Social positions of group members (as defined by spatial and social interaction measures) can be correlated with *ecosocial* activity profiles (of the same group members) in natural or captivity environmnents by means of different methods. A more controlled analysis of this type is possible by recording the ecosocial interactions of two group members of known relationship in a representative range of experimental Z_a/Z_b-combinations; when conducted comparatively with a number of individuals and of species, such an analysis would allow conclusions about the ecosocial qualities of different types of relationships to be drawn.

Another approach to the problem is formulated in the following questions: how are stable social relationships affected by changes in the frequency and/or type of ecosocial interactions following experimental manipulation of Z-conditions?

The problems mentioned here are not new, of course. Many studies have dealt with them, starting from different theoretical viewpoints, for example role theory. Applying the concept of ecosocial behavior allows a systematization of this type of experiment and makes comparisons between species easier. Furthermore, with the systems approach, the emphasis will be on analyzing patterns of interanimal control and information flow. Some of the questions which could only be briefly mentioned here will be discussed in the following sections.

Ecosocial Interaction Patterns as a Basis for Adaptive Group Organization

Members of a primate group do not interact at random. They maintain relationships, which can be defined as the content, quality, and patterning of their *interactions;* the content, quality, and patterning of the *relationships* between the group members, in turn, make up the social structure or social organization of the group (Hinde, 1975). For a first approach to the problem of adaptedness, I will concentrate on relationships: the dyad is conceived as a two-member system. Even with this restriction only a few aspects of the problem can be discussed. Hinde (1976) has reviewed the conceptual tools presently used for describing relationships in the social sciences and in primatology without considering the question of adaptedness. He has also proposed a tentative classification of the dimension of relationships. One possible dimension is the characterization of relationships according to the predominant patterns of interaction they contain. This leads to the formulation of several questions. Could particular ecosocial interaction patterns predominate in a relationship? What are the conditions which favor such types of relationship? What is the range of adaptive responses of a given type of relationship? These questions will now be discussed for two basic patterns of ecosocial interaction, the "dominance" pattern (patterns 1.5 and 2.5 in Table 2) and the "partner-action soliciting" pattern (patterns 1.8 and 2.8 in Table 2). Of all basic patterns in our catalog (Table 2), these two represent the main modes of (partial) control of M_i over $M_j \rightarrow$Z-behavior. A control hierarchy with a decider subsystem and executive subsystems, as postulated for a generalized living system (see p. 317 and Miller, 1966b), therefore would be expected to develop from relationships in which these two patterns of interaction predominate. This will be discussed in the next section.

It is conventional practice to characterize relationships among primates in terms of *dominance*. From our point of view, however, the "dominance" pattern appears as only one out of a variety of ecosocial interaction patterns, which may become characteristic of a relationship.

The conditions which favor this particular pattern are well known: these are situations where access to a desired resource (especially food) is limited in space and time, forcing M_i and M_j into competition. Wrangham (1974) has demonstrated that increasing the frequency of such Z-situations by artificial feeding resulted in a significant increase of aggressiveness and dominance interactions in free-ranging chimpanzees, and thereby changed the quality of the relationships. A similar increase in food-competitive interactions results from artificial feeding in many rhesus monkey groups in India (Southwick *et al.*, 1976). Singh's (1969) comparison of town and forest rhesus monkeys has revealed persistent differences of aggressiveness and competitiveness under laboratory test conditions, which were clearly correlated with the different feeding conditions in their rearing environments. In captive groups, this feeding effect is combined with the effects of restricted space and lack of adequate cover. In many captive groups the dominance-rank order is more pronounced than in wild groups of the same species in that dominance interactions are more frequent and behavioral and morphological differences between high- and low-ranking animals are more pronounced in captive groups (for example, Rowell, 1967, 1971).

An additional explanation for this phenomenon was suggested to me from my observations on a captive group of sykes monkeys. It is obvious that a variety of Z-situations, which are normal in the wild, usually do not occur in captivity (for example, predator attacks, coordinated group movements and communal searching for food, encounters with neighboring groups). The corresponding ecosocial behaviors are (in most cases) not displayed under the restricted conditions; the relationships of captive primate are, thus, lacking important ecosocial qualities. Our hypothesis is that such a decrease in frequency (or even disappearance) of ecosocial interaction patterns, which under natural conditions are associated with the dominance pattern (compare Table 3), is correlated—*ceteris paribus*—with an increase in the frequency of dominance interactions. (Several mechanisms could produce such a negative correlation, but this problem is beyond the scope of this paper.) Observations of a sykes monkey group during several (independent) changes in cage conditions provided grounds for postulating such a correlation, but the hypothesis could not be adequately tested with the same group (compare Nagel, 1979).

Let us briefly look at the ecosocial qualities of a dominance relationship in free-ranging primates. Several ecosocial interaction patterns are often associated with the dominance pattern in such a relationship. These associated patterns and the roles possibly resulting from them, are listed on Table 3. There are, of course, differences among primate species (or even groups) as to which and how many associated patterns occur in a dominance relationship. This interesting problem cannot, however, be treated here. A look at the limitations of a dominance relationship is now needed. Which pattern of differentiation among members of a primate group cannot be attained through the dominance pattern? Generally speaking, a dominant position does not enable a monkey or ape to force its subordinate partner to carry out particular (desired) activities toward Z. Through aggressive intervention the dominant M_i can teach M_j what it should *not do*, but not what it *should* do (at least not any better than the subordinate partner could teach the dominant by rewarding). Theoretically, M_i could teach M_j by punishing every undesired behavior of M_j towards Z, thus singling out the desired behavior. This complicated procedure could only work if Z is available long enough for M_j to learn. In any case, the expenditure of time and energy for M_i may be too high in relation to the possible benefits. A dominant male anubis baboon weighing more than 30 kg, for example, cannot force a juvenile male to climb into a small-branched tree and bring down favorite food; he can only stay under the tree and profit from the falling fruits, when the juvenile himself decides to feed in this tree. Similarly,

Table 3. Ecosocial Qualities of Dominance Relationships in Freeranging Nonhuman Primates.

Ecosocial Interaction Patterns (Pattern Numbers Are Taken from Table 2, pp. 324–325)	*Role Activities (cf. p. 320) which May Result from These Ecosocial Interaction Patterns*
Patterns 1.5 and 2.5 ("dominance interactions") are conventionally used to characterize these relationships. Subordinates employ various behavioral strategies to diminish their exclusion from Z.	Possible function of the dominant: prevent the subordinate from doing something dangerous (by aggressive intervention). Possible function of the subordinate: peripheralization of the subordinate results in patterns of situation 2, which may have role character (see 2.8 and 2.9).
Pattern 1.7. ("observing partner-action")	The subordinate may open up new resources for the dominant or vice versa.
Pattern 1.8 ("soliciting partner-action"); M_i = subordinate partner	Possible functions of the dominant: defense of subordinates and intervention against potentially harmful Z (cf. p. 331) including intergroup defense ("control animal role").
Pattern 1.9 ("imitation"); M_i = dominant partner (in most cases)	The dominant (M_i) transmits information about how to render $M_j \rightarrow Z$ more efficient. (The reverse is possible, too.)
Pattern 2.3 ("concealing"); M_i = subordinate partner	No roles.
Pattern 2.7 ("attention focus"); M_i = dominant partner	Possible functions of the dominant: group leadership and transmission of useful information.
Pattern 2.8 ("soliciting partner-action"); M_i = subordinate partner *Pattern 2.9* ("arousing interest"); M_i = subordinate partner	"Peripheral animal role": detection and notification of potential dangers, new resources and novel Z-conditions.

the dominant animal cannot force a subordinate to give him information about Z, which the latter does not want to give (pattern 2.3). Finally, there is evidence that in several monkey species the transmission of information (patterns 1.9, 1.10, and 2.9) is impeded from subordinate to dominant animals, but not in the opposite direction (sykes monkeys: personal observations; Japanese macaques: Kawai, 1965; hamadryas baboons: Sigg, personal communication). The question of how far the "power" of a dominant primate reaches will be discussed further in terms of control in the next section.

Unlike the dominance pattern, the "*partner-action-soliciting*" pattern is not normally used to characterize relationships among primates. Are there any relationships in which this pattern predominates and which therefore could be labeled partner-action soliciting relationships? To answer this question, we shall first look at a few examples of this type of ecosocial interaction; then, we shall try to extrapolate the general conditions which favor the establishment of such a pattern. (If no reference is given, the following examples are taken from personal observations.)

When a sykes infant gives one of two intensive distress calls ("lost call" or "screaming," calls of Type 3 and Type 7 respectively according to Gautier's (1975) inventory of *Cercopithecus* vocalizations), the mother approaches and protects and/or reassures the infant. High intensity calling also alerts the adult male, soliciting him to attack any intruder (or even group members) potentially harming the infant. Similar protective responses are known from many other monkey species (for example, Bernstein, 1966a, b; Rowell, 1974) and are familiar to every monkey keeper. An interesting case is reported from *Cebus* monkeys by Zola (1968). A strange two-month-old *Cebus* was adopted by juveniles of this species in a short time and defended against a human intruder only after the residents had become familiar with him; the response seemed automatic. Captive-reared juveniles of either sex showed the protective response from an age of about 10 months on, without ever having seen a model animal performing this activity. The information for signalling (giving the soliciting or appeal signal) and for responding to the signal seems to be innate.

When one animal first detects a source of potential menace, its alarm calls elicit specific responses from the other group members, mostly alertness and flight, adapted to the specific Z-situation (pattern 2.8 in Table 2). Frequently, the alarmed animals respond by automatically rushing a short distance away from the potential source of danger—off the ground or to the ground, depending on the species—and then stop to accurately assess the Z-situation. A sender animal may learn to use this effect tactically in situations of social tension, as observed in two females of my captive sykes group. Generally, any situation where a group member, through his special signals, regularly triggers a specific ecosocial behavior of a partner, can be classified as partner-action soliciting interaction (PAS-interaction). Conditions which favor the PAS-interaction pattern in primate groups are:

1. The soliciting animal (M_i) is threatened or otherwise impaired by Z, M_j's response may bring protection or help
2. It is difficult for M_i to attain a desire resource or goal, M_j's response may bring help or increase M_i's chances of achieving his goal
3. A special case are situations in which immediate responses to Z are advantageous; automatic responding of M_j to Z-related signals of M_i is then favored. In the majority of known examples, M_i is subordinate to and/or weaker than M_j.

We are now in a position to answer the question about PAS-relationships among primates: mother-infant relationships come closest to the "pure" PAS-relationship-type. In terms of ultimate causation, development of PAS-relationships is most likely in kinship networks (compare Wilson, 1975) or in social networks where conditions of reciprocity of soliciting are advantageous (compare Trivers, 1971).

What is the range of adaptive responses of a PAS-relationship? In primates, it is not the strong and dominant animals who are equipped with the signals of appealing and soliciting. Thus, in situations of direct conflict, the dominant partner in a PAS-relationship may assert himself at the cost of his counterpart. Furthermore, if *M_i's signals are to solicit specific,* Z-directed behaviors of M_j, they must furnish M_j with some information about Z, especially if Z is not present to M_j's senses. To a limited extent, this is possible even without a means of symbolic representation of Z as developed in human language. Menzel and Halperin (1975) have demonstrated that, among chimpanzees, purposive movements (that is, movements which clearly have some external referent), and especially locomotion, can have sign value for social partners transmitting to them information about the environment. They state that "the ability of nonverbal animals to 'tell' each other the precise nature and location of their goals is limited only by the richness of the signaler's purposive movements and the receiver's knowledge of the signaler and the environment in which he is operating" (p. 652). The ecosocial interactions involved in the communicatory processes in Menzel's experiments correspond to patterns 2.6,

2.7 and 2.9 of Table 2. In a similar way, the soliciting signals emitted by M_i in patterns 2.8 and 1.8 (which may sometimes be supplemented with purposive movements) can transmit information about the nature and location of Z_a by expressing M_i's emotional response to Z_a. How much information about Z is transmitted to Z_b in this way (again) depends on the richness of the signaler's expressions and the receiver's knowledge about the signaler and the common environment. The soliciting signals in the infant-mother relationship contain information about the emotional state of the infant (distressed, anxious, satiated, and so on) and the mother has to find out herself from observation of the context, how far these emotions are induced by and related to specific Z-situations or not. This is a simple communication system, but it has its limitations: it only works adaptively if the mother is able to interpret the infant's appeals adequately, in other words to assess the infant's emotional state and goal in relation to the environment. One can hypothesize that a mother will not be able to do this, when she is severely stressed and/or largely controlled in her behavior by group members other than her infant, or when she is not experienced and/or intelligent enough to assess the significance of Z-situations for the infant realistically.

Evolution of Control in Primate Group Systems: The Emergence of Language

The question of evolutionary perspective brings us back to the question of selection. Is there direct selection on certain ecosocial behavior patterns? Are relationships shaped by such a process? As mentioned on page 318, there are reasons for assuming that selection in primate groups works mainly at the individual level, supplemented by kin selection in groups or subgroups of about ten animals.

For species with relatively large and/or open groups (baboons, macaques, chimpanzees), therefore, the assumption can be made that for each individual, most members of its group are part of the selecting environment, the animals being comparable to resources. Ontogenetically, the group is the mother-infant-dyad's primary environment; and for an adult primate, the direct interactions within the group make up an important (and more or less stable) portion of his activities on any one day. Thus, one can assume that mechanisms of *triadic* interactions, formed within this social framework of selection and modification, and generalized to the "outgroup environment," may provide an important basis for an animal's *ecosocial* interactions and relationships. In part, ecosocial interactions, adapted to particular pressures only present in the outgroup environment, result in additional differentiation mechanisms, which may (or may not) be generalized and transferred to triadic ("ingroup") situations. At present, the study of the abilities of different primate species of transferring ingroup adaptations to the outgroup environment and vice versa has hardly begun. The dominance pattern is an example of a mechanism obviously working in both contexts; rank orders in regard to food are often, but not always, correlated with rank orders in regard to "social resources," for example, estrous females (compare Jolly, 1972:185). In hamadryas baboons, it has been experimentally proven that a mechanism of triadic differentiation in the group context—with the function of protecting pair bonds (Kummer *et al.*, 1974)—also determines analogous *ecosocial* interactions, thereby protecting the animal with prior access to food (compare pattern 1.6, Table 2) against a dominant partner (Falett, 1975). Even in this case, there is no proof that the differentiation mechanism evolved in the triadic context first. In chimpanzees, however, this same pattern 1.6 is observed in the ecosocial context of predatory behavior—among adult males, carcass "ownership rights" are mutually recognized once a captor has managed to hoard a carcass for several minutes—but not (so far) in the triadic social context (Teleki, 1973). Other mechanisms and other cases can only be qualitatively compared. At the present stage of research, it seems to be difficult to make general statements about how and where selection pressures operate on the differentiation mechanisms underlying relationships.

However, when we approach this problem as a question of control within the group system, one further step can be made. In a previous section I have argued that the primate group should not be conceived as a hierarchically organized system of the organism type. It will be pointed out in this section that in primate evolution, the group system has undergone a major transformation into a *potentially hierarchical* system, through the emergence of language in the phylogeny of man. The problem of how language has evolved in the human line will not be discussed here. Rather, we ask how the ecosocial interactions in a primate group are changed (in terms of control) when language is introduced as a medium of communication and of reasoning. To recognize the change clearly, we must look at the situation in nonhuman primates once more. Let me briefly recapitulate and complement what was written in the previous section about the means to control the Z-directed activities of the partner in dominance relationships and in PAS-relationships.

In a PAS-relationship, M_i is able to elicit from M_j specific responses to Z. M_i thereby exerts potential control over a given range of $M_j \rightarrow Z$-responses; but this control potential is limited, since M_j normally regulates its responses according to its *own* assessment of the M_i–Z-context.

The control of a dominant primate over a subordinate one is limited to (1) aggressive interventions against any undesired M_j-Z-interaction, (2) aggressive interactions forcing M_j into a particular, mostly peripheral group position where it will encounter specific Z-conditions, and (3) leading M_j to specific Z-conditions and/or serving as a model for M_j–Z (on the basis of the "attention structure"; patterns 2.7, 2.9, and 1.9). In this relationship, control is mainly indirect and, again, far from complete.

Additionally, limited, indirect control of M_i over $M_j \rightarrow Z$ is possible in the following ecosocial interaction patterns. M_i may selectively conceal (pattern 2.3) or by active interference transmit (pattern 1.10) information about Z to M_j; M_i and M_j may mutually stimulate each other (pattern 1.1) and/or coordinate (pattern 1.2) various activities towards Z.

Why is the potential of ecosocial control so limited in primate groups? It seems to me that the "missing link" must be sought in a mechanism which allows the combination (in one animal) of the ability or "power" to *stimulate* specific $M_j \rightarrow Z$ activities (PAS-pattern) with the ability or "power" to *inhibit* specific $M_j \rightarrow Z$ activities (dominance pattern). In nonhuman primates, apparently, these two "powers" are not combined to any significant extent. (This is a postulate which remains to be checked.)

The "missing link" is—language! Language enables a dominant group member to give precise instructions and orders for a subordinate's Z-directed activities, and to punish every instance of disobedience (or reward obedience) by the subordinate. The dominant male anubis baboon of our example (p. 329) could demand some of the small tree's delicate fruits from the juvenile male—if he only were able to speak! In a group of speaking hominids complete role differentiation and specialization on the basis of a dominance-rank order is *possible*. This group can be described as a *potentially hierarchical system*. Language provides the basis on which a decider subsystem (probably a dominant clan or kin) can develop: (1) a new medium for the storage of a "program" of supraindividual norms (cultural norms), which provide the criteria for the decisions, and (2) the effective means to control the regulative behavior of the executive subsystems.

Let me briefly formulate this step in terms of interindividual "exploitation." Exploitation is defined here as a process whereby M_i satisfies his needs by making use of M_j's ecosocial behavior and capacities. The possibilities of nonhuman primates to control their partners' ecosocial behavior only give limited potential for interindividual exploitation, the main mechanism being partner-action soliciting (which is typical of infant-mother relationships) and the main context being protection and defense against external dangers. Language, by enabling a dominant M_i to give directive instructions for M_j's ecosocial behavior, provides an important additional potential for interindividual exploitation in humans. (Further steps are tied to economic developments and are not considered here.) The extent to which this potential is used ultimately depends on the consequences of this use for the (inclusive) fitness of the involved members. This problem has been extensively discussed in recent years under the label "evolution of altruism."

Besides the spectacular expansion of the functional range of the dominance pattern to potentially every Z-situation, the emergence of language also has important consequences for the other ecosocial interaction patterns of the primates. By means of the systems approach, it will be possible to investigate comparatively alterations in these patterns and in the mechanisms of regulation and control within the group.

Acknowledgments

I gratefully acknowledge grants from the Julius Klaus-Stiftung, Zürich, and the Zürcher Hochschulverein, which financed the keeping of a group of sykes monkeys (*Cercopithecus mitis albogularis*) at the University of Zurich. Observations and experiments for my Ph.D. thesis were made on this group, and many of the ideas and concepts presented here were developed from these observations. I thank Professor Hans Kummer, my colleagues Hansueli Müller, Hans Sigg, and Alexander Stolba, and my wife Trudi for their help. Many thanks also for the valuable comments and criticism of the symposium participants. None of them are, of course, responsible for the conclusions presented in this paper.

Notes

1. Alternative terms for "object with attributes" or "element" are "component" or "member" (Miller, 1965a); in this paper the term "member" will be used.
2. Kummer (1967) defines tripartite relations as "... composed of sequences in which three individuals simultaneously interact in three essentially different roles and each of them aims its behavior at both of its partners" (p. 64).
3. Remember Miller's functional concept of subsystems as the totality of the members which carry out a particular process for their system. In Miller's terms "role" can be defined as "subsystem process" (1975:354).

References

Bernstein, I. S. 1966a. Analysis of a key role in a capuchin (*Cebus albifrons*) group. *Tulane Stud. in Zool.* 13:49–54.

Bernstein, I. S. 1966b. An investigation of the organization of pigtail monkey groups through the use of challenges. *Primates* 7:471–480.

Bertalanffy, L. von 1968. *General systems theory: Foundations, development, applications*. New York: Braziller.

Clutton-Brock, T. H. 1974. Primate social organization and ecology. *Nature, Lond.* 250:539–542.

Clutton-Brock, T. H., and Harvey, P. H. 1976. Evolutionary rules and primate societies. Pages 195–237 in *Growing points in ethology*. Eds. P. P. G. Bateson and R. A. Hinde. Cambridge: Cambridge University Press.

Clutton-Brock, T. H., and Harvey, P. H. 1977. Primate ecology and social organization. *J. Zool. Lond.* 183:1–39.

Crook, J. H ., and Gartlan, J . S. 1966. Evolution of primate societies. *Nature, Lond.* 210:1200–1203.

Falett, J. 1975. Experimente zur Besitzrespektierung beim Mantelpavian (*Papio hamadryas*). Unpublished master thesis. University of Zurich.

Fedigan, L. M. 1976. A study of roles in the Arashiyama West troop of Japanese monkeys (*Macaca fuscata*). *Contrib. Primat.* 9:1–95.

Gautier, J. P. 1975. Etude comparée des systèmes d'intercommunication sonore chez quelques cercopithécines forestiers africains—Mise en évidence de correlations phylogénétiques et socio-écologiques. Thèse de Docteur d'Etat. (Ph.D. thesis.) Université de Rennes.

Hall, A. D., and Fagen, R. E. 1956. Definition of system. *Yrbk. Soc. Gen. Syst. Res.* 1:18–28.

Hinde, R. A. 1975. Interactions, relationships and social structure in nonhuman primates. Pages 13–24 in *Proc. Symp. 5th Cong. Int.*

Primat. Soc. (1974). Eds. S. Kondo, M. Kawai, A. Ehara, and S. Kawamura. Tokyo: Japan Science Press.

Hinde, R. A. 1976. On describing relationships. *J. Child Psychol. Psychiat.* 17:1–19.

Hladik, C. M. 1975. Ecology, diet and social patterning in Old and New World primates. Pages 3–36 in *Socioecology and psychology of primates*. Ed. R. H . Tuttle. The Hague: Mouton.

Jolly, A. 1972. *The evolution of primate behavior*. New York and London: Macmillan.

Kawai, M. 1965. Newly acquired precultural behavior of the natural troop of Japanese monkeys on Koshima Island. *Primates* 6:1–30.

Kummer, H. 1967. Tripartite relations in hamadryas baboons. Pages 63–72 in *Social communication among primates*. Ed. S. A. Altmann. Chicago: University of Chicago Press.

Kummer, H ., Götz, W., and Agnst, W. 1974. Triadic differentiation: An inhibitory process protecting pair bonds in baboons. *Behaviour* 49:62–87.

Lewontin, R. C. 1970. The units of selection. *Ann. Rev. Ecol. Syst.* 1: 1–18.

Maynard Smith, J . 1976. Group selection. *Q. Rev. Biol.* 51:277–283.

Menzel, E. W., and Halperin, S. 1975. Purposive behavior as a basis for objective communication between chimpanzees. *Science* 189:652–654.

Miller, J. G. 1965a. Living systems: Basic concepts. *Behav. Sci.* 10:193–237.

Miller, J. G. 1965b. Living systems: Structure and process. *Behav. Sci.* 10:337–379.

Miller, J. G. 1971. Living systems: the group. *Behav. Sci.* 16:302–398.

Miller, J. G. 1975. The nature of living systems. *Behav. Sci.* 20:343–365.

Nagel, U. 1973. A comparison of anubis baboons, hamadryas baboons and their hybrids at a species border in Ethiopia. *Folia. Primat.* 19: 104–165.

Nagel, U. 1979. Beiträge zur Biologie der funktionellen Verhaltensdifferenzierung in Primatengruppen. Unpublished Ph.D. thesis. University of Zurich.

Rowell, T. E. 1967. A quantitative comparison of the behaviour of a wild and a caged baboon group. *Anim. Behav.* 15:499–509.

Rowell, T. E. 1971. Organization of caged groups of cercopithecus monkeys. *Anim. Behav.* 19:625–645.

Rowell, T. E. 1974. Contrasting adult male roles in different species of nonhuman primates. *Arch. Sex. Behav.* 3:143–149.

Singh, S. D. 1969. Urban monkeys. *Sci. Amer.* 221:108–115.

Southwick, C. H., Siddiqi, M. F., Farooqui, M. Y., and Pal, B. C. 1976. Effects of artificial feeding on aggressive behaviour of rhesus monkeys in India. *Anim. Behav.* 24:11–15.

Teleki, G. 1973. *The predatory behavior of wild chimpanzees*. Lewisburg, Pa.: Bucknell University Press.

Trivers, R. L. 1971. The evolution of reciprocal altruism. *Quart. Rev. Biol.* 46:35–57.

van Hooff, J. A. R. A. M., and de Waal, F. 1975. Aspects of an ethological analysis of polyadic agonistic interactions in a captive group of *Macaca fascicularis.* Pages 269–274 in *Contemporary primatology.* Eds. S. Kondo, M. Kawai, and A. Ehara. Basel: Karger.

Watzlawick, P., Beavin, J. H ., and Jackson, D. D. 1967. *Pragmatics of human communication.* New York: Norton.

Weiss, P. A. 1969. The living system: determinism stratified. Pages 3–55 in *Beyond reductionism.* Eds. A. Koestler and J. R. Smythies. Boston: Beacon Press.

Wilson, E. O. 1975. *Sociobiology: The new synthesis.* Cambridge, Mass.: Belknap–Harvard University Press.

Wrangham, R. W. 1974. Artificial feeding of chimpanzees and baboons in their natural habitat. *Anim. Behav.* 22:83–93.

Zola, S. 1968. Protective-adoptive aggression in young *Cebus* monkeys. *Lab. Primate Newsl.* 7:17–19.

Chapter 14

In Summary

Irwin S. Bernstein
Euclid O. Smith

In the introduction, we presented the viewpoint which we hoped to take in examining the relationship between primate social organization and ecology. We expressed confidence in the scientific method and outlined the premises which have traditionally been used in developing models or theories relating social organization to ecology. In the chapters that followed each of the authors developed specific theoretical approaches to the problem. Some developed the framework of a theory, defining terms and hypothesizing relationships among the constituent elements; others tried specific applications of various theories, whereas others examined relevant data in an attempt to test the explanatory power of a theory, or to suggest additional theoretical elements.

Thelma Rowell, in the first chapter, set the stage by challenging explanations and predictions based on presumed adaptive functions. She strongly suggests that this particular approach would be incapable of identifying data conditions under which the null hypothesis should be accepted, that is, those conditions under which we would accept the null that a behavior was not adaptive. Since we are limited only by our imaginations in the postulation of possible adaptive functions, support of a theoretical formulation on the basis of its "adaptive value" is useless. She concludes that a new look at our theoretical assumptions is required and begins the task.

Our other contributors take up the same task from multiple perspectives and, in the last chapter, Ueli Nagel proposes a new conceptualization of the relationship between social organization and ecology. He uses a systems approach to develop a method for identifying the nature of interrelationships between behavior and ecology. This produces then not a new theory summarizing existing knowledge, nor an explanatory theory to account for what we know, but rather a new tool which can be used to obtain data relevant to the development of such theories.

This is perhaps where we should be: collecting data and testing hypotheses. It is far too early for any grand scheme, for we really know very little about: the elements of social organization or ecology which relate to one another, the causes of variation in social behavior, the evolutionary mechanisms which bring about such changes, the ontogenetic processes which influence social behavior, and the functions of the social mechanisms which we do see. Several authors in this book have examined the sources of variability which influence the expression of social behavior, which is also presumably directed at coping with a particular ecological pressure. Both human and nonhuman societies were examined, and multiple contexts were considered in attempting to gain an appreciation of the extent to which behavioral adaptations are in direct response to ecology. Phylogenetic heritage has to be considered, of course, but in an order noted for its generality and its plasticity of response, problem solving, learning abilities, information acquisition, and retention mechanisms must be closely attended to.

Behavioral adaptation thus may be a result of genetic mechanisms expressing themselves in a species typical environment, or a result of responses modified as a function of experience in an environment. Obviously these two sources of adaptation will interact with one another, and there are several variations of each. In terms of behavioral adaptations acquired during the lifetime of an individual, however, in socially living organisms it is useful to distinguish between responses acquired by direct problem solving experiences and those acquired by some form of social learning. Whether we call the information transmitted from one generation to another traditions, culture, or some less specific term, it is clear that we must consider the mechanisms of such information transfer and compare it to genetic mechanisms of information transmission in order to understand how social behavior may be responsive to ecological situations. The social organization in which social behavior is expressed is itself an outcome of behavioral interaction patterns. Social organization thus not only influences social behavior, but is defined by the pattern of expression of such behavior. The possible patterns of expression are limited by the characteristics of the members of the social unit, including their age, sex, and physical condition, and their relative numbers. The composition of the social group, its social structure, thus strongly influences the expression of social behavior and the organization of the unit. As such, we must consider these variables which influence social structure when we consider how ecology may influence social organization.

What began as an examination of existing theories relating ecology and social organization, ended as a major reevaluation of the entire approach to understanding such relationships. It was not a matter of testing, refining, and extending existing models, but rather an examination of basic assumptions. We discarded some premises and specified others and recognized that oversimplifications would be useless in understanding the outcome of the complex interface between behavior and environment.

No one needs to be told that the world is complex. What we need is identification and understanding of the principles which produce regularities consistently enough to permit useful prediction. Simplified approaches accept a level of variability in predicted outcomes, but in the present case, the level of variability remnant after accounting for one or two basic factors makes predictions useless. We shall have to develop more basic theories encompassing several of the major sources of variability.

In asking how ecology influences social organization, Bourlière begins by asking what are the relevant features in the environment which influence the individual. Parameters relating to energetic and nutrient sources, resource distribution and requirements, toxins, predator pressures, temperature limits, and so on are surely all relevant, but they cannot be studied in isolation from the individuals they impact upon. The quality of the habitat is assessed by the individual inhabitant who must be able to recognize and respond to resources. Mere presence is not enough to insure availability to the inhabitant. The individual's perception of the environment is thus crucial, rather than our "objective" measurement of the environment. It matters less what is theoretically available than what the individual accepts and utilizes. Our task in identifying the relevant ecological parameters is thus twofold, identification of the requirements of the individuals and assessment of the individual's responses to the available resource parameters. Neither task will be easy, but it is how the individual perceives and responds to the habitat which will determine its success or failure in that environment.

Altmann and Altmann then remind us that ecological constraints will act through differential selection on life history processes. The outcome of reproduction in a single year will tell us little about differential genetic fitness of primates. A particular breeding strategy can only be evaluated in terms of reproductive success calculated over the lifetime of an individual. Selection operates on life history processes to influence ontogenetic rates, longevity, fertility, and reproductive strategies. It is these processes which account for demographic rates, such as age specific mortality and fertility, which are determinant of population compositions. The information contained in a life table reflects the product of demographic rate information and represents the population composition which may be organized into social units. The actual structure of a social unit influences the types and frequencies of social behavior and the pattern of social behavior expressed defines the social organization. Selection for social organization will thus be indirect and there will be a long feedback loop from social behavior to the influence on life history processes which will then feedback to influence social behavior.

The identification of the patterns of social behavior which may be influenced by ecology will, in and of itself, be insufficient to explain any particular expressed instance of social organization or social structure, for many variables may influence life history processes, in addition to the selection operating on social organization. Dunbar presents us with a specific example of a social organization influenced by demography. The social structure of gelada units strongly influences social behavior, but social structure is influenced by several random processes in addition to any selective pressures which modify life history processes. In relatively small populations the number of individuals of each sex born in any particular year may depart from mean equlbrium rates as a result of random events. The effects of a simple departure from equilibrium in a single year, however, will persist for many years into the future. Given differential ontogenetic rates for the two sexes, even if the ratio of births in the following year were exactly opposite, the long-term effects on social structure would not be balanced. Dunbar shows how departures from equilibrium may persist in a population for more than a generation, even if all but the first year were idealized equilibrium years. Random processes thus can have profound consequences for demography.

Perturbations in age-sex ratios can be demonstrated to have significant consequences on the social organization of gelada groups and appear to be responsible for some major shifts in the social behavior patterns of adult males. Moreover, the frequency of perturbations from mean rates in small populations, such as most primate social units, will be of sufficient frequency due to random processes such that idealized stationary life tables may never be achieved. As a consequence, we may expect social structure to vary over time in primate groups, and to the extent that social structure limits and influences social organization, we may find different social patterns described not only for different troops of the same species but also for the same troop at different times.

The implications of this conclusion for primate field work are profound. Not only will we have to contend with variability due to species differences and habitat differences, but we will have to recognize that the usual tenure of field studies, a year or two, is insufficient to describe the "average" social patterns, even for our sample of one. It is hard enough to gain recognition for the fact that each troop, no matter how large, is still a sample of one in studying social organization. Now we will also have to consider the variability of the obtained measure even in that sample of one.

Such considerations make attempts to relate the social behavior of a troop of one taxon in one ecology to the social behavior of a troop of another taxon in another ecology seem naive indeed. Baldwin and Baldwin, however, state that the process may not be the primary influence on primate social organization since, in the order Primates, ontogenetic processes may have enormous impacts on social relationships. In accordance with Rowell, and several other contributors, the Baldwins note the importance of traditions in primate social organization, but rather than considering tradition drift per se, they look at the mechanism of transmission itself. Socialization and other learning processes may be seen in terms of differential reinforcement of variable response systems. With the extreme plasticity of primate behavior, the Baldwins suggest that ecological influences on social behavior will most importantly be a function of individual histories in response to the environment, with only a small genetic component influencing behavioral adaptations to ecology. Phylogenetic inputs may therefore be virtually totally overridden by ontogenetic processes, and the impact of ecology upon primate social organization may best be studied in terms of direct reinforcement for particular responses.

From this perspective, primate social organizations may be very finely attuned to ecological situations. One might expect that responses patterns would be direct expressions of the differential reinforcement individuals received for responding to ecological pressures with various available responses. In considering the primate with the greatest ability to modify behavior as a function of experience, Birdsell indicates that this is not so. A great deal of the learning that takes place is not a result of individual problem solving, but rather a consequence of the transmission of traditions. Traditional life-styles can be maintained even in the face of massive changes in the habitat; in fact, traditions are sustained not because they are optimal solutions to ecological problems, but because they are tolerable. Thus, no matter if social organization is adapted to ecology through genetic or traditional mechanisms, it is unlikely that we will see either one particular solution, or that we will see optimal solutions for any situation. A range of variability is to be expected with multiple tolerable patterns in any environment and the most prevalent and successful patterns will be those that are tolerable under the widest range of conditions. So much for the precision of adaptation.

Suzuki demonstrates this principle nicely in describing multiple nonhuman primate taxa living in the same habitat and the social organization of chimpanzees living in multiple habitats. The chimpanzee social organization can best be understood not by how it adapts the unit to any particular environment, but rather by how it succeeds in functioning across the range of environments in which the chimpanzee may be found. Even the same troop may drastically alter its environmental setting across the time span of a single year, and the social organization of the chimpanzee must include the flexibility to make the absolute changes required while still retaining the continuity of social organization under all conditions, such that troops may be organized and reorganized periodically without major disruptions.

Coelho *et al.* further elaborate on the theme of the tolerable in primate social organization by examining two New World primate species with dramatically different social organizations that live in the same habitat. Not only are the howler and spider monkeys at Tikal sympatric, but they apparently exploit much the same food resources and display only minor differences in econiches. Coelho *et al.* demonstrate that this condition can persist at Tikal, despite theoretical restrictions on competing species in the same niche, by showing that the aspects of habitat attended to normally, food distribution and abundance, are not the limiting factors at Tikal. In the face of superabundant food resources, the manner of exploitation of this resource matters little so long as it permits a sufficiency of resource to be collected and used. Whatever determines the carrying capacity at Tikal, it is apparently not caloric availability. Whatever it is though, it will be the primary selective pressure and we might expect niche differentiation between sympatric species only in regard to critical resources in short supply.

At the moment, however, it is not clear what the limiting factors are at Tikal and we can only speculate as to how the two different social organizations seen relate to features of the ecology. Clutton-Brock and Harvey develop a list of attributes which relate to ecology and hypothesize that where the distribution and density of food is crucial, certain consequences should be forthcoming. If food is generally the limit on primate carrying capacity, then home ranges, day ranges, population densities, and biomass should be lawfully related to food availability. They test this hypothesis by running correlations of these measures with all of the available data they could collect from the literature on all known primates. Their analysis is, by and large, restricted to the generic level and many of the values used are admittedly estimates, but they nonetheless find evidence supporting their hypotheses. Thus, food availability may be the major factor limiting the carrying capacity for primates in most habitats. Dietary strategies and phylogeny also influence the correlations, but the authors are encouraged by the results of this correlational approach.

Eisenberg looks at some of these same factors but develops the most all-encompassing approach to an understanding of primate social organization that we have seen. His perspective includes phylogeny, geology, and distributional data as well as the more usual measures. He covers the New World species in depth with a scope that is perhaps unique and no doubt reflects his interests in mammalogy as well as his special concern for broad principles of New World ecology. A massive amount of data from diverse sources is summarized to yield a comprehension of the problem which perhaps foreshadows the efforts that must be made to eventually understand the relationship of social organization to ecology.

If all of this information is needed to relate ecology to social organization, then what can we hope to achieve in understanding how a fossil taxon might have responded socially in the face of its now unobservable ecological situation? The first problem will be to discover what the behavior of the extinct taxon was, and the next to discover what the ecological pressures were for that taxon. Relating these two to each other will be no simple task, as indicated in the previous chapters, but the discovery and description of the data themselves will be the more serious obstacle to understanding in the case of extinct species.

If it were not for our interest in understanding our own origins, this task might be regarded as merely a formal exercise in extrapolating from models. As it is, however, the application of techniques is of intense interest to students and investigators trying to piece together the evolution of our own species. Loring Brace begins with detailed measures of a single anatomical feature, the cross-sectional area of the dentition, and shows us how much information can be obtained from this one measure and how this can be used in conjunction with other less exhaustive measures to piece together the story of our own origins. Although much ingenious speculation is required, and many of the details of social organization may never be known, certain broad limits on the social organization of early hominids can be deduced. These deductions are based on both measured data and theoretical premises. The basis for these premises lies in our understanding of the broad principles which govern the relationship between primate social organization and ecology. Tests of these models using extant forms thus strongly influence the premises found to be acceptable in deducing the social behavior of early hominids.

A more extensive model based on presumed ecological inputs to social behavior is provided by Campbell when he considers more recent hominids for which much more detailed information is available concerning both anatomy and the artifacts that relate to their activities. The existence of tools, shelters, and similar objects, consequent to activities, may be regarded in this case as "fossil behavior." When we understand the functions of artifacts we can reconstruct the behavior of their makers to a large degree, and thus bridge the time gap that might ordinarily separate us from a study of the impact of ecology upon the social organization or social behavior of extinct forms.

In conclusion, then, this conference may be regarded as laying the foundation for future studies relating social organization to ecology, including considerations of multiple factors. The mechanisms by which such influences may be expressed are presented and the expected limitations of methodology are discussed. Two major themes are developed by the contributors.

First, we stress the importance of considering life history processes as they influence demography and the impact of demography upon social structure and ultimately social organization. Any narrower perspective is considered likely to have serious shortcomings. The implications for field studies should be clear. No single study of even two or three years duration will produce data which will characterize a species nor necessarily reveal a social organization which will be typical of the study group through time. Due to the time duration of primate life history processes, and the time course of demographic perturbations, the same society may show alternative social patterns as social structure changes. In the absence of stationary life tables, one must expect social change.

The second major point we should like to make is that evolutionary processes do not invariably produce optimal adaptations. Even where selection may operate to favor the best available social adjustment to a particular ecology, the critical word is "available." Neither mutations nor social innovations are produced on demand. The biological and social processes which produce variability in social organization are essentially random processes. As environments change, selective pressures change. Social organizations which are tolerably adapted to the new conditions may survive. Although the best of these may have some advantage, none may be optimal since neither biological nor social evolution is teleological. In fact, both genetic and ontogentic processes may be remarkably conservative in the face of changing conditions.

The major dichotomy of genetic and ontogenetic inputs to social behavior was examined in detail and the primacy of "traditional" input in ontogenetic processes was noted for socially living primates. The incredible similarity of function and consequence of mechanisms of genetic transmission and tradition transmission was revealed in detailed presentations of field studies of human and nonhuman primates. The mechanisms which modify and transmit information using the two modalities are essentially parallel and the same processes apply, although the specific mechanisms differ, and the time scales of effect are at least theoretically dissimilar. The process of change in both is one essentially of evolution, with genetic behavioral propensities and traditional modes persisting in so long as they are tolerable solutions to the problems of survival and reproduction in any particular environment.

Index